# A Level Advancing Physics for OCR

## Year 1 and AS

**Third Edition**

B


**Series Editor**
Lawrence Herklots

**Authors**
Lawrence Herklots
John Miller
Helen Reynolds

Great Clarendon Street, Oxford, OX2 6DP, United Kingdom

Oxford University Press is a department of the University of Oxford. It furthers the University's objective of excellence in research, scholarship, and education by publishing worldwide. Oxford is a registered trade mark of Oxford University Press in the UK and in certain other countries



First published in 2015



British Library Cataloguing in Publication Data
Data available

978-0-19-834093-5

10 9 8 7 6 5 4 3

Paper used in the production of this book is a natural, recyclable product made from wood grown in sustainable forests. The manufacturing process conforms to the environmental regulations of the country of origin.

Printed in China by Leo Paper Products Ltd

This resource is endorsed by OCR for use with the specification H157 AS Level Physics B (Advancing Physics) and with year 1 of the specification H557 A Level Physics B (Advancing Physics). In order to gain endorsement this resource has undergone an independent quality check. OCR has not paid for the production of this resource, nor does OCR receive any royalties from its sale. For more information about the endorsement process please visit the OCR website www.ocr.org.uk

Index compiled by INDEXING SPECIALISTS (UK) Ltd., Indexing House, 306A Portland Road, Hove, East Sussex BN3 5LP United Kingdom

This book has been written to support students studying for OCR AS Physics B and for students in their first year of studying for OCR A Level Physics B. It covers the AS content from the specification, the content of which will also be examined at A Level. The content covered is shown in the contents list, which also shows you the page numbers for the main topics within each chapter. There is also an index at the back to help you find what you are looking for. If you are studying for OCR AS Physics B, you will only need to know the content in the blue box.

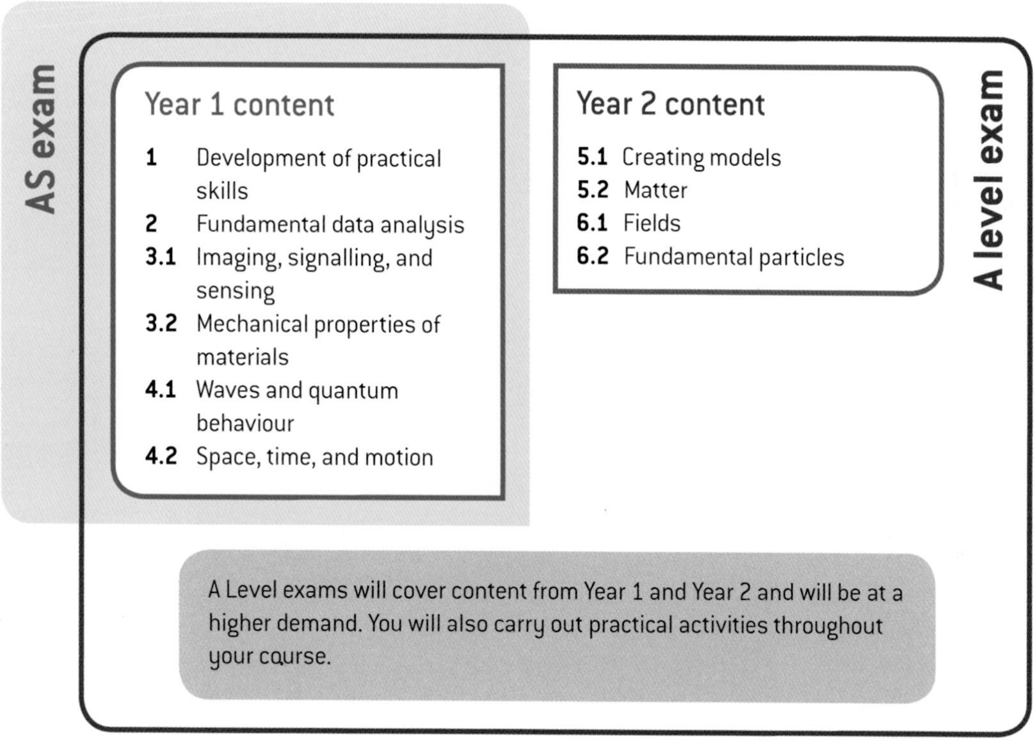

**AS exam**

**Year 1 content**

**1** Development of practical skills
**2** Fundamental data analysis
**3.1** Imaging, signalling, and sensing
**3.2** Mechanical properties of materials
**4.1** Waves and quantum behaviour
**4.2** Space, time, and motion

**A level exam**

**Year 2 content**

**5.1** Creating models
**5.2** Matter
**6.1** Fields
**6.2** Fundamental particles

A Level exams will cover content from Year 1 and Year 2 and will be at a higher demand. You will also carry out practical activities throughout your course.

This book contains many different features. Each feature is designed to support and develop the skills you will need for your examinations, as well as foster and stimulate your interest in physics.

Terms that you will need to be able to define and understand are highlighted by **bold text**.

## Learning outcomes

- → At the beginning of each topic, there is a list of learning outcomes.
- → These are matched to the specification and allow you to monitor your progress.
- → A specification reference is also included in the topic title. *Specification reference: 2.1.3*

### Practical features

These features support further development of your practical skills, and cover the required practicals for this course.

### Extension features

These features contain material that is beyond the specification. They are designed to stretch and provide you with a broader knowledge and understanding and lead the way into the types of thinking and areas you might study in further education. As such, neither the detail nor the depth of questioning will be required for the examinations. But this book is about more than getting through the examinations.

1 Extension features also contain questions that link the off-specification material back to your course.

### Study Tips

Study tips contain prompts to help you with your understanding and revision.

### Synoptic link

These highlight the key areas where topics relate to each other. As you go through your course, knowing how to link different areas of physics together becomes increasingly important. Many exam questions, particularly at A Level, will require you to bring together your knowledge from different areas.

### Summary Questions

1 These are short questions at the end of each topic.

2 They test your understanding of the topic and allow you to apply the knowledge and skills you have acquired.

3 The questions are ramped in order of difficulty. Lower-demand questions have a paler background, with the higher-demand questions having a darker background. Try to attempt every question you can, to help you achieve your best in the exams.

# MODULE 4.1

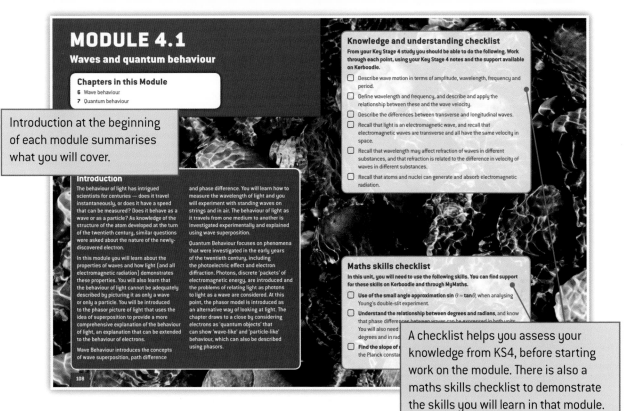

## Waves and quantum behaviour

### Chapters in this Module

**6** Wave behaviour
**7** Quantum behaviour

Introduction at the beginning of each module summarises what you will cover.

### Introduction

The behaviour of light has intrigued scientists for centuries — does it travel instantaneously, or does it have a speed that can be measured? Does it behave as a wave or as a particle? As knowledge of the structure of the atom developed at the turn of the twentieth century, similar questions were asked about the nature of the newly-discovered electron.

In this module you will learn about the properties of waves and how light (and all electromagnetic radiation) demonstrates these properties. You will also learn that the behaviour of light cannot be adequately described by picturing it as only a wave or only a particle. You will be introduced to the phasor picture of light that uses the idea of superposition to provide a more comprehensive explanation of the behaviour of light, an explanation that can be extended to the behaviour of electrons.

Wave Behaviour introduces the concepts of wave superposition, path difference and phase difference. You will learn how to measure the wavelength of light and you will experiment with standing waves on strings and in air. The behaviour of light as it travels from one medium to another is investigated experimentally and explained using wave superposition.

Quantum Behaviour focuses on phenomena that were investigated in the early years of the twentieth century, including the photoelectric effect and electron diffraction. Photons, discrete 'packets' of electromagnetic energy, are introduced and the problems of relating light as photons to light as a wave are considered. At this point, the phasor model is introduced as an alternative way of looking at light. The chapter draws to a close by considering electrons as 'quantum objects' that can show 'wave-like' and 'particle-like' behaviour, which can also be described using phasors.

108

### Knowledge and understanding checklist

**From your Key Stage 4 study you should be able to do the following. Work through each point, using your Key Stage 4 notes and the support available on Kerboodle.**

- ☐ Describe wave motion in terms of amplitude, wavelength, frequency and period.
- ☐ Define wavelength and frequency, and describe and apply the relationship between these and the wave velocity.
- ☐ Describe the differences between transverse and longitudinal waves.
- ☐ Recall that light is an electromagnetic wave, and recall that electromagnetic waves are transverse and all have the same velocity in space.
- ☐ Recall that wavelength may affect refraction of waves in different substances, and that refraction is related to the difference in velocity of waves in different substances.
- ☐ Recall that atoms and nuclei can generate and absorb electromagnetic radiation.

### Maths skills checklist

**In this unit, you will need to use the following skills. You can find support for these skills on Kerboodle and through MyMaths.**

- ☐ **Use of the small angle approximation sin $\theta \approx$ tan$\theta$,** when analysing Young's double-slit experiment.
- ☐ **Understand the relationship between degrees and radians,** and know that phase differences between waves can be expressed in both units. You will also need ... degrees and in rad...
- ☐ **Find the slope of a** ... the Planck constan...

A checklist helps you assess your knowledge from KS4, before starting work on the module. There is also a maths skills checklist to demonstrate the skills you will learn in that module.

Module summaries highlight the key concepts of each module. They break down the important facts and formulas by chapter and topic from across the module.

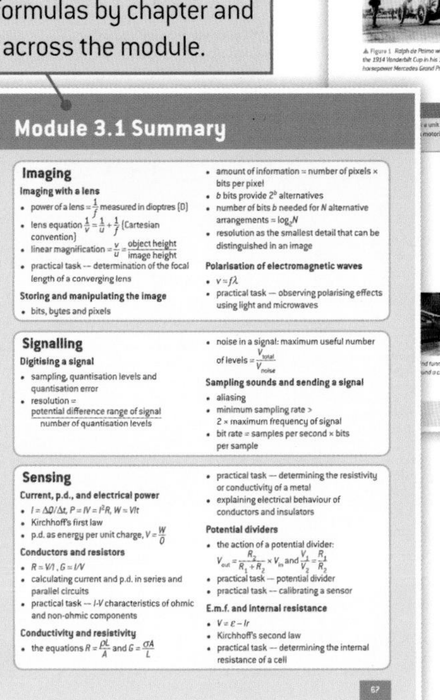

## Module 3.1 Summary

### Imaging

**Imaging with a lens**

- power of a lens $= \frac{1}{f}$ measured in dioptres (D)
- lens equation $\frac{1}{v} = \frac{1}{u} + \frac{1}{f}$ (Cartesian convention)
- linear magnification $= \frac{v}{u} = \frac{\text{object height}}{\text{image height}}$
- practical task — determination of the focal length of a converging lens

**Storing and manipulating the image**

- bits, bytes and pixels
- amount of information = number of pixels × bits per pixel
- $b$ bits provide $2^b$ alternatives
- number of bits $b$ needed for $N$ alternative arrangements $= \log_2 N$
- resolution as the smallest detail that can be distinguished in an image

**Polarisation of electromagnetic waves**

- $v = f\lambda$
- practical task — observing polarising effects using light and microwaves

### Signalling

**Digitising a signal**

- sampling, quantisation levels and quantisation error
- resolution $= \frac{\text{potential difference range of signal}}{\text{number of quantisation levels}}$
- noise in a signal: maximum useful number of levels $= \frac{V_{total}}{V_{noise}}$

**Sampling sounds and sending a signal**

- aliasing
- minimum sampling rate > 2 × maximum frequency of signal
- bit rate = samples per second × bits per sample

### Sensing

**Current, p.d., and electrical power**

- $I = \Delta Q/\Delta t$, $P = IV = I^2R$, $W = VIt$
- Kirchhoff's first law
- p.d. as energy per unit charge, $V = \frac{W}{Q}$

**Conductors and resistors**

- $R = V/I$, $G = I/V$
- calculating current and p.d. in series and parallel circuits
- practical task — $I$-$V$ characteristics of ohmic and non-ohmic components

**Conductivity and resistivity**

- the equations $R = \frac{\rho L}{A}$ and $G = \frac{\sigma A}{L}$
- practical task — determining the resistivity or conductivity of a metal
- explaining electrical behaviour of conductors and insulators

**Potential dividers**

- the action of a potential divider: $V_{out} = \frac{R_2}{R_1 + R_2} \times V_{in}$ and $\frac{V_1}{V_2} = \frac{R_1}{R_2}$
- practical task — potential divider
- practical task — calibrating a sensor

**E.m.f. and internal resistance**

- $V = \varepsilon - Ir$
- Kirchhoff's second law
- practical task — determining the internal resistance of a cell

67

## Physics in perspective

### Racing car design

▲ Figure 1 Ralph de Palma winning the 1914 Vanderbilt Cup in his 115 brake horsepower Mercedes Grand Prix race car

When motor racing started, the aim of car designers was to use ever more powerful engines. Figure 1 shows a 1914 Mercedes, which had a mass of about a tonne. It was fast, with a top speed of 115.0 mph and an engine power of 115 brake horsepower. It won many races at that time, but it was hardly what you would call streamlined.

**Drag and supercharging**

As speeds increased in the 1920s and 1930s, designers realised that they needed to tackle drag force. This force is generally described by the equation

$$F_D = \frac{1}{2}\rho C_D A v^2$$

where $\rho$ is the density of fluid through which the object is moving (in this case, air), and $v$ and $A$ are the velocity and cross-sectional area of the moving object. $C_D$, the drag coefficient, is a dimensionless number, usually less than 1. Increasing $v$ greatly increases the drag, but increasing $v$ is what car designers wanted — thankfully, both $C_D$ and $A$ can be reduced. $A$ is reduced by lowering the profile of the car, and $C_D$ is reduced by finding ways to streamline the vehicles so that the air flows smoothly over them.

Automobile designers also introduced supercharging, which involved pumping extra air into the engine so that it could burn fuel faster and generate greater power, up to 180 kW. A modern development of this, which uses waste heat and so takes less energy from the engine, is called turbocharging.

**Acceleration and stability**

Acceleration is change in velocity. In motor racing, it comes up in 3 main situations.

- Speeding up — the easiest way to improve this is increasing the engine power. This increases the forces involved in driving the car, and so the driver must be protected against the force with which the seat pushes him or her forwards during acceleration.
- Slowing down — the braking system must be able to safely bring the car, along with the driver, to a stop in a reasonable time.
- Cornering may not involve a change of speed, but it does involve a change in direction and so also a change in velocity. The force needed to push the car into a new direction has to be provided by friction with the ground. So the wheel size and tyres used must maximise the contact between the vehicle and the ground.

As cars became more and more streamlined they became less stable. Just as aircraft reaching a certain speed take off, so too do a car's wheels leave the ground if it is travelling fast enough. Furthermore, race tracks became smaller in size, resulting in tighter bends and more laps to be driven, so more corners had to be negotiated at speed. In the 1960s and

Physics in perspective pages go into detail on some real-life applications of physics — a defining feature of this context-led specification.

smooth, soft tyres called 'slicks'.

**Safety and Formula 1**

Formula 1 refers to the highest class of single-seat car racing, and the 'formula' is the set of regulations which must apply to all cars. By the 1970s, speeds had increased dramatically but safety had not, and deaths were common. As a result, drivers campaigned for improved safety, which gave rise to modern Formula 1 Regulations.

Besides the use of protective clothing and regular inspections of vehicles and track, a number of safety measures were integrated into the cars. These included crumple zones, automatic cut-off of fuel lines and a 'survival cell' in which the driver's cockpit was built. Public safety was also considered, with the introduction of double guard rails and wide verges to the track.

Other regulations restrict the performance of vehicles — these include increasing the minimum mass of vehicles, restricting the engine capacity (and so keeping the power down), and banning some aerodynamic developments which increased the downforce. These restrictions make the race fairer and also less risky for the drivers. As car designers continually come up with technological improvements, the Formula 1 regulations have to be refined each year.

**Summary questions**

1 Ralph de Palma won the 1914 Vanderbilt Cup in a car of mass 1080 kg with a top speed of 51.5 m s$^{-1}$ and an engine output power of 85.8 kW. The race consisted of 35 laps on a circuit of length 13.6 km.
   a Calculate the shortest time that this race could have taken and explain why the race would have taken longer than this.
   b After the race, de Palma demonstrated the car's maximum speed by driving in a straight line on horizontal ground. Draw a labelled diagram showing all the forces acting on the car at this steady speed. You can assume that the rear wheels are providing the driving force.
   c Assuming that 80% of the car's output power is delivered to work done against resistive forces, calculate the value of the resultant resistive force when driving at a constant 51.5 m s$^{-1}$.

2 The drag force acting on a car is given by $F_D = \frac{1}{2}\rho C_D A v^2$. Assuming that a family car has a drag coefficient of 0.32 and a maximum cross-sectional area of 2.1 m$^2$, calculate the drag force while driving down the motorway at 70 mph (31.3 m s$^{-1}$). The density of air, $\rho = 1.2$ kg m$^{-3}$.

217

Physics in perspective pages contain summary questions to further test your knowledge and understanding of the science behind the applications.

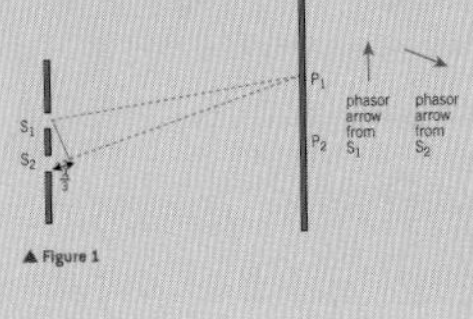

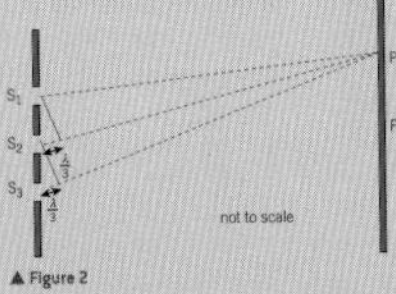

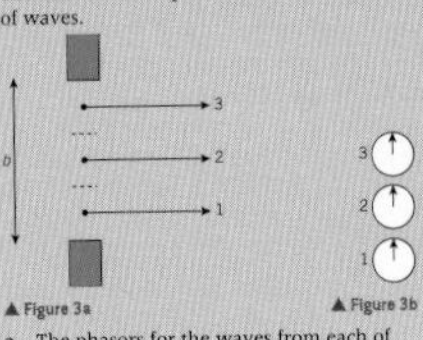

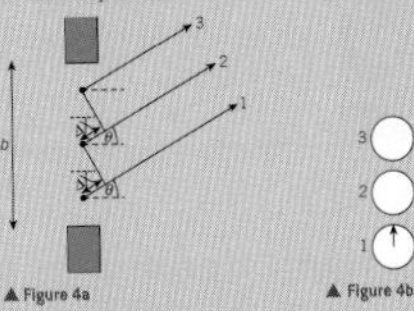

Practice questions at the end of each chapter including questions that cover practical and maths skills.

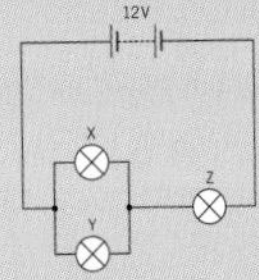

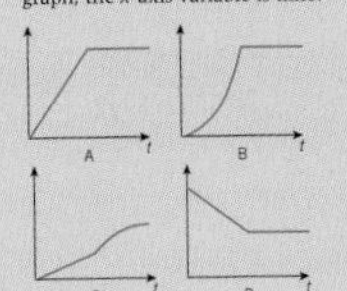

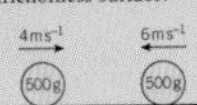

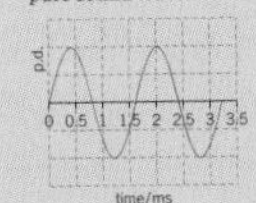

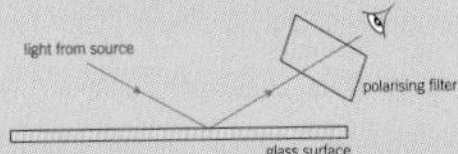

Practice questions at the end of the book, with multiple choice questions and synoptic style questions, also covering the practical and math skills.

This book is supported by next generation Kerboodle, offering unrivalled digital support for independent study, context, differentiation, assessment, and the new practical endorsement.

If your school subscribes to Kerboodle, you will also find a wealth of additional resources to help you with your studies and with revision.

- Study guides
- Maths skills boosters and calculation worksheets
- Practicals activities to support the practical endorsement
- Interactive progress quizzes that give question-by-question feedback
- Animations and revision podcasts
- Self-assessment checklists

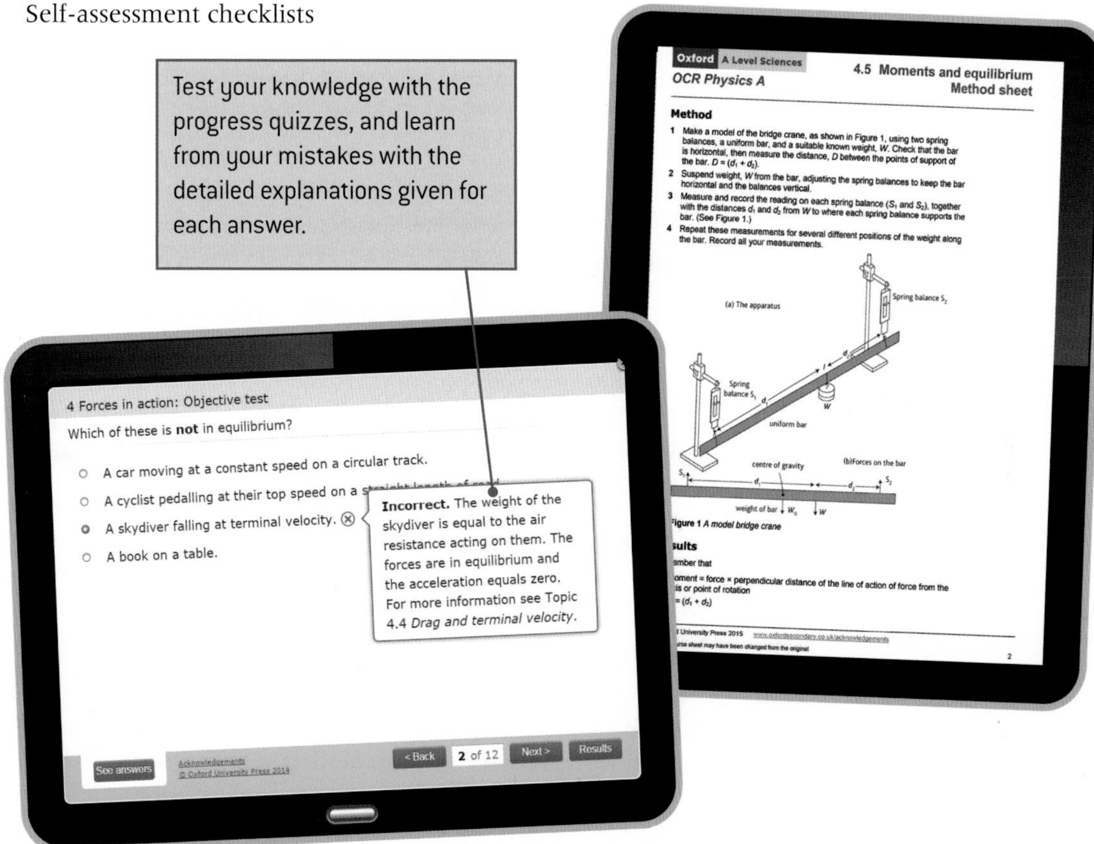


For teachers, Kerboodle also has plenty of further support, including answers to the questions in the book and a digital markbook. There are also full teacher notes for the activities and worksheets, which include suggestions on how to support students and engage them in their own learning. All of the resources are pulled together into teacher guides that suggest a route through each chapter.

# Studying Advancing Physics

▲ **Figure 1** *Rutherford, although critical of stamp collectors, actually appeared on stamps. This stamp shows a representation of the alpha-particle scattering experiment that led to the nuclear model of the atom*

Advancing Physics is about *understanding* the world around us, and appreciating how it is revealed by the experiments we perform and the explanations we study. Ernest Rutherford, one of the finest experimental physicists of all time, is often quoted as saying

*All science is either physics or stamp-collecting.*

Whether or not he actually said this, the statement can be interpreted as suggesting that physics is more than just collecting facts. That is not to say that facts aren't important — it is still useful to *know* fundamental facts when studying physics, such as the fact that an atom is around $1 \times 10^{-10}$ m in diameter, or that the Universe is about 14 billion years old.

Remembering facts is one thing, but understanding ideas is quite another. To use the example of Rutherford, whilst it is useful to remember the diameter of an atom, it is far more interesting to understand how we measured this value. As you progress through the Advancing Physics course you will be encouraged to seek explanations for the statements you are given and to use your understanding of basic physics in a range of different contexts.

## Experiments

Facts do not begin in books: they begin in the world around us. The study of what we now call science really got going when people began to perform controlled experiments and make carefully recorded observations. Experimental work remains a very important part of physics and you will be given many opportunities to develop your practical skills.

Experiments come in a variety of forms — some demonstrate effects or processes, whilst others require careful measurements to investigate relationships. Some experiments aim to find values for fundamental characteristics, such as the refractive index of a transparent material.

Careful experimentation is extremely important. Many advances in science have stemmed from experimental results not quite matching with the expected theoretical value. You should always try to make your practical work as good as it can be — it's vital to be self-critical and estimate the uncertainty in your measurements. Don't think to yourself that *this is good enough*, but rather think *how can this experiment be improved?*

▲ **Figure 2** *The advances in communication technology since the last century are such that the old phone is barely even recognisable anymore*

## Technology

We are in a time of rapid technological change. This is perhaps most clearly seen in the fields of electronics and communication, but technological developments are having an impact on different areas of life on a much wider scale than just mobile phones and tablet computers. Automotive design and digital imaging in the health sciences are just two more examples of the positive impact of technology on our lives.

Much of this development relies on fundamental physics, and this course will highlight some of the links between physics and developing technology.

## Mathematics

Physics is a science that relies upon mathematical reasoning, so much so that many explanations in physics are actually mathematical relationships written in word form. You will be introduced to any new mathematical ideas you need as you meet them. The technique of modelling physical processes such as accelerating bodies, or the decay of radioisotopes, using repetitive or 'iterative' calculations is particularly powerful.

Graphical representations of relationships are extremely useful. Choosing the correct graph to draw and techniques such as estimating the area beneath a graph line or calculating its gradient are skills that will be used and developed throughout the course.

## Asking Questions

Perhaps a more suitable heading would be asking the *right* questions. Simple questions can lead to great developments in our understanding of the world around us. For example, the question 'why does light travel in a straight line?' appears to be so simple that it is hardly worth asking, but answering this question requires a theory (or model) of how light behaves — something that physicists debated for centuries.

▲ **Figure 3** *Light beams travelling in straight lines*

To take another example, Albert Einstein asked himself the question, 'if I jump off a roof why do I feel no force, even though I am accelerating towards the ground?' He later called this the happiest thought of his life and the answer led to his greatest work — the General Theory of Relativity.

## Excitement

Physics is an exciting and challenging subject. Some of the excitement comes from new discoveries or technological developments arising from fundamental physics. We cannot tell how important the development of graphene and carbon-nanotube technologies will be in the future economy, just as we cannot predict what future space missions will reveal about our local Solar System or the greater reaches of space. Then there is another form of excitement — the enjoyment of finding things out yourself, of deepening your understanding of the world around you. The more you understand, the more you will appreciate the new discoveries, and studying physics is a very good place to start.

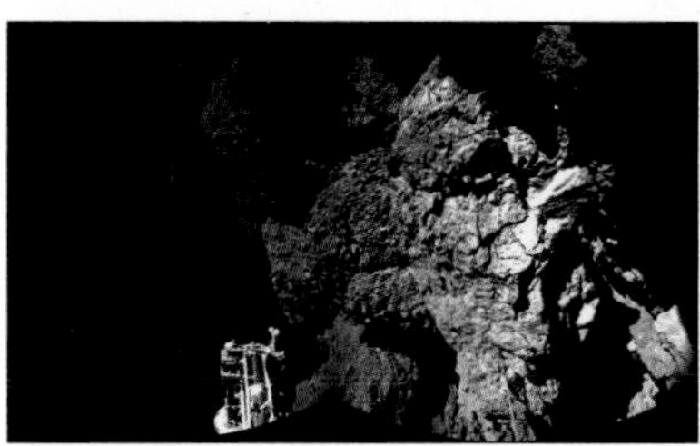

▲ **Figure 4** *Photo of comet Churyumov-Gerasimenko taken in 2014 by Philae, the first human-built spacecraft to successfully land on a comet*

# MODULE 1

## Development of practical skills in physics

Physics is a practical subject and experimental work provides you with important practical skills, as well as enhancing your understanding of physical theory. You will be developing practical skills by carrying out practical and investigative work in the laboratory throughout both the AS and the A Level Physics course. You will be assessed on your practical skills in two different ways:

- written examinations (AS and A level)
- practical endorsement (A level only)

### Practical coverage throughout this book

Practical skills are a fundamental part of a complete education in science, and you are advised to keep a record of your practical work from the start of your A level course that you can later use as part of your practical endorsement. You can find more details of the practical endorsement from your teacher or from the specification.

In this book and its supporting materials practical skills are covered in a number of ways. By studying Application boxes and practice questions in this student book, and by using the Practical activities and Skills sheets in Kerboodle, you will have many opportunities to learn about the scientific method and carry out practical activities.

### 1.1 Practical skills assessed in written examinations

The practical skills of planning, implementing, analysis, and evaluation will be assessed in all the written papers at both AS and A Level. The A level examination paper 3, *Practical skills in physics*, will include longer questions focusing on practical skills and analysis. This section summarises the skills that you need to develop when answering questions in the written papers.

### 1.1.1 Planning

- Designing experiments
- Identifying variables to be controlled
- Evaluating the experimental method

### Skills checklist

- ☐ Selecting apparatus and equipment
- ☐ Selecting appropriate techniques
- ☐ Selecting appropriate quantities of materials and substances and scale of working
- ☐ Solving physical problems in a practical context
- ☐ Applying physics concepts to practical problems
- ☐ Identifying and controlling variables (where appropriate)

## 1.1.2 Implementing

- Using a range of practical apparatus
- Carrying out a range of techniques
- Using appropriate units for measurements
- Recording data and observations in an appropriate format

### Skills checklist

- ☐ Understanding practical techniques and processes
- ☐ Identifying hazards and safe procedures
- ☐ Using SI units
- ☐ Recording qualitative observations accurately
- ☐ Recording a range of quantitative measurements
- ☐ Using the appropriate precision for apparatus

## 1.1.3 Analysis

- Processing, analysing, and interpreting results
- Analysing data using appropriate mathematical skills
- Using significant figures appropriately
- Plotting and interpreting graphs

### Skills checklist

- ☐ Analysing qualitative observations
- ☐ Analysing quantitative experimental data, including
  - calculation of means
  - amount of substance and equations
- ☐ For graphs,
  - selecting and labelling axes with appropriate scales, quantities, and units
  - drawing tangents and measuring gradients
  - representing uncertainty in graphs correctly

## 1.1.4 Evaluation

- Evaluating results to draw conclusions
- Identify anomalies
- Explain limitations in method
- Identifying uncertainties and errors
- Suggesting improvements

### Skills checklist

- ☐ Reaching conclusions from qualitative observations
- ☐ Identifying uncertainties and calculating percentage errors
- ☐ Identifying procedural and measurement errors
- ☐ Refining procedures and measurements to suggest improvements
- ☐ Understanding accuracy and precision

## 1.2 Practical skills assessed in practical endorsement

You will also be assessed on how well you carry out a wide range of practical work and how to record the results of this work. These hands-on skills are divided into 12 categories and form the practical endorsement. This is assessed for the A Level Physics qualification only.

The endorsement requires a range of practical skills from both years of your course. If you are taking only AS Physics, you will not be assessed through the practical endorsement but the written AS examinations will include questions that relate to the skills that naturally form part of the AS common content to the A level course.

The practicals you do as part of the endorsement will not contribute to your final grade awarded to you. However, these practicals must be covered and your teacher will go through how this is to be done in class. It is important that you are actively involved in practical work because it will help you with understanding the theory and also how to effectively answer some of the questions in the written papers.

The practical activities you will carry out in class are divided into Practical Activity Group (PAGs). PAG1 to PAG 6 will be undertaken in Year 1, PAG7 to PAG 10 in Year 2, and PAG11 to PAG 12 throughout the two-year course.

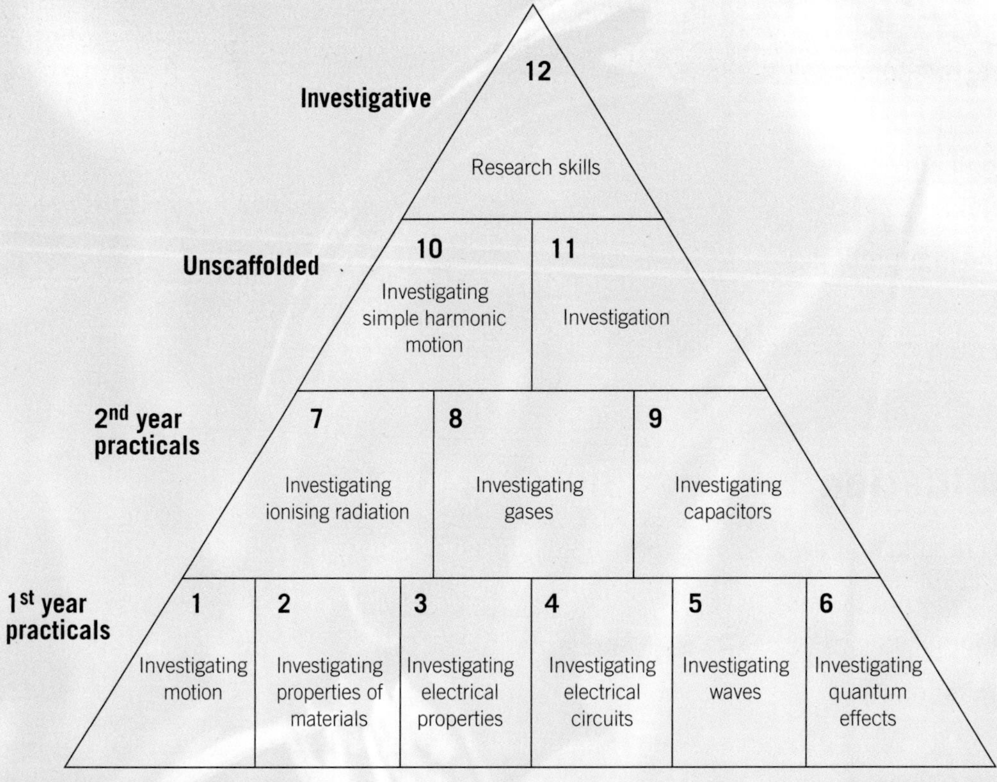

The PAGs are summarised below, together with the topic references in the book that relate to the specific PAG.

| PAG | Topic reference |
|---|---|
| 1 Investigating motion | 8.1 |
| 2 Investigating properties of materials | 4.3 |
| 3 Investigating electrical properties | 3.2, 3.3 |
| 4 Investigating electrical circuits | 3.5 |
| 5 Investigating waves | 6.1, 6.3, 6.4 |
| 6 Investigating quantum effects | 7.1 |
| 7 Investigating ionising radiation | 10.1 |
| 8 Investigating gases | 14.1 |
| 9 Investigating capacitors | 10.4 |
| 10 Investigating simple harmonic motion | 11.3, 11.4 |
| 11 Investigation | Throughout |
| 12 Research skills | Throughout |

## Maths skills and How Science Works across Module 1

In order to develop your knowledge and understanding in A Level Physics, it is important to have specific skills in mathematics. All the mathematical skills you will need during your physics course have been embedded into the individual topics for you to learn as you meet them. An overview is available in each of the module openers and these skills are further supported by the worked examples, summary questions, practice questions and the Maths appendix.

How Science Works (HSW) is another area required for success in A Level Physics, and helps you to put science in a wider context, helping you to develop your critical and creative thinking skills in order to solve problems in a variety of contexts. Once again, this has been embedded into the individual topics covered in the books, particularly in application boxes and examination-style questions. The application and extension boxes cover some of the HSW elements.

You can find further support for maths skills and HSW on Kerboodle.

## Base units

When you make a measurement, or calculate a physical quantity, you need both a number *and* a unit. In the **SI system** of units (SI comes from *Le Système International d'Unités*), there are seven **base units** from which all the others are derived. The base units are the metre (m), kilogram (kg), second (s), ampere (A), kelvin (K), candela (cd), and mole (mol). They are the units of distance, mass, time, electric current, temperature, luminosity, and amount of substance, respectively.

The units of other quantities can all be expressed in base units, though we tend to use **derived units**.

▼ **Table 1** *Equivalent and derived units*

| Quantity | Can be calculated with | Equivalent to | Derived unit |
|---|---|---|---|
| force | mass × acceleration | $kg\,m\,s^{-2}$ | newton N |
| energy | force × distance | N m | joule J |
| power | energy per second | $J\,s^{-1}$ | watt W |
| charge | current × time | A s | coulomb C |
| potential difference | energy per coulomb | $J\,C^{-1}$ | volt V |

Note that m/s and $m\,s^{-1}$ are equivalent, but it is better to write $m\,s^{-1}$ at A Level.

## Checking equations with units

You can use units to check that equations make physical sense. The units on both sides of an equation must match. If they do not, the equation is wrong.

### Hint

When you put quantities into an equation, write the number *and* the unit. This can help you to work out the unit of the quantity that you are calculating.

### Worked example: Power = *IV*

Power is measured in watts and $1\,W = 1\,J\,s^{-1}$. Show that the units of $IV$ are also $J\,s^{-1}$.

**Step 1:** Lay out the units for $I \times V$.

Units of $IV = A\,J\,C^{-1}$

**Step 2:** Since the coulomb C = A s, this can be substituted into the equation above.

Units of $IV = A\,J\,A^{-1}\,s^{-1}$

**Step 3:** Cancel out terms where possible.

Units of $IV = J\,s^{-1}$ This is the same as the unit of power.

## Radian measure

Although you can define an angle using degrees (°), where there are 360° in a circle, you can also define an angle by the ratio of the arc length to the radius. If the arc length is equal to the radius then the angle at the centre is equal to one **radian** ($^c$ or rad). When the arc length equals the circumference then the angle at the centre is 360°, or $2\pi^c$. So $180° = \pi$ radians, and $90° = \frac{\pi}{2}$ radians. This means that 1 radian $= \frac{180}{\pi}$.

$$\text{angle (radians)} = \frac{\text{arc length (m)}}{\text{radius (m)}}$$

### Synoptic link

The use of radians is important when discussing phasors, which you will meet in Topic 6.1, Superposition of waves.

### Worked example: Radians and degrees

A circle has a radius of 0.5 m. Calculate the angle at the centre if the arc length is 1.0 m. Convert the angle to degrees.

**Step 1:** Use the equation to calculate the angle.

$$\text{angle (radians)} = \frac{\text{arc length (m)}}{\text{radius (m)}}$$

$$\text{angle} = \frac{1.0\,\text{m}}{0.5\,\text{m}} = 2 \text{ radians}$$

**Step 2:** Convert from radians to degrees using 1 radian $= \frac{180}{\pi}$.

$$2 \text{ radians} = 2 \times \frac{180}{\pi} = 110° \text{ (2 s.f.)}$$

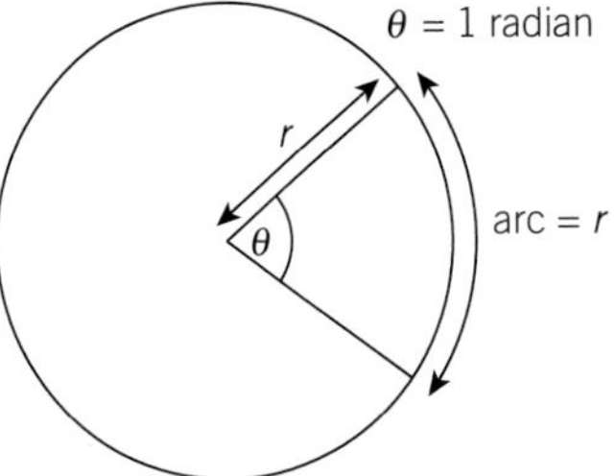


▲ **Figure 1** *An arc length equal to r means the angle is one radian*

## Standard form

In physics, numbers can be very large, like the distance from the Earth to the Sun, or very small, like the size of an atom. In **standard form** (also called scientific notation) you can write numbers with one digit in front of the decimal place, multiplied by the appropriate power of ten, positive or negative.

For example, 1000 m can be written $1.0 \times 10^3$ m, 0.1 s = $1.0 \times 10^{-1}$ s, 20 000 Hz = $2 \times 10^4$ Hz, and 0.0005 kg = $5 \times 10^{-4}$ kg. Note that $1.0 \times 10^3$ m is the same as $10^3$ m. This means that:

- the distance from the Earth to the Sun = 150 000 000 000 m = $1.5 \times 10^{11}$ m
- the diameter of an atom is 0.000 000 000 1 m = $1 \times 10^{-10}$ m.

You will see many physical constants written in standard form, for example, the speed of light = $3 \times 10^8\,\text{m}\,\text{s}^{-1}$.

▲ **Figure 2** *The Sun is $1.5 \times 10^{11}$ m from the Earth*

### Multiplying numbers in standard form

When you multiply two numbers in standard form together you *add* the powers of ten. When you divide two numbers you *subtract* the powers of ten of the second number from that of the first.

- If the powers *add* when you multiply then $10^2 \times 10^3 = 10^5$ or 100 000.
- If the powers *subtract* when you divide then $10^2 \div 10^4 = 10^{-2}$.

▲ **Figure 3** *You need a scientific calculator to do calculations involving standard form*

## Hint

When you enter a number in standard form into your calculator, such as $3 \times 10^{-3}$, you just need to press 3, EXP, +/–, 3. You do not need to press the 1 or 0 buttons.

Scientific calculators have a button that you use when you are calculating with numbers in standard form. Work out which button you need to use (it could be EE, EXP, $10^x$, $\times 10^x$).

### Worked example: How many?

A full library contains 200 000 books. Each book contains 400 pages. Calculate the total number of pages. The total length of the shelves in the library = 4800 m. Calculate the thickness of a page.

**Step 1:** Convert the numbers to standard form.

$$200\,000 = 2 \times 10^5,\ 400 = 4 \times 10^2$$

**Step 2:** Multiply the numbers and work out the answer by adding the powers.

Number of pages = number of books × pages per book

$$= 2 \times 10^5 \times 4 \times 10^2 = (2 \times 4) \times (10^5 \times 10^2)$$
$$= 8 \times 10^7 \text{ pages}$$

**Step 3:** Divide the numbers and work out the answer by subtracting the powers.

$$\text{Width of a page} = \frac{\text{width of all the books}}{\text{number of pages in all the books}} = \frac{4.8 \times 10^3}{8 \times 10^7}$$

$= \frac{4.8}{8} \times \frac{10^3}{10^7} = 0.6 \times 10^{-4}$ m because you subtract the powers: 3 – 7 = –4. This can be written as $= 6 \times 10^{-5}$ m.

## Metric prefixes

You can use various **metric prefixes** to show large or small multiples of a particular unit. Here the unit is the metre. For example, 3 nanometres $= 3 \times 10^{-9}$ m. Most of the prefixes that you will use in physics involve multiples of $10^3$.

▼ **Table 2** *The most common prefixes*

| T | G | M | k | c | m | µ | n | p | f |
|---|---|---|---|---|---|---|---|---|---|
| tera | giga | mega | kilo | centi | milli | micro | nano | pico | femto |
| $10^{12}$ | $10^{9}$ | $10^{6}$ | $10^{3}$ | $10^{-2}$ | $10^{-3}$ | $10^{-6}$ | $10^{-9}$ | $10^{-12}$ | $10^{-15}$ |

### Converting between units

It is helpful to use standard form when you are converting between units. You need to think about how many of the smaller units are in the bigger one.

- There are 1000 mm in 1 m. So 1 mm $= \frac{1}{1000}$ m $= 10^{-3}$ m.
- There are 1000 mm × 1000 mm = $10^6$ mm$^2$ in 1 m$^2$. So 1 mm$^2$ = $10^{-6}$ m$^2$.
- There are 100 cm × 100 cm × 100 cm = $10^6$ cm$^3$ in 1 m$^3$. So 1 cm$^3$ = $10^{-6}$ m$^3$.

### Worked example: Cross-sectional area of a wire

A wire has a diameter of 0.5 mm. Calculate the cross-sectional area using $A = \pi r^2$.

**Step 1:** Convert mm to m using standard form.

$$\text{Diameter } d = 0.5 \times 10^{-3}\,\text{m} = 5 \times 10^{-4}\,\text{m}$$

Radius $r = 2.5 \times 10^{-4}$ m, or you could simply use $0.25 \times 10^{-3}$ m

**Step 2:** Substitute the number in standard form into the equation.

$$A = \pi r^2 = \pi \times (2.5 \times 10^{-4}\,\text{m})^2 = 2 \times 10^{-7}\,\text{m}^2 \text{ (1 s.f.)}$$

#### Hint

When carrying out multiplications or divisions using standard form, add or subtract the powers of ten to work out roughly what you expect the answer to be. This will help you to avoid mistakes.

## Uncertainty

There are two main ways to estimate the **uncertainty** of a measurement:

- repeat it many times and make an estimate from the variation you get
- look at the process of measurement used, and inspect and test the instruments used.

Usually, you should focus mainly on the second way, which is the process of measuring, and on the qualities of the instruments you have. The main reason for being interested in the quality of a measurement is to try to do better.

### Properties of instruments

The essential qualities and limitations of measuring instruments are:

- **resolution** – the smallest detectable change in input, for example, 1 mm on a standard ruler
- **sensitivity** – the ratio of output to input, for example, the change in potential difference across a thermistor when the temperature changes by 1 °C
- stability (**repeatability, reproducibility**) – the extent to which repeated measurements give the same result, including gradual change with time (drift)
- **response time** – the time interval between a change in input and the corresponding change in output, for example, how long it takes a temperature sensor to respond when you put it in hot water
- **zero error** – the output for zero input, for example, a newtonmeter that reads 0.1 N when there is no force acting
- **noise** – variations, which may be random, superimposed on a signal, for example, changes to a reading on a temperature sensor as someone opens a door
- **calibration** – determining the relation between output and true input value, including linearity of the relationship, for example, the relationship between the resistance of a thermistor (in Ω) and the temperature (in °C).

▲ **Figure 4** *The resolution of this multimeter is ±0.01 V*

## Estimating uncertainty

Uncertainty can be estimated in several ways:

- from the resolution of the instrument concerned. For example, the readout of a digital instrument ought not to be trusted to better than ±1 in the last digit
- from the stability of the instrument, or by making deliberate small changes in conditions (a tap on the bench, maybe) that might anyway occur, to see what difference they make
- by trying another instrument, even if supposedly identical, to see how the values they give compare
- from the range of some repeated measurements.

When comparing uncertainties in different quantities, it is the **percentage uncertainties** that need to be compared, to identify the largest. You can make the biggest improvement to your measurements by trying to reduce the largest uncertainty.

### Worked example: Calculating percentage uncertainty

The meter is reading 12.51 V, and the uncertainty in the reading is ±0.01 V (±1 in the last digit). Calculate the percentage uncertainty of this reading.

**Step 1:** Identify the equation for percentage uncertainty.

$$\text{The percentage uncertainty} = \frac{\text{uncertainty in reading} \times 100\%}{\text{actual reading}}$$

**Step 2:** Substitute values.

$$\text{Percentage uncertainty} = \frac{0.01\,\text{V} \times 100\%}{12.51\,\text{V}} = 0.08\%\ (1\ \text{s.f.})$$

## Analysing uncertainties

The final uncertainty in an answer depends on how quantities are combined. Here are three important rules about the way uncertainties propagate.

1 **Adding or subtracting quantities**

When you add or subtract quantities in an equation, you add the absolute uncertainties for each value.

### What is the extension?

The original length of a spring is 2.5 ± 0.1 cm and the final length is 15.0 ± 0.2 cm. Calculate the extension of the spring and the absolute uncertainty.

**Step 1:** Calculate the extension by subtracting the lengths.

extension = 15.0 – 2.5 = 12.5 cm

**Step 2:** Add the absolute uncertainties.

absolute uncertainty = 0.1 + 0.2 = 0.3

**Step 3:** Write the answer in the normal convention.

extension = 12.5 ± 0.3 mm

**2 Multiplying or dividing quantities**

When you multiply or divide quantities, you add the percentage uncertainties for each value.

## What is the resistance?

The current $I$ in a resistor is 1.60 ± 0.02 A and the potential difference $V$ across the resistor is 6.00 ± 0.20 V. Calculate the resistance and the absolute uncertainty.

**Step 1:** Calculate the resistance $R$ of the resistor.

$$R = \frac{V}{I} = \frac{6.00}{1.60} = 3.75\,\Omega$$

**Step 2:** Calculate the percentage uncertainty in each measurement.

$$\%\text{ uncertainty in } I = \frac{0.02}{1.60} \times 100 = 1.25\%$$

$$\%\text{ uncertainty in } V = \frac{0.20}{6.00} \times 100 = 3.33\%$$

**Step 3:** Add the percentage uncertainties.

$$\%\text{ uncertainty in } R = 1.25 + 3.33 = 4.58\%$$

**Step 4:** Calculate the absolute uncertainty in $R$.

absolute uncertainty in $R = 0.0458 \times 3.75 = 0.2\,\Omega$ (1 s.f.)

**Step 5:** The uncertainty in $R$ limits the precision with which the final answer for resistance can be given. The final answer should be rounded to the same precision, which is one decimal place.

$R = 3.8 \pm 0.2\,\Omega$ (1 d.p.)

**3 Raising a quantity to a power**

When a measurement in a calculation is raised to a power $n$, your percentage uncertainty is increased $n$ times. The power $n$ can be an integer or a fraction.

## Cross-sectional area of a wire

The diameter of a wire is recorded as 0.51 ± 0.02 mm. Calculate the cross-sectional area of the wire and the absolute uncertainty.

**Step 1:** Calculate the cross-sectional area $A$ of the wire.

$$A = \frac{\pi d^2}{4} = \frac{\pi \times (0.51 \times 10^{-3})^2}{4} = 2.04 \times 10^{-7}\,\text{m}^2$$

**Step 2:** The percentage uncertainty in $A$ is equal to 2 times the percentage uncertainty in $d$.
(The $\pi$ and the 4 are numbers and therefore have no uncertainty associated with them.)

$$\%\text{ uncertainty in } A = 2 \times \left(\frac{0.02}{0.51} \times 100\right) = 7.84\%$$

**Step 3:** Calculate the absolute uncertainty in $A$.

$$\text{absolute uncertainty in } A = 0.0784 \times 2.04 \times 10^{-7} = 0.16 \times 10^{-7}\,\text{m}^2$$

**Step 4:** The diameter of the wire is quoted to 2 significant figures, therefore the final answer for the cross-sectional area must also be written to 2 significant figures.

$$A = (2.0 \pm 0.2) \times 10^{-7}\,\text{m}^2 \text{ (2 s.f.)}$$

## Why results vary

There are different kinds of *variation, uncertainty,* or *error*:

- inherent *variation* in the measured quantity, for example, variation in the value amongst a set of 'identical' commercial resistors
- small (maybe random) uncontrollable variations in conditions, including noise, leading to *uncertainty*
- simple mistakes, for example, misreading a scale, or a *one-off* accidental *error*, which needs to be detected and removed – **outliers** often turn out to be due to such mistakes
- systematic *error* or bias, for example, reading a scale at an angle – this may show up as an intercept on a suitable graph
- a genuine outlying value, which may be an *error*, or may be significant for another reason.

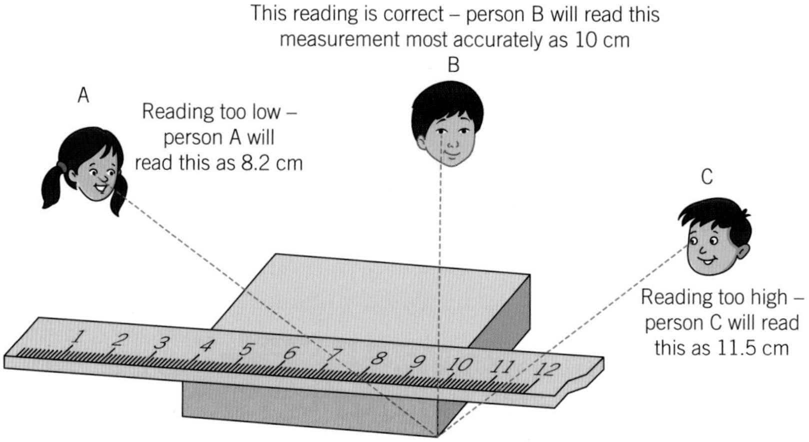

This reading is correct – person B will read this measurement most accurately as 10 cm

B

A

Reading too low – person A will read this as 8.2 cm

C

Reading too high – person C will read this as 11.5 cm

1 2 3 4 5 6 7 8 9 10 11 12

▲ **Figure 5** *If you always read a scale at an angle you can introduce a systematic error to your measurements*

### Dot plots

Suppose you measure the resistance of resistors that all come out of a packet labelled 47 Ω. There will be a variation in the value of resistance that you measure. You can display the values on a simple **dot plot**, such as the one in Figure 6.

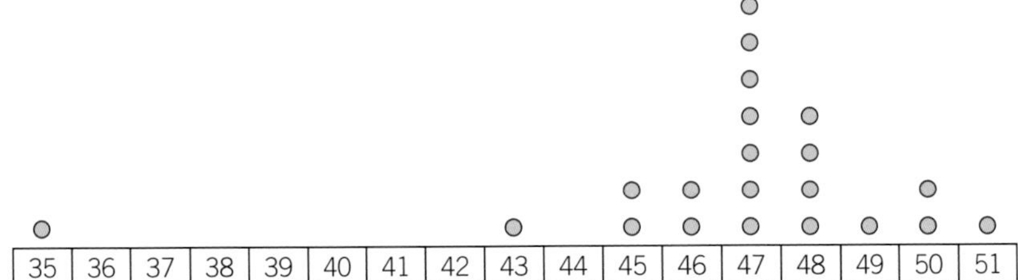


▲ **Figure 6** *A dot plot can show the spread of results*

You can estimate the uncertainty of a measurement in terms of the **range** of typical values, excluding outliers. Here the value of 35 Ω is probably an outlier. The **spread** is then

$$\text{spread} = \pm\frac{1}{2}\text{range}$$

A good rule of thumb is that a value is likely to be an outlier if it lies more than 2 × spread from the **mean**, but this is only a rule of thumb. There may be a reason that you have that reading.

To calculate the mean of a set of numbers you need to add up all the values and divide by the number of values.

#### Worked example: Estimating uncertainty from the range

Use the dot plot in Figure 6 to find the mean resistance, the spread, and the percentage uncertainty.

**Step 1:** Identify any outliers.

35 Ω is an outlier.

**Step 2:** Calculate the mean.

$$\text{Mean} = (43 + 2 \times 45 + 2 \times 46 + 7 \times 47 + 4 \times 48 + 49 + 2 \times 50 + 51)\,\Omega \div 20 = \frac{946}{20}\,\Omega = 47\,\Omega \text{ (2 s.f.)}$$

**Step 3:** Calculate the spread.

$$\text{Spread} = \pm\frac{1}{2}(51\,\Omega - 43\,\Omega) = \pm\,4\,\Omega$$

**Step 4:** Calculate the percentage uncertainty.

Measurement of resistance = 47 Ω ±4 Ω

$$\text{Percentage uncertainty} = \frac{\text{uncertainty} \times 100\%}{\text{reading}} = \frac{4\,\Omega \times 100\%}{47\,\Omega} = 9\% \text{ (1 s.f)}$$

**Hint**

You usually use the mean of a set of results, but there may be occasions when the mean is not representative – for example, when there are results that are far apart from the rest but not clearly outliers.

**Hint**

Don't just completely discard outliers – you should investigate them carefully and see if there's a reason for them.

### Graphs and uncertainty

The uncertainty in a measurement can be used to give a small range or **uncertainty bar** for each measurement. Instead of plotting just the points on a graph, you can plot an uncertainty bar for all of your measurements.

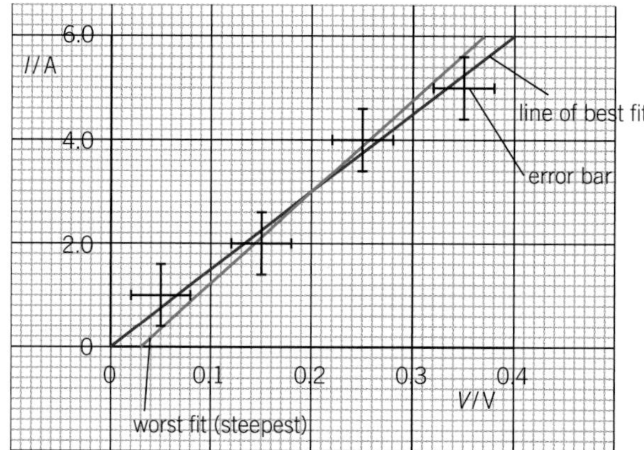


▲ **Figure 7** *Uncertainty bars are useful when you draw the line of best fit and the worst line for your measurements*

Your straight best fit line must pass through all the uncertainty bars (Figure 7). You would use this line to determine the value of the gradient. How can you determine an approximate value for the uncertainty in the gradient? You would draw the **line of worst fit** – the least acceptable straight line through the data points – this can either be the steepest or the shallowest line.

The absolute uncertainty in the gradient is the positive difference between the gradient of the line of best fit and the gradient of the line of worst fit.

The percentage uncertainty in the gradient can be calculated as

$$\% \text{ uncertainty in gradient} = \frac{\text{absolute uncertainty}}{\text{gradient best fit line}} \times 100\%$$

## Precision and accuracy

In everyday life, people often use the words *precise* and *accurate* to mean the same thing – this is not the case in physics.

- **Accuracy** is to do with how close a measurement result is to the true value – the closer it is, the more **accurate** it is.
- **Precision** is to do with how close repeated measurements are to each other – the closer they are to each other, the more **precise** the measurement is.

Figure 8 is a visual way of appreciating the terms accuracy and precision using a dartboard as an example. You are aiming for bullseye – it represents the true value.

not accurate
not precise

accurate
not precise

not accurate
precise

accurate
precise

▲ **Figure 8** *Accuracy and precision*

### Estimations

Sometimes an exact answer is absolutely necessary, such as when calculating the speed of a spacecraft at various stages on a trip to the

Moon. Sometimes a rough estimate is helpful. It can help you to work out if a calculation is correct, or where you expect an answer to be. A simple estimate is an **order of magnitude** estimate, which is an estimate to the nearest power of 10. For example, to the nearest power of 10 you are 1 m tall, and can run $10\,m\,s^{-1}$. You may also simplify the question by imagining objects as squares or cubes, for example, when estimating how many atoms there are in a given volume.

You, your desk, and your chair are all of the order of 1 m tall.

▲ **Figure 9** *Order of magnitude of 1 m*

## Worked example: Uncertainties and significant figures

The radius of a circle $r = 1.3 \pm 0.1$ m. Estimate the area of the circle. Calculate the area of the circle and estimate the uncertainty in the area. Justify the number of significant figures in your answer.

**Step 1:** Estimate the area by using approximations.

The area = $\pi r^2$. Assume that $\pi$ approximately equals 3. So the area lies between $3 \times (1\,m)^2 = 3\,m^2$ and $3 \times (2\,m)^2 = 12\,m^2$. 1.3 m is closer to 1 than to 2, so the area is closer to $3\,m^2$ than $12\,m^2$. Therefore, $5\,m^2$ is a reasonable estimate.

**Step 2:** Calculate the area of the circle using the measurement.

The area of the circle = $\pi r^2 = \pi(1.3\,m)^2 = 5.30929218...\,m^2$

**Step 3:** Use the uncertainty to identify how big or small the measurement could be.

The radius could be as big as 1.3 m + 0.1 m = 1.4 m, or as small as 1.3 m − 0.1 m = 1.2 m.

**Step 4:** Calculate the area using these numbers.

The area of the circle = $\pi r^2 = \pi(1.4\,m)^2 = 6.157521601...m^2$. This is bigger by about $0.8\,m^2$. (Note that you would not write down a number with this number of significant figures for any answer.)

The area of the circle = $\pi r^2 = \pi(1.2\,m)^2 = 4.523893421...m^2$. This is smaller by just under $0.8\,m^2$.

**Step 5:** Use the largest and smallest values to estimate the uncertainty.

The area lies somewhere between $4.5\,m^2$ and $6.1\,m^2$, so the uncertainly = $\pm 0.8\,m^2$ and area = $5.3 \pm 0.8\,m^2$.

**Step 6:** Justify the number of significant figures in your answer.

We are effectively using two significant figures for the answer, which is what we were given in the question. All the numbers after the 3 are not significant because of the value of the uncertainty.

### Hint

You should only write your final answer using a justifiable number of significant figures. If you need to use one part of a question to calculate something in another part of the question, you should use the unrounded number in your calculator, and then round the answer to the same number of significant figures as before.

# MODULE 3.1

## Imaging, signalling and sensing

### Chapters in this Module



### Introduction

This module introduces you to some fundamental ideas in physics and shows how they are used in a variety of situations. It gives you the first taste of a very useful rule — learn the fundamental ideas in detail and you will be able to apply them to all kinds of situations or contexts. For example, the physics of the simple lens can lead to a better understanding of imaging by satellite. Similarly, understanding the humble potential divider can help you appreciate how all kinds of sensor systems work. You will also be introduced to some of the mathematical and experimental skills you will use throughout the course.

**Imaging** begins by looking at a simple converging lens, the sort that has been used for hundreds of years in reading glasses and magnifying glasses. We then consider how the image is captured as an array of numbers. This is a very important area of technology to understand as nearly all images are now stored digitally as an array of numbers. This allows manipulation and editing of the images, a feature useful in the word of medical imaging as well as simple photo manipulation and editing.

**Signalling** looks at how information is sampled and represented by a chain of digital data. This means information from images, text, music, and more can be transmitted quickly across the room, across the country or across space. This will give you an appreciation of the basic principles used in video and music streaming, digital radio, and digital television. Nearly all modern communication technology depends on digital transmission — imagine the world without the Internet.

**Sensing** introduces you to electrical circuits and some basic ideas such as internal resistance. You will learn to use potential dividers to make simple sensors for detecting light level, temperature, and other variables. You will investigate the sensitivity and resolution of sensors, which are important features when choosing a sensor for a particular task. These basic sensors are at the core of many, more complex circuits, and often produce data that is stored and transmitted digitally.

## Knowledge and understanding checklist

**From your Key Stage 4 study you should be able to do the following. Work through each point, using your Key Stage 4 notes and the support available on Kerboodle.**

- [ ] Describe the difference between a transverse and a longitudinal wave and recall that electromagnetic waves are transverse.
- [ ] Recall that different substances may refract waves.
- [ ] Recall that current is a rate of flow of charge.
- [ ] Recall the units for current, resistance and potential difference and apply the relationship $V = IR$.
- [ ] Describe the difference between series and parallel circuits.
- [ ] Calculate the current, potential difference and resistance in d.c. series circuits.
- [ ] Recognise and use common circuit symbols including diodes, LDRs and thermistors.

## Maths skills checklist

**Maths is a vitally important aspect of Physics. In this unit, you will need to use the following maths skills. You can find support for these skills on Kerboodle and through MyMaths.**

- [ ] **Change the subject of non-linear equations** when using the lens equation in this chapter.
- [ ] **Use a calculator to find logs** in calculations involving the number of bits of data required to code images and signals.
- [ ] **Sketching graphs of relationships** in studying the relationship between resistance and temperature for components such as thermistors.
- [ ] **Calculate cross-sectional areas** of wires when investigating conductivity and resistivity.
- [ ] **Use appropriate numbers of significant figures** such as in experimental work in Sensing — you will need to think carefully about how many significant figures to use in your calculations.
- [ ] **Make calculations using the appropriate units** such as in calculations involving quantities like coulombs, newtons or metres. Using units correctly is an important part of physics.
- [ ] **Rearranging and solving algebraic equations** in many situations throughout the course

MyMaths.co.uk
Bringing Maths Alive

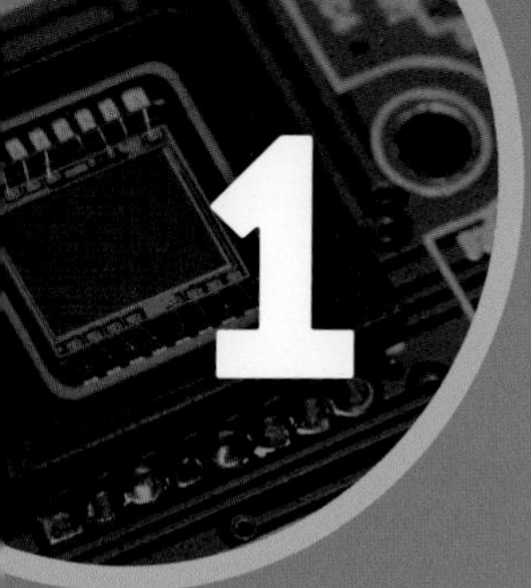

# IMAGING

## 1.1 Bending light with lenses

Specification references: 3.1.1a(i), 3.1.1b(i), 3.1.1b(ii), 3.1.1c(ii)

**Learning outcomes**

Describe, explain, and apply:

- how a lens changes the curvature of the incident wave-front
- the terms: focal length, power
- the equation for power of a converging lens $P = \frac{1}{f}$
- diagrams of the passage of light through a converging lens.

What do you notice when you throw a stone into the calm water of a pond? Circular ripples spread out from the point where the stone enters the water. These ripples are examples of **wave-fronts.** A wave-front can be thought of as a line of disturbance moving through a material or through space. In the case of the stone in a pond, the wave-front is a ripple moving through the water. The distance between two consecutive wave-fronts is the wavelength.

Wave-fronts in water give us a useful way of thinking about light, as light spreads out from a small source in a similar manner to the ripples spreading out on a pond (Figure 1) – although light wave-fronts will spread out as three-dimensional spheres.

▲ **Figure 1** *The curvature of ripples and wave-fronts becomes less as the distance from the source increases*

### Wave-fronts and curvature

The curvature of the circumference of a circle is the reciprocal of its radius. For a circle of radius $r$ the curvature is $\frac{1}{r}$.

As $r \rightarrow \infty$, $\frac{1}{r} \rightarrow 0$. This means that, as the distance from the source increases, the curvature of the wave-fronts decreases. Wave-fronts from a very distant source will appear to not curve at all. These are often called **plane wave-fronts.**

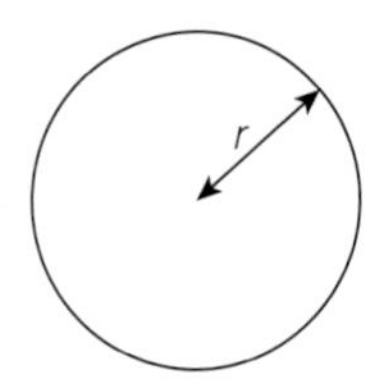


▲ **Figure 2** *The curvature of the circumference of this circle is $\frac{1}{r}$*

**Worked example: Curvature**

Calculate the curvature of a wave emitted from a point 1.5 m away.

**Step 1:** Select the appropriate equation.

$$\text{curvature of wave} = \frac{1}{r}$$

**Step 2:** Substitute values and evaluate.

$$\text{curvature of wave} = \frac{1}{1.5\,\text{m}} = 0.67\,\text{m}^{-1}\ (2\ \text{s.f.})$$

**Study tip**

Answers cannot be more precise than the data provided in a question. If the distance from the point source in this example is given as 1.50 m to three significant figures (s.f.), your answer would be 0.667 $\text{m}^{-1}$ (also to three s.f.).

### Light as rays and wave-fronts

Light can be thought of as wave-fronts or rays (Figure 3). These ways of thinking about light are connected. A ray of light points along the direction of motion of the wave-front and is always at right angles to the wave-front.

**Ray point of view**

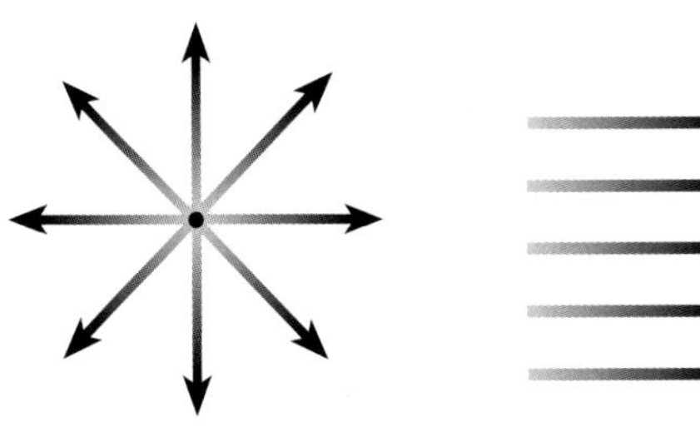

Light travels in straight lines from a small source.

Light in a parallel beam, or from a very distant light source, has rays (approximately) parallel to one another.

**Wave point of view**

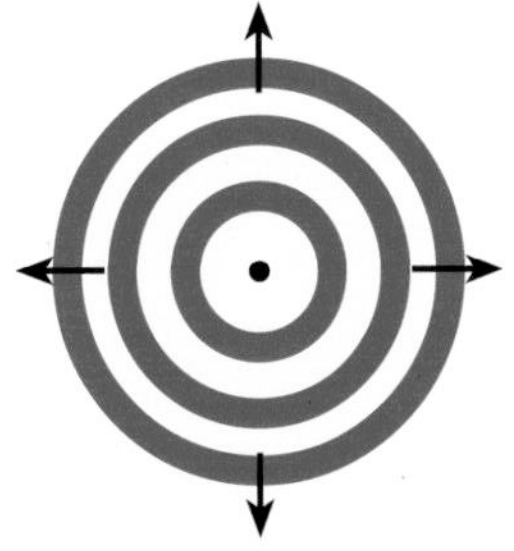

Light spreads out in spherical wave-fronts from a small source.

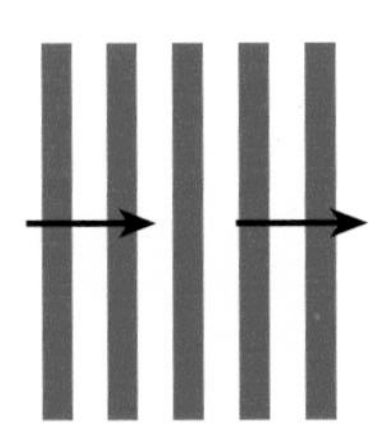

Wave-fronts in a parallel beam, or from a very distant light source, are straight (not curved) and parallel to one another.

▲ **Figure 3** *Rays and waves – two ways of picturing light*

## Light through a converging lens

A converging lens is a lens that focuses light to a point behind it. It can be used to concentrate light rays as a *burning glass*. This concentrates sunlight striking the surface of the lens into a tiny image of the Sun. The intensity of the light at this point, the **focus** of the lens, can be great enough to start a fire.

In terms of rays, the burning glass works by bending the parallel rays from the Sun and focusing them together at the focus. In wave terms, the lens works by altering the curvature of the waves, changing the plane waves that strike it from the Sun into spherical ripples that converge on a focus (see Figure 4).

The distance from the lens to the focus is the radius of the wave-fronts just after passing through the lens. This is the focal length, $f$. Powerful lenses have small focal lengths, for example, lenses in smartphone cameras have focal lengths of about 4 mm.

A converging lens adds a curvature of $\frac{1}{f}$ to wave-fronts passing through the lens.

$\frac{1}{f}$ is the **power** of the lens. This is measured in dioptres (D).

$$\text{Lens power (D)} = \frac{1}{\text{focal length } f \text{ (m)}}$$

**Wave point of view:**
The lens adds curvature to the waves, centering them on the focus.

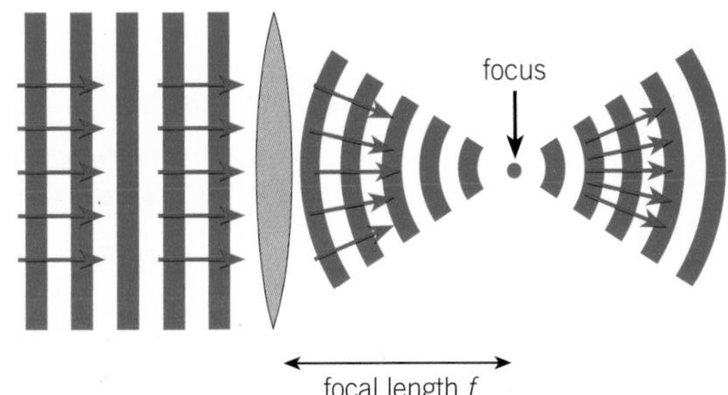


**Ray point of view:**
The lens bends the rays, bringing them to a focus.

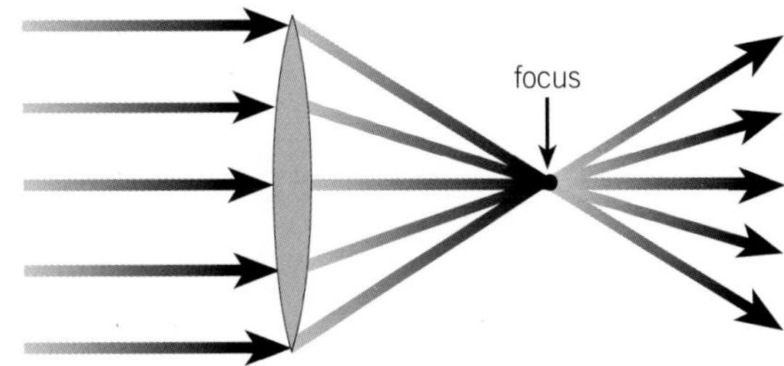


▲ **Figure 4** *Rays and waves – two ways of representing the burning glass*

### Worked example: Calculating lens power

Calculate the power of a lens with a focal length of 4.0 mm.

**Step 1:** Identify the equation to use.

$$\text{Lens power} = \frac{1}{f}$$

**Step 2:** Convert focal length into metres.

$$4.0\,\text{mm} = 4.0 \times 10^{-3}\,\text{m}$$

**Step 3:** Substitute value into the equation in Step 1, and evaluate.

$$\text{Lens power} = \frac{1}{4.0 \times 10^{-3}} = 250\,\text{D}$$

### Study tip

1 D is equivalent to 1 $m^{-1}$

### Study tip

Remember to always check the units before performing a calculation. The formula for curvature uses radius in metres.

## Summary questions

**1** A wave-front has a radius of 32 cm. Calculate the curvature of the wave-front. *(1 mark)*

**2** A camera lens has a focal length of 50 mm.

**a** Calculate the lens power in dioptres. *(2 marks)*

**b** State the additional curvature the lens adds to a wave-front as it passes through the lens. *(1 mark)*

**3** A lens has a power of 7.4 D. Calculate its focal length in millimetres. *(2 marks)*

**4** **a** Draw a diagram of plane wave-fronts passing through a converging lens towards the focal point. *(2 marks)*

**b** Draw a diagram showing plane wave-fronts of the same wavelength (spacing between wave-fronts) passing through a lens of half the power of the lens in **a**. *(2 marks)*

# 1.2 Finding the image

## Specification references: 3.1.1a(i), 3.1.1b(i), 3.1.1c(iii), 3.1.1c(iv), 3.1.1d(i)

A converging lens makes an image of the source of light (the object) because light from every part of the object goes through the lens. The cornea and the lens of the eye focus light from an object onto the retina. The image on the retina is upside down as rays from the top of the object are focused at the bottom of the image and vice versa (Figure 1). The brain flips the image so you *see* the world the right way up.

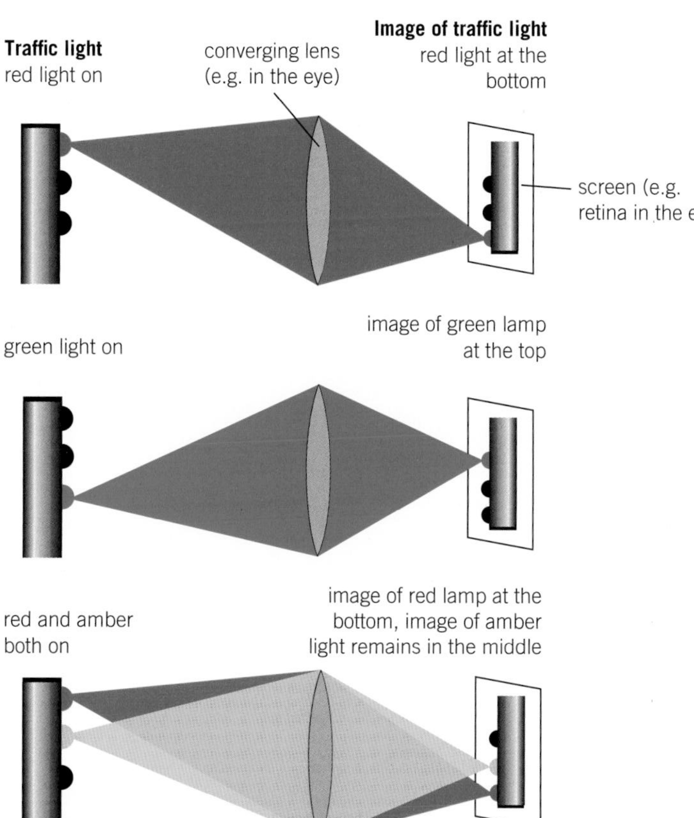


▲ **Figure 1** *How a lens makes an image*

### Learning outcomes

Describe, explain, and apply:

- → the formation of a real image by a thin converging lens
- → the term: magnification
- → the equation $\frac{1}{v} = \frac{1}{u} + \frac{1}{f}$
- → the equation for linear magnification $m = \frac{\text{image height}}{\text{object height}} = \frac{v}{u}$
- → an experiment to determine the power and focal length of a converging lens.

### Synoptic link

You have met how the curvature of a wave can be calculated in Topic 1.1, Bending light with lenses.

## Making images

The rule for how a lens shapes light is simple.

**Curvature of waves leaving the lens** = **curvature of waves entering the lens** + **curvature added by the lens**

The distance from the lens to the image of the source (the image distance) is represented by $v$. The distance from the lens to the source (the object distance) is represented by $u$.

After passing through the lens, the waves form part of spheres centred on a point at distance $v$. So the curvature of the waves leaving the lens $= \frac{1}{v}$. The curvature of the waves entering the lens $= \frac{1}{u}$.

**Study tip**

When using this equation, take care with negative signs. Always measure from the lens, counting distances to the right of the lens (or the side of the image) as *positive*, and distances to the left (or the side of the object) as *negative*.

We can translate the rule for shaping light into a rule for the distances of the object and image from the lens:

$$\frac{1}{v} = \frac{1}{u} + \frac{1}{f}$$

## Worked example: Using the lens equation

**a** Calculate the curvature of the wave-fronts before and after passing through the lens, and use your results to calculate the power of the lens.

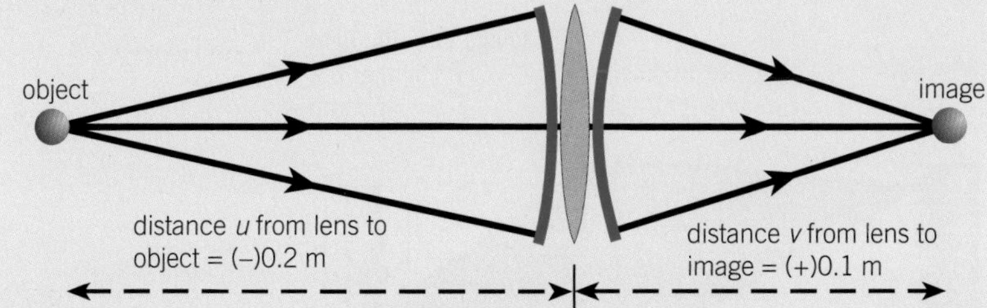


▲ Figure 2

**Step 1:** Calculate the curvature of the wave-fronts after passing through the lens:

$$\text{curvature} = \frac{1}{v} = \frac{1}{0.1\,\text{m}} = 10\,\text{D}$$

**Step 2:** Calculate the curvature of the wave-fronts reaching the lens:

$$\text{curvature} = \frac{1}{u} = \frac{1}{-0.2\,\text{m}} = -5\,\text{D}$$

**Step 3:** Curvature added by lens = power of lens = curvature after – curvature before

$$= 10\,\text{D} - (-5)\,\text{D} = 15\,\text{D}$$

**b** Calculate the focal length of the lens.

**Step 1:** Select the appropriate equation: focal length $= \frac{1}{\text{power}}$

**Step 2:** Substitute and evaluate: focal length $= \frac{1}{15} = 0.067\,\text{m}$ or 67 mm (2 s.f.)

The image distance $v$ is different from the focal length $f$, because the focal point is the point at which parallel waves from a very distant object are brought to focus. The focal length is the distance from the centre of the lens to the focal point, and is a constant for a particular lens with a fixed shape. Waves from a nearer object are brought together beyond the focal point. The image distance $v$ is greater than the focal length $f$, except for very distant objects.

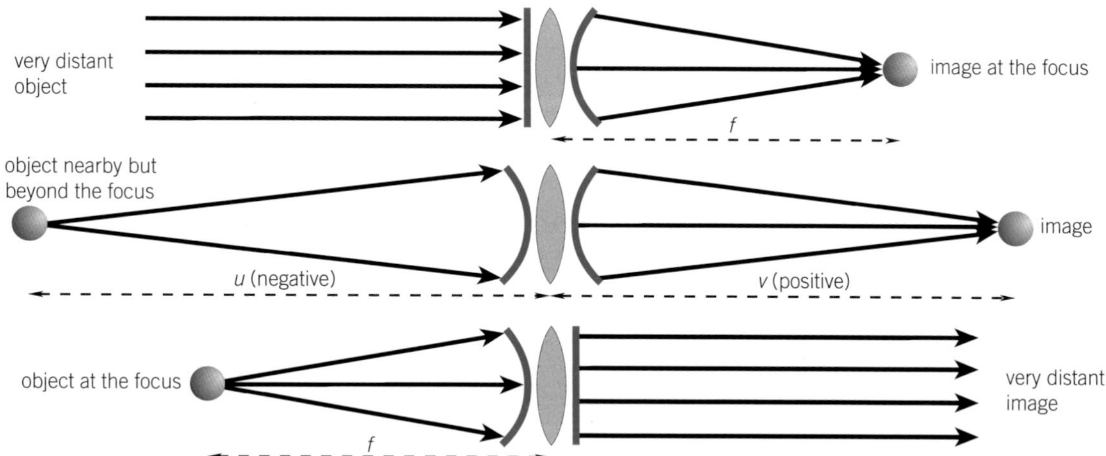


▲ **Figure 3** *Effect of a converging lens on light from a distant object, a nearby object, and an object at the focus*

### Practical 3.1.1d(i): Determining power and focal length of a converging lens

A small filament lamp is used as an object and a lens is used to project its image onto an opaque screen. The object distance *u* is changed several times, and the screen is moved each time so that the image is in focus. At each object distance, the values of *u* and *v* (the image distance) are measured.

The values of $\frac{1}{v}$ (*y*-axis) and $\frac{1}{u}$ (*x*-axis) are plotted on a graph. This will give a straight line. The intercepts on the axes both give $\frac{1}{f}$, the power of the lens. The reciprocal of this value gives the focal length of the lens.

**Study tip**

Give careful thought to the uncertainty in the measurements for experiments like this. The uncertainty in the image distance measurement is likely to contribute the most to the overall uncertainty of the value for the focal length and power. This is because it's difficult to tell exactly when the image is perfectly in focus.

## Magnification

Lenses can also be used to **magnify** the image of an object. This means that the size of the image appears larger than the original object. The linear magnification of an image is defined as:

$$\text{linear magnification } m = \frac{\text{image height (m)}}{\text{object height (m)}}$$

Linear magnification is also related to the object distance and the image distance by:

$$\text{linear magnification } m = \frac{\text{image distance } v \text{ (m)}}{\text{object distance } u \text{ (m)}}$$

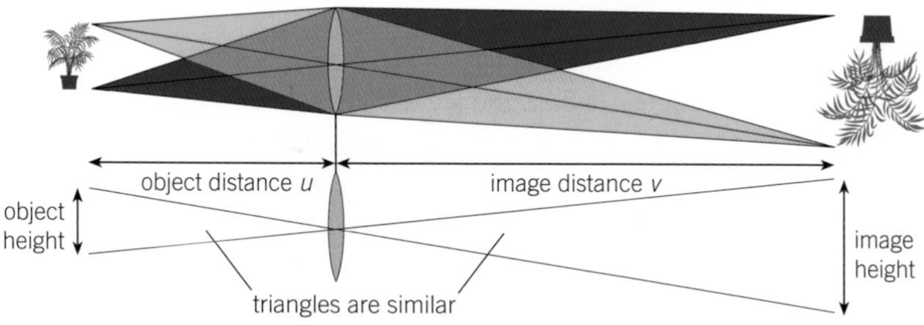


▲ **Figure 4** *Magnification by a lens*

**Study tip**

A negative *m* occurs when the image is inverted (upside down). When *m* is less than one, the image appears smaller than its source – magnification can also refer to a reduction in size.

### Worked example: Magnification using the lens equation

**a** A lens produces an image of a lamp. The distance from the lamp to the lens is −0.40 m. The distance from the lens to the focused image is 0.95 m. Calculate the magnification of the image.

**Step 1:** Choose the appropriate equation: $m = \frac{v}{u}$

**Step 2:** Substitute and evaluate. $m = \frac{0.95\,\text{m}}{-0.40\,\text{m}} = (-)2.4$ (2 s.f.).

**b** Calculate the focal length of the lens.

**Step 1:** Choose the appropriate equation: $\frac{1}{v} = \frac{1}{u} + \frac{1}{f}$

**Step 2:** Substitute and evaluate to find $\frac{1}{f}$:

$$\frac{1}{0.95\,\text{m}} = \frac{1}{-0.40\,\text{m}} + \frac{1}{f}$$

$$\therefore \quad \frac{1}{f} = \frac{1}{0.95\,\text{m}} + \frac{1}{0.40\,\text{m}}$$

$$\frac{1}{f} = 3.553\,\text{m}^{-1}$$

**Step 3:** Take the reciprocal to find $f$:

$$\therefore \quad f = 0.28\,\text{m (2 s.f.)}$$

### Study tip

The negative sign has been put in brackets because it is usually not included in the answer – but don't let negative magnifications worry you.

▲ **Figure 5** *Painting from 1466 AD, showing the use of reading glasses*

### Reading glasses

As we get older, the muscles in our eyes weaken. Close objects can no longer be focused. From about the year 1300, those who needed to work with close objects (e.g. reading a book) began to wear reading glasses, often with quartz lenses. The lenses added curvature to the light waves from the object, helping the eye focus near objects. The same principle was employed in the development of magnifying glasses around the same time.

1 The invention of reading glasses has been called one of the greatest inventions of the Medieval period. Why do you think this is? What effect might the development of reading glasses and magnifying glasses have had on the spread of learning and knowledge? Explain your answer.

## Summary questions

1 A converging camera lens with a focal length of 50 mm produces a focused image of a face. The distance between the lens and the face is 1 m. Calculate the image distance. *(2 marks)*

2 An object is placed 1.5 m in front of a converging lens. A focused image is formed on a screen 2.5 m from the lens. Calculate the magnification of the image. *(2 marks)*

3 You tell a colleague that the magnification of a particular system is 0.27. You are told that this must be wrong because images can't be magnified *and* be smaller than the object. Explain what is meant by 'magnification' in this context. *(2 marks)*

4 A lamp 400 mm from a lens is in focus on a screen 400 mm from the other side of the lens.

**a** Calculate the power of the lens. *(2 marks)*

**b** Use the lens equation to explain why the lamp must be moved nearer to the lens to project its image further away. *(3 marks)*

# 1.3 Storing and manipulating the image

Specification reference: 3.1.1a(ii), 3.1.1b(i), 3.1.1c(i), 3.1.1c(vi)

Digital cameras are getting ever more compact. Figure 1 shows a tiny camera that can be swallowed. It transmits video images as a stream of binary numbers as it passes through the digestive system. White LEDs (light-emitting diodes) provide the light source for this camera.

## CCDs and pixels

In a digital camera, the image from the lens is focused onto a light-sensitive microchip called a charge-coupled device or CCD. This is a screen covered by millions of tiny 'picture elements' or **pixels** (Figures 2 and 3). Each pixel stores electric charge when light falls on it – the brighter the light falling on the pixel, the greater the charge stored on it. The image becomes an array of numbers, which can then be manipulated to edit the image.

## Bits and bytes

Numbers in computers, such as the numbers representing the image in a digital camera, are stored as *on* or *off* values. 'On' may be a high potential difference, 'off' may be low. This is a **binary** system. The two values can be thought of as two digits, 1 and 0.

If one pixel simply records 'high' or 'low', then only one memory location storing a 1 or 0 is needed. This is called one **bit** (binary digit) of information. In practice, pixels do not simply record dark = black and bright = white. Pixels can often have 256 shades of grey, with each shade recorded as a number from 0 to 255.

### Learning outcomes

Describe, explain, and apply:

- the storing of images in a computer as an array of numbers that may be manipulated to enhance the image
- the terms: pixel, bit, byte, resolution, noise
- the equation: amount of information in an image = no. of pixels × bits per pixel
- the equation for the number of bits $b$, using $b = \log_2 N$.

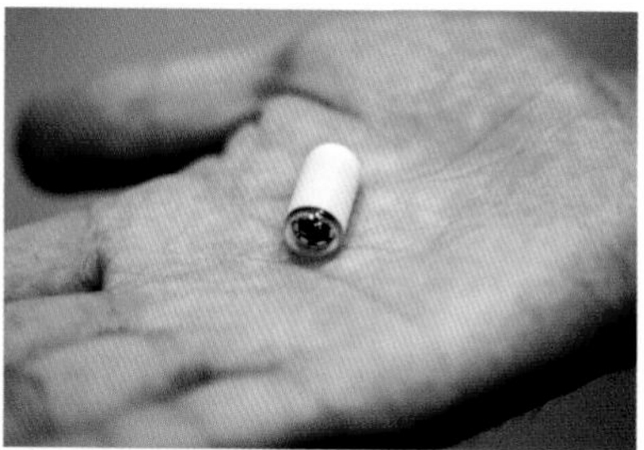

▲ **Figure 1** *A capsule endoscope*

▲ **Figure 2** *A typical CCD. In modern cameras a CCD can have many millions of pixels*

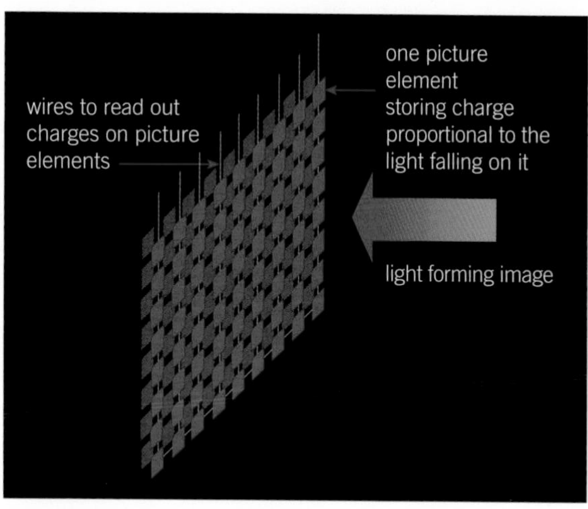


▲ **Figure 3** *Structure of a CCD*

Surprisingly, the number of bits required to store 256 alternatives is only eight. A group of eight bits is called a **byte**, so one byte can store 256 alternatives.

Table 1 shows you the basic idea. Consider the decimal number 5. This is represented by a 1 in the 1s column and a 1 in the 4s column, that is, 5 = 1 + 4, represented in binary as 101.

How about the decimal number 13? That is 8 + 4 + 1, or 1101 in binary.

▼ **Table 1** *Bit representations of decimal numbers*

| Bit D (8s) | Bit C (4s) | Bit B (2s) | Bit A (1s) | Decimal value |
|---|---|---|---|---|
| 0 | 0 | 0 | 0 | 0 |
| 0 | 0 | 0 | 1 | 1 |
| 0 | 0 | 1 | 0 | 2 |
| 0 | 0 | 1 | 1 | 3 |
| 0 | 1 | 0 | 0 | 4 |
| 0 | 1 | 0 | 1 | 5 |
| 0 | 1 | 1 | 0 | 6 |
| 0 | 1 | 1 | 1 | 7 |

The number of arrangements of bits $N$ can also be calculated using the equation

$$N = 2^b \text{ or } b = \log_2 N$$

where $b$ is the number of bits available.

### Worked example: Calculations using bits

**a** Show that 8 bits can have 256 arrangements.

**Step 1:** Identify the equation.

$$N = 2^b$$

**Step 2:** Substitute values and evaluate.

$$N = 2^8 = 256$$

**b** Calculate the number of bits required to give 65 536 possible arrangements.

**Step 1:** Identify the equation required.

$$b = \log_2 N$$

**Step 2:** Substitute values and evaluate.

$$b = \log_2 65\,536 = 16$$

**Study tip**

To use the equation $b = \log_2 N$, you will need the button on your calculator that looks like . You might need to change your calculator to maths mode.

## Image resolution

The **resolution** of a digital image is the scale of the smallest detail that can be distinguished. If an image is about 20 mm across, with approximately 40 pixels across, this means that the resolution of the image is 0.5 mm per pixel ($0.5\,\text{mm}\,\text{pixel}^{-1}$). To calculate the resolution, divide the distance represented in the image by the number of pixels in that distance.

### Worked example: Resolution of an image

Figure 4 shows the surface of Europa, a satellite of Jupiter. It represents an area of 30 × 70 km. There are 1300 pixels across the image. Calculate the resolution of the image.

**Step 1:** Identify the correct equation to use.

$$\text{resolution} = \frac{\text{width of object in image}}{\text{number of pixels across object}}$$

**Step 2:** Substitute and evaluate.

$$\text{resolution} = \frac{70\,000\,\text{m}}{1300} = 54\,\text{m}\,\text{pixel}^{-1}\ (2\ \text{s.f.})$$

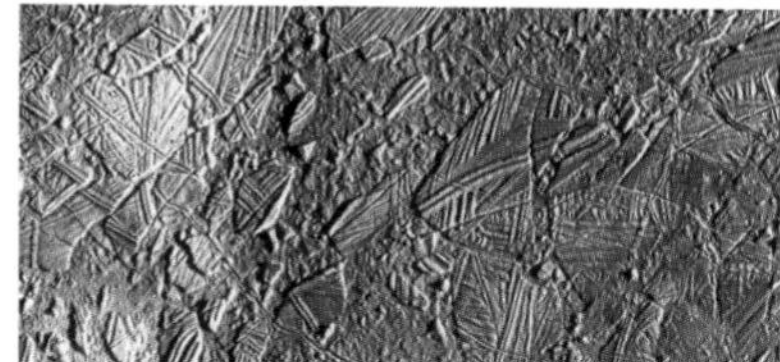

▲ **Figure 4** *Surface of Europa imaged by the Galileo spacecraft*

## Information in an image

The amount of information stored in an image can be calculated using the equation

$$\text{amount of information in image} = \text{no. of pixels} \times \text{bits per pixel}$$

### Worked example: Information in an image

Calculate the storage required for a six megapixel camera that uses three bytes to encode colour information for each pixel.

**Step 1:** Identify the correct equation required.

the amount of information in an image
= no. of pixels × bits per pixel

**Step 2:** Convert values given so that they may be substituted into the equation in Step 1.

6 megapixels = $6 \times 10^6$ pixels

3 bytes = 3 × 8 = 24 bits

**Step 3:** Substitute values into the equation in Step 1 and evaluate.

$$\text{amount of storage required} = 6 \times 10^6 \times 24 = 144\ \text{Mbits}\ (18\ \text{Mbytes})$$

## Image processing

In a digital image each pixel is coded by a number. An 8 bit (1 byte) pixel will usually be coded so that the darkest value is represented by 0 and the brightest by 255. Changing the number will change the appearance of the pixel in the image. This means that mathematical operations can be used to edit or process the images. Examples of how digital images can be processed are shown in Table 2.

▼ **Table 2** *Image processing*

| | |
|---|---|
| Changing brightness | A dim image can be brightened by increasing the value on each pixel by the same amount until the brightest pixel in the image is coded at 255. |
| Removing noise | Noise in images refers to the random speckles across the image. This can be reduced by smoothing, where the value of each pixel is replaced with the median or mean of its value and those around it. |
| Edge detection | To enhance edges in an image the average value of the pixel's neighbours is subtracted from each pixel. This removes uniform areas of brightness and picks out the places where the gradient of the brightness changes abruptly – at the edges. |
| Changing contrast | An image with little contrast will not use the full range of pixel values. An image may only use the values between 75 and 150. To improve the contrast this range is stretched across the 256 possible values so the value 75 becomes 0 and 150 becomes 255. |

### Study tip

Remember that the median is the middle value in a series of values arranged in an ordered list, whereas the mean is the sum of all the values divided by the number of values.

### Summary questions

**1** Replace the pixel in the middle of Table 3 with

**a** the mean of all the values, *(2 marks)*

**b** the median of all the values. *(2 marks)*

**c** Compare the processes above for eliminating possible noise. Suggest which is the better choice. Explain your answer. *(3 marks)*

▼ **Table 3**

| | | |
|---|---|---|
| 100 | 100 | 100 |
| 100 | 200 | 100 |
| 100 | 100 | 100 |

**2** Calculate the number of bits required to code for 4096 alternative values. Express this as a number of bytes. *(3 marks)*

**3** A satellite system to image the Earth's surface is designed to have a resolution of 10 m $pixel^{-1}$ and to cover an area of 100 $km^2$ in each image.

**a** State what the term 'resolution' means in this context. *(1 mark)*

**b** Calculate the number of pixels required to achieve this resolution. *(2 marks)*

**c** Each pixel requires 3 bytes. Calculate the amount of memory in Mbytes that each image requires. *(2 marks)*

# 1.4 Polarisation of electromagnetic waves

Specification references: 3.1.1a(iv), 3.1.1b(i), 3.1.1b(iii), 3.1.1c(v), 3.1.1d(ii)

Some snow goggles and sunglasses have polarising lenses or filters in them to dramatically cut down the glare in bright environments. These are particularly useful in the mountains on sunny days. For this reason polarising lenses (or filters) are used by sportspeople and photographers. They are also used in microscopy.

## Frequency, speed, and wavelength

Light can be thought of as a wave. Three important characteristics of waves are: speed $v$, frequency $f$, and wavelength $\lambda$. These are related by the equation

$$\text{speed } v\ (\text{m s}^{-1}) = \text{frequency } f\ (\text{Hz}) \times \text{wavelength } \lambda\ (\text{m})$$

Frequency is the inverse of the time period $T$ of the wave – the time for one complete oscillation of the wave. This is expressed as $f = \frac{1}{T}$.

**Time picture**

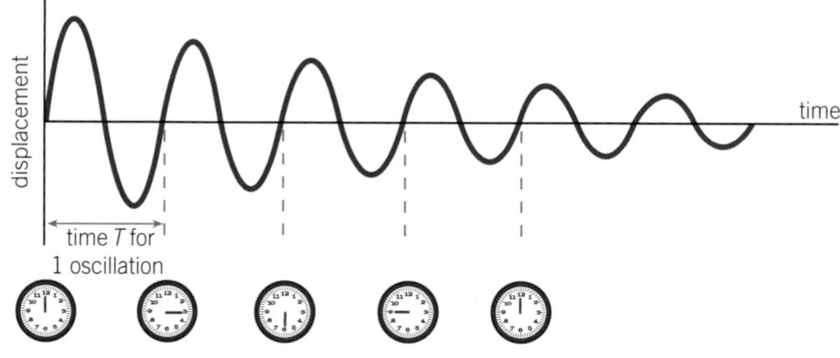


**Position picture: two different wave speeds compared (same frequency)**

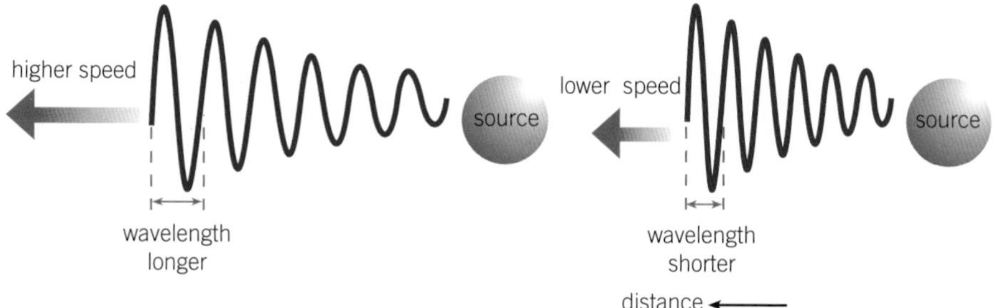


▲ **Figure 2** *Frequency, speed, and wavelength*

### Learning outcomes

Describe, explain, and apply:

- → evidence of the polarisation of electromagnetic waves
- → the term: polarisation
- → diagrams of wave-forms
- → the equations $v = f\lambda$ and $f = \frac{1}{T}$
- → an experiment to observe the polarising effects of using microwaves and light.

▲ **Figure 1** *Polarising filter in ski goggles*

### Synoptic link

You will meet further uses of polarisation in Chapter 5, Looking inside materials.

## Synoptic link

Waves will be discussed in more detail in Chapter 6, Wave behaviour.

## Study tip

The speed of light $c$ is sometimes used in equations instead of $v$, so for example the equation $v = f\lambda$ becomes $c = f\lambda$.

## Study tip

It is a good idea to memorise approximate values of wavelengths on the electromagnetic spectrum. This means that you can check you have obtained the correct order of magnitude in a calculation.

### Worked example: Frequency and wavelength of ultrasound

The time period of an ultrasound signal is 0.1 μs. It travels at $2000\,\text{m}\,\text{s}^{-1}$.

**a** Calculate the frequency and wavelength of this signal.

**Step 1:** Identify the correct equation to use for frequency.

$$\text{frequency} = \frac{1}{T}$$

**Step 2:** Substitute values into the equation and evaluate.

$$\text{frequency} = \frac{1}{1 \times 10^{-7}\,\text{s}} = 1 \times 10^{7}\,\text{Hz or } 10\,\text{MHz}$$

**Step 3:** Identify an equation that links frequency to wavelength.

$$v = f\lambda$$

**Step 4:** Rearrange $v = f\lambda$ to make $\lambda$ the subject.

$$\lambda = \frac{v}{f}$$

**Step 5:** Substitue values and evaluate.

$$\lambda = \frac{2000\,\text{m}\,\text{s}^{-1}}{1 \times 10^{7}\,\text{Hz}} = 2 \times 10^{-4}\,\text{m (or 0.2 mm)}$$

## The Electromagnetic spectrum

Visible light is part of the electromagnetic spectrum. All electromagnetic waves are transverse waves and have the same speed of propagation $c$ in empty space. $c = 3.00 \times 10^{8}\,\text{m}\,\text{s}^{-1}$.

You can see from Figure 3 that visible light is a narrow band of wavelengths. The range of wavelengths of visible light is roughly $4 \times 10^{-7}$ m (violet) to $7 \times 10^{-7}$ m (red).

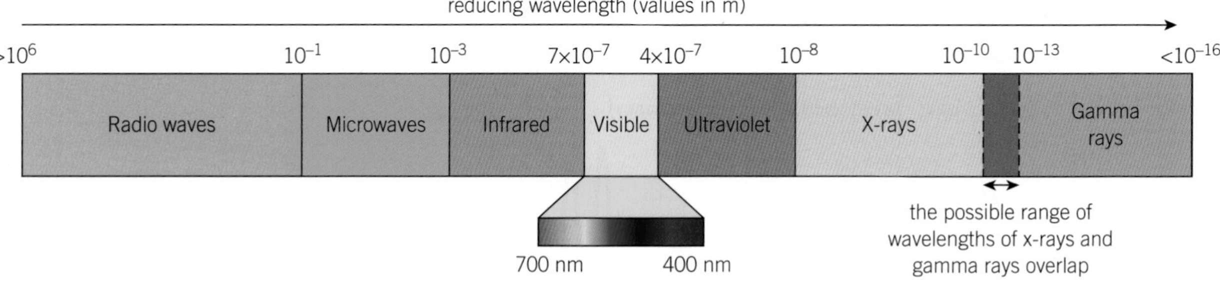


▲ **Figure 3** *The electromagnetic spectrum*

## Polarisation

Electromagnetic waves can be **polarised**. This is a property of transverse waves. Transverse waves are polarised if they vibrate in one plane only. Unpolarised transverse waves vibrate in a randomly changing plane.

Electromagnetic waves are waves of oscillating magnetic and electric fields. The two fields are at right angles to each other and to the direction of travel of the wave, as shown in Figure 4. You can see that the electric field is oscillating in the vertical plane (up and down) and the magnetic field is oscillating in the horizontal plane. All electromagnetic waves can be polarised.

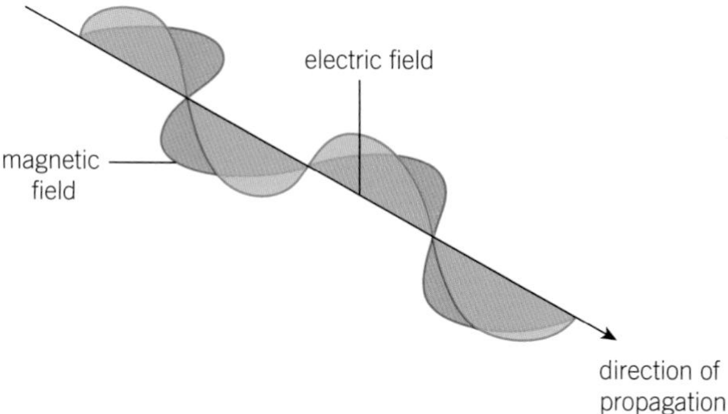


▲ **Figure 4** *An electromagnetic wave*

If a wave is plane-polarised, the direction of oscillation remains fixed. In our example, the electric field strength will always be oscillating in the vertical plane. If the wave is unpolarised the direction of the oscillation will not be fixed. This is represented in Figure 5.

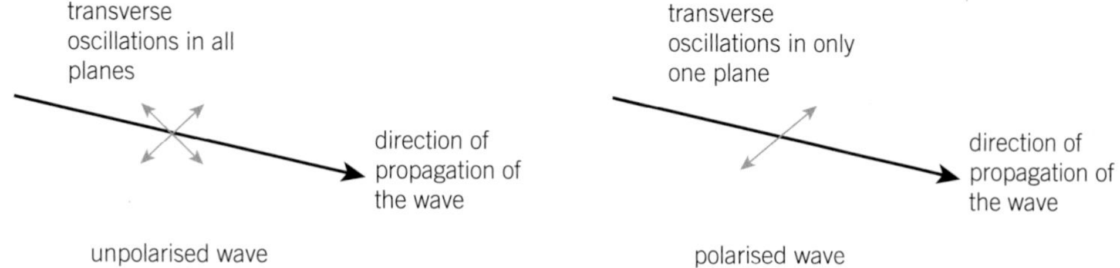


▲ **Figure 5** *Polarised and unpolarised waves*

## Demonstrating polarisation

Unpolarised light is polarised when it passes through a polarising filter. Figure 6 represents unpolarised light passing through a pair of filters. The first filter only allows through the vertical component of the wave, and the second filter only allows through the horizontal component of the wave. No light is transmitted through the second (horizontal) filter because the light reaching it after passing through the first filter is fully polarised in the vertical direction.

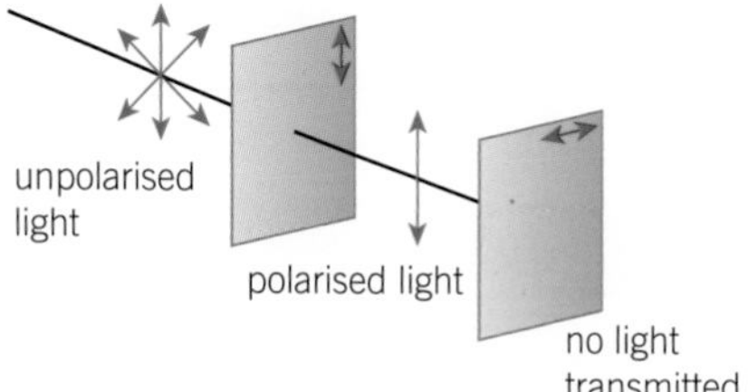


▲ **Figure 6** *The effect of filters on polarised light*

You can detect whether light is polarised by observing it through a single polarising filter and rotating the filter. If the intensity of the light remains constant, the light source is emitting unpolarised light. If the intensity varies as the filter rotates, the source is emitting polarised light. The polarisation of radio waves can be detected by rotating the receiving aerial. The aerial will pick up the strongest signal when it is set up parallel to the plane of polarisation of the radio waves.

## Practical 3.1.1 d(ii): Observing the polarisation of light and microwaves

Light can be polarised by shining a narrow beam of light through a tank of water that contains a few drops of milk (Figure 7).

By looking through a polarising filter at light scattered vertically upwards (looking down at the tank from above), and rotating the filter slowly, the light observed will vary in intensity, showing that the light has been polarised. A similar effect is seen when you view the tank side on.

If light is completely cut out at one angle you will know that the polarised light at that point is vibrating perpendicular to the filter.

A similar experiment can be performed with microwaves (Figure 8). If a metal grille is placed between a transmitter of polarised waves and a receiver, the strength of the signal at the detector can be changed by rotating the grille in the same way as rotating a polarising filter for the light experiment above.

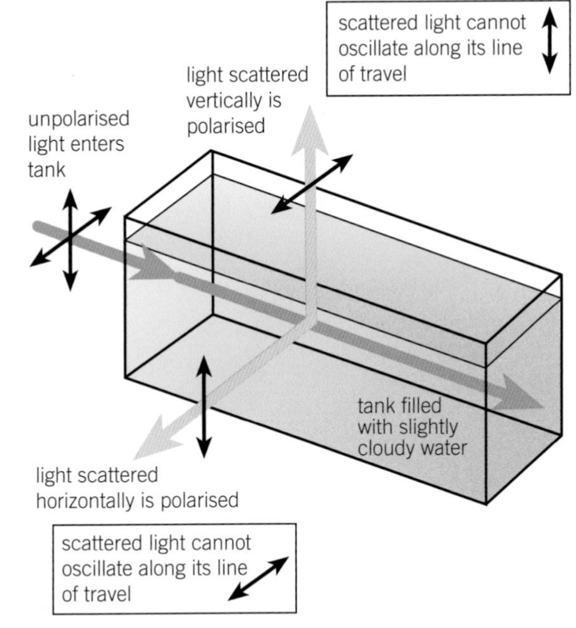


▲ **Figure 7** *Polarisation of light by scattering*

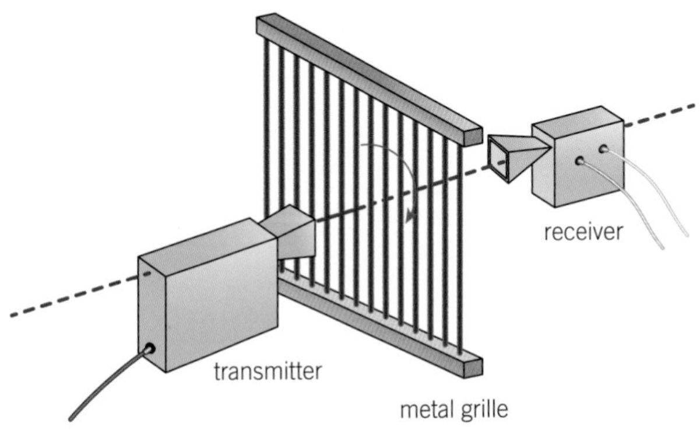


▲ **Figure 8** *An experiment to show the scattering of microwaves*

### Hint

Radio waves and microwaves are both electromagnetic waves. The same principles apply to both.

## Summary questions

1 A radio station broadcasts at a frequency of 91 MHz. Calculate the wavelength of the waves assuming the speed of the waves is $c = 3.0 \times 10^8\,\mathrm{m\,s^{-1}}$. Explain how you decided how many significant figures to use in your answer. *(2 marks)*

2 **a** Draw a displacement–time graph showing a wave with time period of 0.1 s. Draw three complete waveforms. *(3 marks)*

**b** Add to your graph a waveform with a frequency of 20 Hz. *(2 marks)*

3 The radio emission from some galaxies is polarised. Suggest how this fact could be confirmed. *(2 marks)*

## Practice questions

1 A digital camera has a 6 megapixel lens. Each pixel is coded by 24 bits. The memory card on the camera can store 160 images. What is the minimum memory required?

**A** 23 kbyte **B** 5 Mbyte

**C** 2.9 Gbyte **D** 23 Gbyte

2 An electronic display uses a grid of 16 × 16 pixels. Each pixel is coded by 4 bits.

**a** State the number of bits the display memory stores for one image. *(1 mark)*

**b** Give your answer to **a** in bytes. *(1 mark)*

3 Here is some data about ultrasound used in medical imaging:

wavelength in soft tissue = $2.9 \times 10^{-4}$ m

speed in soft tissue = $1450\,\mathrm{m\,s^{-1}}$

**a** Calculate the frequency of the ultrasound. *(1 mark)*

**b** The speed of ultrasound in air is about $340\,\mathrm{m\,s^{-1}}$. State what happens to the frequency and wavelength of ultrasound as it travels from soft tissue into air. *(2 marks)*

4 Radio waves from a transmitter are vertically polarised. A teacher demonstrates this by positioning a radio with its aerial vertical. The signal is at a maximum with the radio in this orientation. She then rotates the radio to position B and then position C.

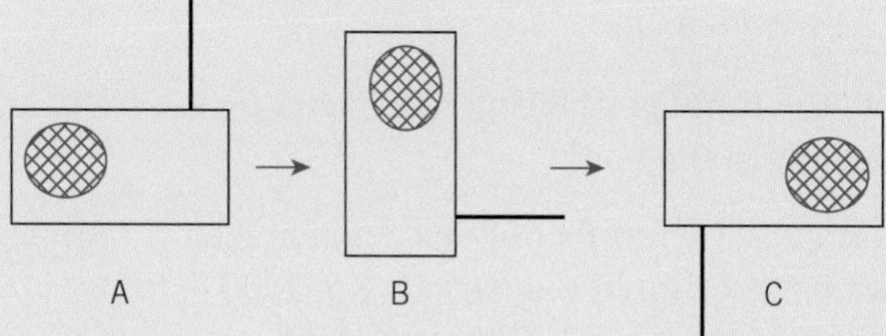


▲ Figure 1

**a** Explain what is meant by polarisation. You may include diagrams in your answer. *(2 marks)*

**b** Describe how the signal will vary as the radio is rotated to position B and then C. *(2 marks)*

5 This question is about finding the focal length of a converging lens.

A student varies the distance between a point source of light and the lens *u* and measures the distance between the lens and the focused image, *v*.

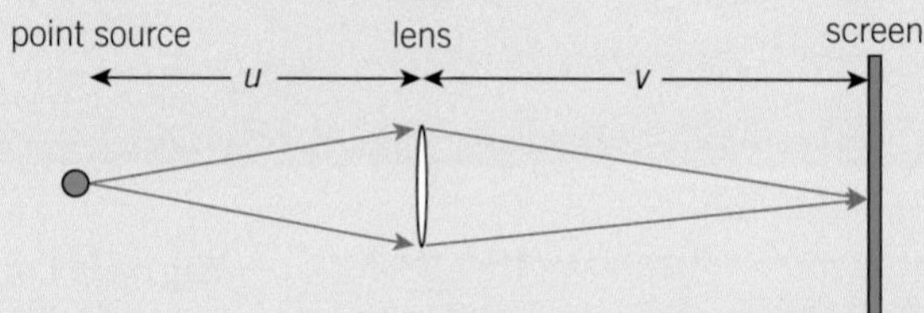


▲ Figure 2

Here is one pair of readings:
object distance = −0.300 m
image distance = 0.145 m

The student suggested that the uncertainty in the image distance is ± 0.01 m

**a** Suggest why there is uncertainty in the image distance measurements, and why the student chooses to ignore the uncertainty in the object distance when calculating the focal length of the lens. *(2 marks)*

**b** Calculate the focal length of the lens, including the uncertainty in the value. *(4 marks)*

The graph shows more data gained from the experiment, plotted as a graph of $\frac{1}{v}$ against $\frac{1}{u}$. The uncertainty bars have not been included.

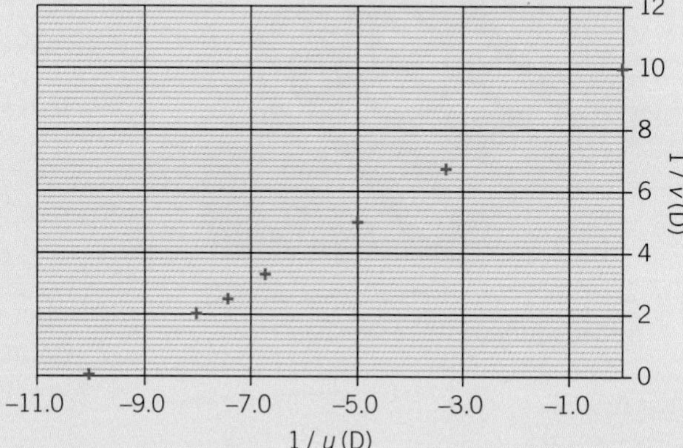


▲ Figure 3

**c** On a copy of the graph, add the uncertainty on the data point (−6.7, 3.3). *(1 mark)*

**d** Use the equation $\frac{1}{v} = \frac{1}{u} + \frac{1}{f}$ to explain why the value of the $\frac{1}{u}$ intercept is equal to the negative of the $\frac{1}{v}$ intercept. *(2 marks)*

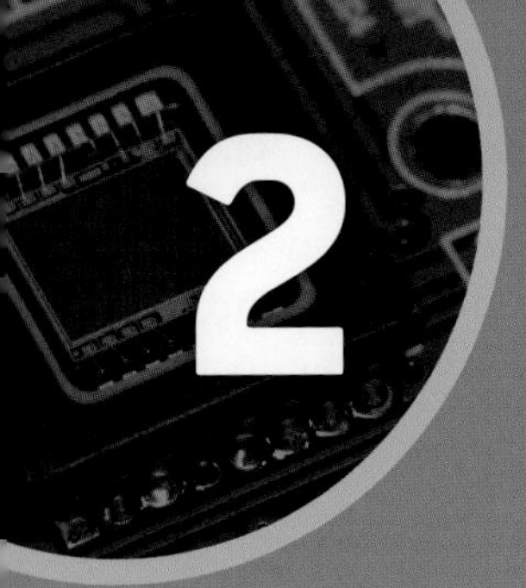

# 2 SIGNALLING

## 2.1 Digitising a signal

Specification references: 3.1.1a(iii), 3.1.1b(i), 3.1.1c(ix), 3.1.1c(x)

### Learning outcomes

Describe, explain, and apply:

- how signals are digitised
- advantages of digital signals
- the terms: sampling, signal, resolution
- the graphical representation of the digitisation of an analogue signal for a given number of levels of resolution
- the equation $b = \log_2 \frac{V_{total}}{V_{noise}}$.

**Signals** come in many different forms. A signal transfers information from one location to another – it can be conveyed through sound or light. A red traffic light is a signal. So is a fire alarm, or even a wink. A signal can be a variation of current in a telephone cable or a stream of binary highs and lows (1s and 0s).

### Digitising signals

We live in a digital world. Information is digitised and sent around the globe at speeds that could not have been imagined 50 years ago. Digitised signals are possible because information can be coded into a string of binary digits that is transmitted (through the air, through space, through wires, or through optical fibres) from the sender to the receiver. For example, emails use digital signals. Each character is sent as a 1 byte number code.

#### Worked example: Encoding characters into digital signals

A book contains 1000 pages. Each page of text contains 500 words that equates to 3000 characters including spaces. Each character is encoded with a 1 byte number.

Calculate the number of bits required to encode the book.

**Step 1:** Convert bytes per character to bits per character.

$$1 \text{ byte} = 8 \text{ bits}$$

**Step 2:** Multiply the number of bits per character by the number of characters in the book.

Number of bits required = number of bits per character × number of characters per page × number of pages = 8 × 3000 × 1000 = $2.4 \times 10^7$ bits or 24 Mbits

▲ **Figure 1** *All aspects of communication technology are rapidly changing*

### Advantages of digital signals

**Analogue** signals vary continuously from one value to the next, without fixed values. For most of the 20th century, long-distance communication systems such as telephone, radio, and television were analogue systems. For example, in an analogue telephone a sound vibration is changed into matching oscillations of potential difference.

One problem with analogue systems is the need for amplification as the signal becomes weaker. If the signal becomes distorted or 'noisy',

#### Hint

Do not confuse Mbit with Mbyte. 1 Mbit = $1 \times 10^6$ bit, whilst 1 Mbyte = $1 \times 10^6$ byte.

the amplification boosts the signal *and* the noise. In this context, noise is the random variation on the signal. For example, holding a conversation in a crowded room is difficult because what your friend is saying (the signal) has to be picked out from the background chatter (the noise). Simply amplifying the sound won't help, because that will also amplify the effect of the background chatter. The hiss on a badly tuned radio is another example of noise – turning the radio up increases the hiss and the signal.

Noise can be filtered out, but this causes a loss of detail in the signal. This is not a problem in a digital signal. It is easy to detect binary 'on/off' signals even when they are weak and noisy. A perfect copy of a message can be regenerated and sent on. Digital signals can also transmit information much faster than analogue signals.

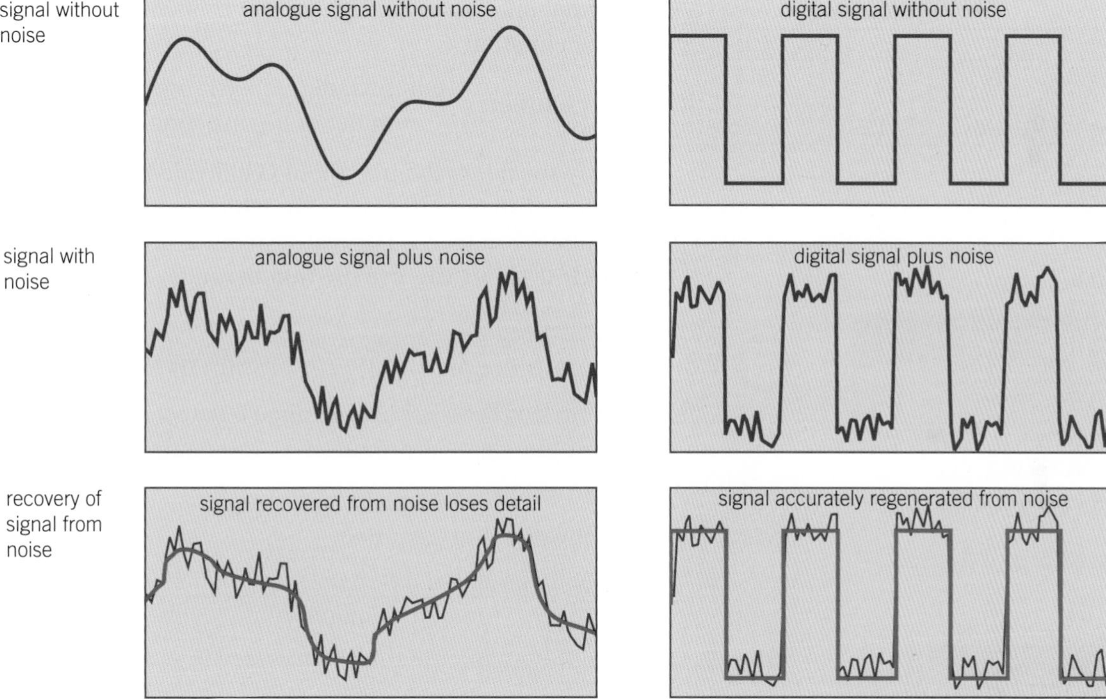


▲ **Figure 2** *Signals and noise*

## Sampling

When musicians refer to sampling, they often mean short extracts of music that are edited and mixed with other recordings. In physics, **sampling** is the process in which the displacement of a continuous (analogue) signal is measured at small time intervals and turned into a digital string of binary numbers (samples). This is shown in Figure 3.

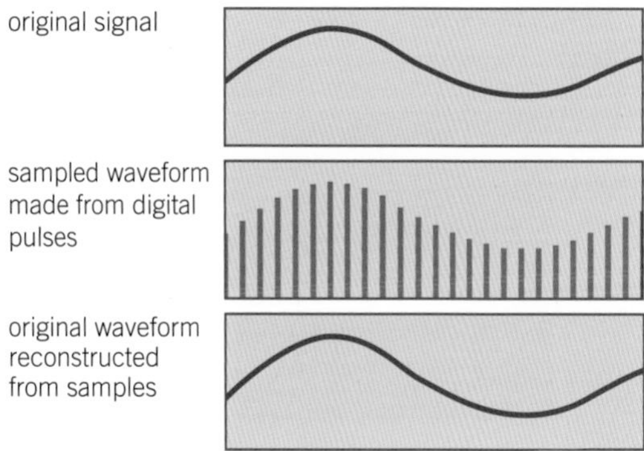


▲ **Figure 3** *Sampling a signal*

### Analogue to digital conversion

Figure 4 (next page) shows the principle of converting a varying analogue signal into a stream of numbers. In this example, each sample is coded with three bits.

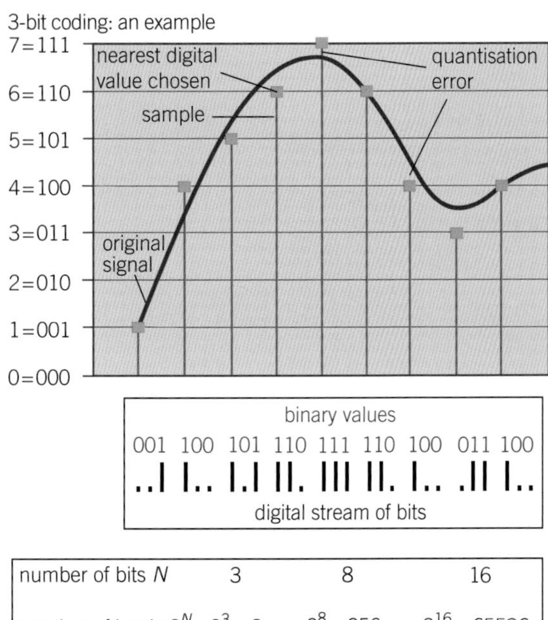


| number of bits $N$ | 3 | 8 | 16 |
|---|---|---|---|
| number of levels $2^N$ | $2^3 = 8$ | $2^8 = 256$ | $2^{16} = 65536$ |

▲ **Figure 4** *Digitising a signal*

This means that there are eight ($2^3$) levels to represent the signal value for any sample. These eight levels, known as quantisation levels, cover the range of signal values from the lowest (000) to the highest (111). You can see that the signal does not always match exactly with a quantisation level. In such cases the system records the nearest level to the signal. The difference between the signal value and the quantisation level is the quantisation error. Increasing the number of quantisation levels produces a better match to the original signal.

## Resolution of a sample

The **resolution** of a sample is the smallest change in potential difference that can be determined. This is given by the equation

$$\text{resolution} = \frac{\text{potential difference range of signal}}{\text{number of quantisation levels}}$$

Strictly speaking, the p.d. should be divided by the number of steps between levels, which is 1 less than the number of levels. For real systems, $2^b$ and $2^b - 1$ are very close so we can use the number of levels.

### Synoptic link

You have met the number of arrangements of bits, $N$, and the available number of bits, $b$, in Topic 1.3, Storing and manipulating the image.

### Worked example: Calculating resolution of a sample

Consider a signal that is detected over a range 0.0 V to 12.0 V. An 8-bit sample of this signal is produced. Calculate the resolution of this sample.

**Step 1:** Identify the equation required.

$$\text{resolution} = \frac{\text{potential difference range of signal}}{\text{number of quantisation levels}}$$

**Step 2:** Calculate the number of quantisation levels possible.

number of possible arrangements $N$ for 8 bits $= 2^b = 2^8 = 256$

**Step 3:** Substitute values into the equation in Step 1 and evaluate.

$$\text{resolution} = \frac{\text{potential difference range of signal}}{\text{number of quantisation levels}}$$

$$= \frac{12.0\,\text{V}}{256} = 0.047\,\text{V (2 s.f.)}$$

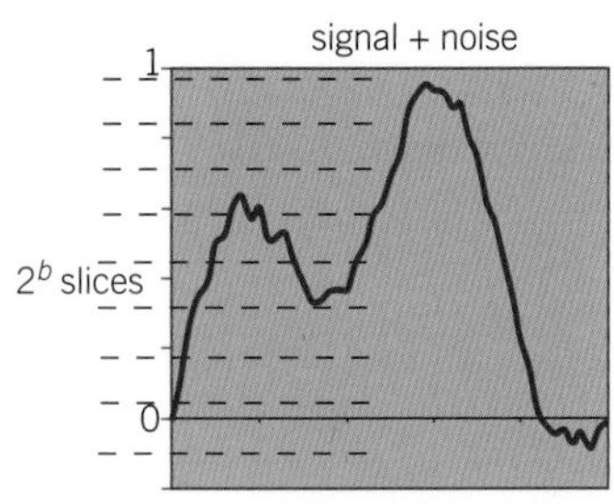


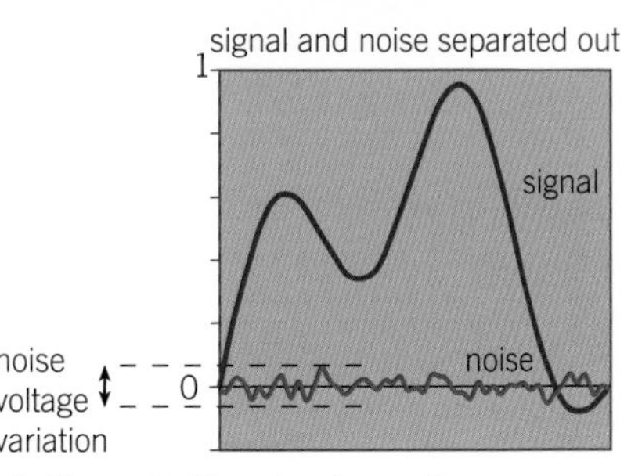


▲ **Figure 5** *Signal, noise and quantisation levels*

## Number of useful quantisation levels

Increasing the number of quantisation levels increases demands on data storage and transmission. Noise also plays a part in the choice of the number of levels to use.

Look at Figure 5. The diagram above shows a noisy signal. The lower diagram shows the signal and noise separated out. It is pointless to have a smaller gap between quantisation levels than the size of the noise variation, as this would mean sampling the noise to great detail, rather than ignoring it.

The maximum number of useful quantisation levels is given by the equation

$$\text{Maximum useful number of levels} = \frac{\text{total noisy signal variation}}{\text{noise variation}} = \frac{V_{total}}{V_{noise}}$$

Since number of possible arrangements (or levels) $N = 2^b$, this equation becomes

$$2^b = \frac{V_{total}}{V_{noise}}$$

Taking $\log_2$ on both sides

$$b = \log_2 \frac{V_{total}}{V_{noise}}$$

### Worked example: Bits and quantisation levels

A signal has a maximum total variation of 200 mV. The noise variation is 5 mV. Calculate the largest number of bits per sample worth using to encode the variation.

**Step 1:** Identify the equation required. $b = \log_2 \frac{V_{total}}{V_{noise}}$

**Step 2:** Substitute values and evaluate.

$$b = \log_2 \frac{V_{total}}{V_{noise}} = \log_2\left(\frac{200}{5}\right) = 5.3, \text{ which rounds } \textit{up} \text{ to 6 bits.}$$

We round up, because 5 bits gives 32 levels and 40 is the maximum useful number of levels.

## Summary questions

1 Calculate the number of bits required to encode this sentence in a signal if each character and space is represented by 8 bits. The previous sentence is 127 characters in length. *(2 marks)*

2 A signal variation is sampled with 6 bits per sample.

a Calculate how many quantisation levels are possible in each sample. *(2 marks)*

b The potential difference is sampled on a scale from 0.00 V to 12.00 V. Calculate the resolution of the sample. *(2 marks)*

3 Give two advantages of digital recording over analogue recording. Explain your answers. *(4 marks)*

4 The ratio $\frac{V_{total}}{V_{noise}}$ in a signal is 75. Calculate the largest number of bits per sample worth using to encode the variation. *(3 marks)*

5 a Draw a diagram showing a waveform sampled with 8 quantisation levels. You should include 6 sampling points at regular intervals on your diagram. *(4 marks)*

b Use your diagram to explain what 'quantisation error' means. *(2 marks)*

# 2.2 Sampling sounds and sending a signal

Specification references: 3.1.1a(iii), 3.1.1c(vii), 3.1.1c(viii)

### Learning outcomes

Describe, explain, and apply:

- → disadvantages of digital signals
- → the equation: minimum rate of sampling > 2 × maximum frequency of signal
- → the equation: rate of transmission of digital information = samples per second × bits per sample.

Have you ever watched old films where the wheels on a speeding car appear to turn slowly or even backwards? This happens because the time interval between individual images (frames) captured by the camera is too long compared to the speed of the rotating wheel. This causes the wheels to appear as though they move backwards as the wheels make nearly a full rotation between each frame. This highlights the problem of not taking enough samples each second. The sampling rate (sampling frequency) is the number of samples taken each second. For a varying signal to be sampled accurately the time interval between samples must be shorter than the time in which important changes in the signal occur. If the time interval is too large the reconstructed signal will lose detail. This is illustrated in Figure 1.

In fact, to ensure that the original signal can be reconstructed accurately two conditions have to be met:

- the signal cannot contain frequencies above a certain maximum
- minimum sampling rate > 2 × highest frequency component.

If the signal contains frequencies above the maximum component the signal will not be reconstructed accurately. Lower frequency signals called *aliases*, which are not present in the original signal, will be generated. The backward-rotating wheels in the film described above are an example of this.

**Sampling too slowly misses high frequency detail in the original signal**

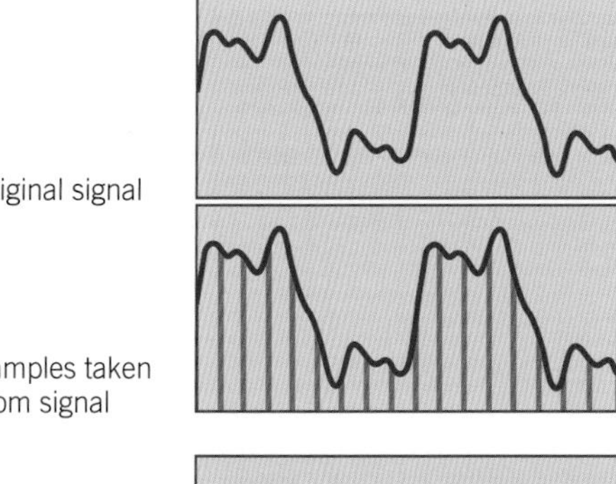


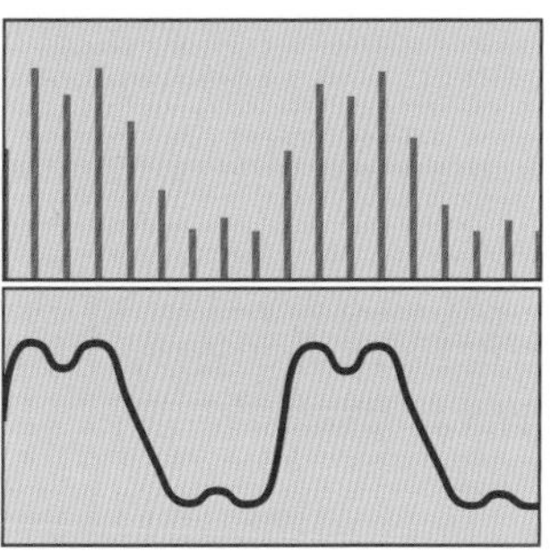


▲ **Figure 1** *Losing high frequency detail*

**Sampling too slowly creates spurious low frequencies (aliases)**

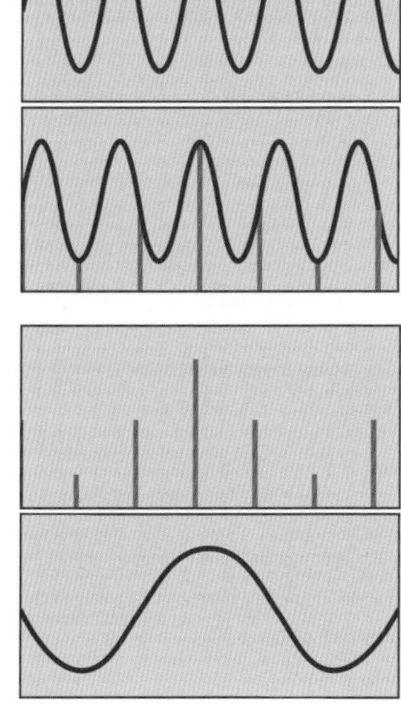

▲ **Figure 2** *Aliasing*

The human ear cannot detect sound above frequencies of about 20 kHz. Therefore, for music to be sampled accurately, the sampling frequency should be greater than 40 kHz. The standard frequency used is 44.1 kHz. Filters remove frequencies above 20 kHz in the original signal so aliasing (the formation of aliases) will not be a problem.

## Bit rate

The **bit rate** is the rate of transmission of digital information. It can be calculated using the equation

$$\text{bit rate (bit s}^{-1}\text{ or Hz)} = \text{samples per second} \times \text{bits per sample}$$

### Worked example: Calculating bit rate

A CD quality sound uses 16 bits per sample at a sample rate of 44.1 kHz for each of the two stereo channels. Calculate the combined bit rate for the two stereo channels.

**Step 1:** Identify the equation required.

bit rate (bit $s^{-1}$ or Hz) = samples per second × bits per sample

**Step 2:** Substitute values and evaluate.

bit rate for one stereo channel (bit $s^{-1}$ or Hz) = $44.1 \times 10^3 \times 16$
$= 7.1 \times 10^5$ bits $s^{-1}$ or $7.1 \times 10^5$ Hz

bit rate for both stereo channels $= 2 \times 7.1 \times 10^5 = 1.4 \times 10^6$ bits $s^{-1}$
$= 1.4$ Mbit $s^{-1}$ or 1.4 MHz

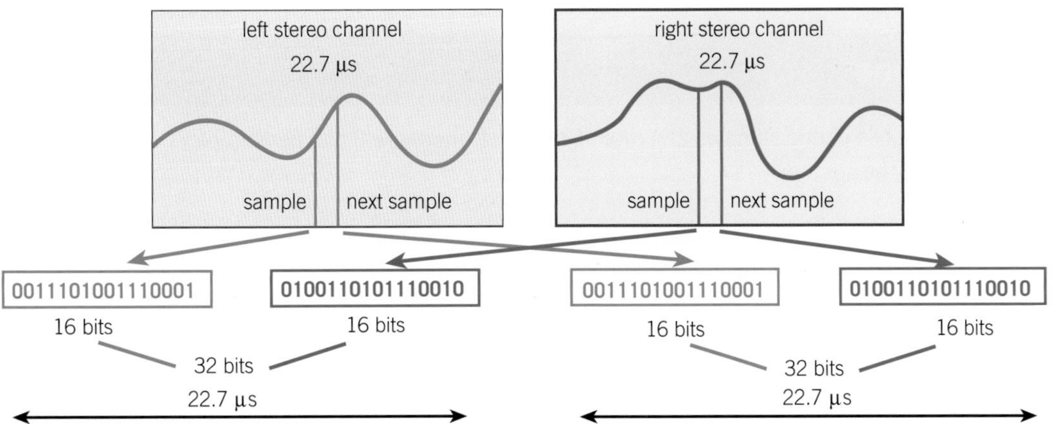


▲ **Figure 3** *Samples and strings of bits*

We can also use the idea of bit rate to calculate the playing time of a recorded sound signal of known size.

$$\text{duration of signal (s)} = \frac{\text{number of bits in signal}}{\text{bit rate (bit s}^{-1}\text{ or Hz)}}$$

## Study tip

Watch out for data given in bytes when the question asks for bits, or the other way round. Remember that 1 byte = 8 bits.

### Worked example: Signal duration from bit rate

A song occupies 28 Mbytes of memory on a mobile phone. If the bit rate of the system is 1.4 Hz, calculate the duration of the song.

**Step 1:** Select the appropriate equation:

$$\text{duration of signal (s)} = \frac{\text{number of bits in signal}}{\text{bit rate (bit s}^{-1}\text{ or Hz)}}$$

**Step 2:** Convert bytes to bits, and MHz to Hz.

$28\,\text{Mbytes} = 8 \times 28\,\text{Mbits} = 224\text{Mbits} = 224 \times 10^6\,\text{bits}$

$1.4\,\text{MHz} = 1.4 \times 10^6\,\text{Hz}$

**Step 3:** Substitute and evaluate.

$$\text{duration of signal (s)} = \frac{224 \times 10^6\,\text{bits}}{1.4 \times 10^6\,\text{Hz}}$$

$= 160\,\text{s} = 2\text{ minutes }40\text{ seconds}$

## Disadvantages to digital signals

How will digital signals affect society in the twenty-first century? Does seeing tragic world events almost as they happen, but being powerless to do anything about them, affect people's sense of responsibility?

Digital signals are numbers and so can be changed or scrambled. Who should have the right to do this? Are online banking details completely secure? How can or should the web be policed? Will cyber-terrorism become an increasing problem? What about cyber-bullying? Images of actors we see on television screens and in cinemas can be digitally enhanced and give the impression of a level of perfection that cannot be met in reality – should this be a cause of concern?

The digital revolution of the twentieth- and twenty-first centuries has produced great benefits, from medicine to entertainment and academic study, but it also raises difficult questions.

## Summary questions

1 A recording app on a mobile phone samples at a rate of 32.0 kHz. There are 8 bits per sample. Calculate the bit rate of the system. *(2 marks)*

2 This topic contains approximately 4500 characters including spaces. If each character is encoded with 8 bits, calculate the time duration of the encoded topic when transmitted at a bit rate of 10 Mbit $s^{-1}$. *(2 marks)*

3 Explain why the sampling rate needs to be greater than 40 000 Hz to encode an audio recording accurately when the highest frequency present is 20 000 Hz. *(3 marks)*

4 If the human ear cannot detect frequencies above 20 000 Hz, explain why it is necessary to filter out frequencies above this maximum before encoding takes place at a sampling rate of 40 000 Hz. *(2 marks)*

5 a Give two disadvantages to society of digital recording of sounds and images. Explain your answers. *(4 marks)*

b Suggest an aspect of digital technology that can be considered as both an advantage and disadvantage to society. Explain your choice. *(3 marks)*

## Practice questions

1 A signal is sampled at a rate of 44.1 kHz. Each sample is coded with 2 bytes.

Which figure gives the bit rate of the sampling?

A 88 bits $s^{-1}$ B 706 bits $s^{-1}$

C 88 200 bits $s^{-1}$ D 705 600 bits $s^{-1}$

2 Here are three statements about hearing and sampling.

1 The limit of human hearing is about 20 kHz.

2 Recordings can use a filter so that no variations above 20 kHz are recorded.

3 Recordings are often sampled at 44.1 kHz, over double the highest frequency that has been recorded.

Explain why it is necessary to limit the frequency range and sample at a rate over double the highest frequency of human hearing to obtain a high-quality recording. *(3 marks)*

3 An analogue signal has a total voltage variation (signal + noise) of 600 mV. The noise variation is 9 mV.

a Calculate the maximum useful number of bits to code the signal. *(2 marks)*

b Explain why it is not worth using more bits to code the signal. *(2 marks)*

4 An analogue signal has a voltage variation of 380 mV. It is sampled with 8 bits per sample. Calculate the resolution of each sample. *(2 marks)*

5 A digital photograph has a file size of 28 Mbyte. Calculate the time it will take the image to download at a download rate of is 1 Mbit $s^{-1}$. *(2 marks)*

6 Music is streamed to a mobile phone at a rate of 96 kilobits per second.

How many kilobytes per second does this represent? *(1 mark)*

7 The graph shows a signal coded with three bits per sample. The blue curved line represents the original waveform.

a How can you tell that there are three bits per sample? *(1 mark)*

b How does the graph show that the original signal cannot be perfectly reconstructed from the digital sample? *(1 mark)*

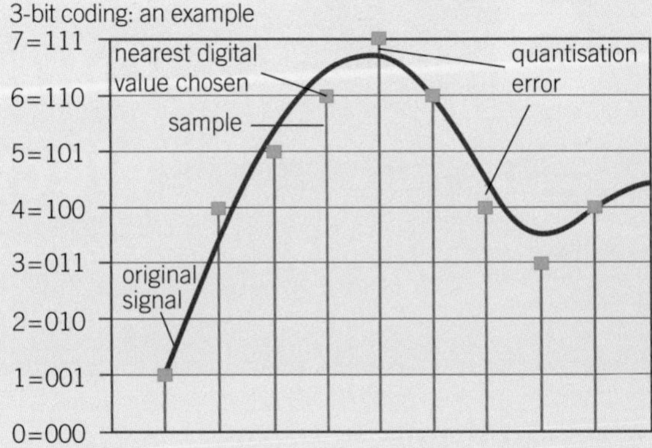


▲ Figure 1

c State two changes to the sampling system that would reduce the error in the reconstructed signal. *(2 marks)*

8 This question is about a sampling system. The graph shows a section of a waveform that has been sampled. Each small triangle represents a sample point.

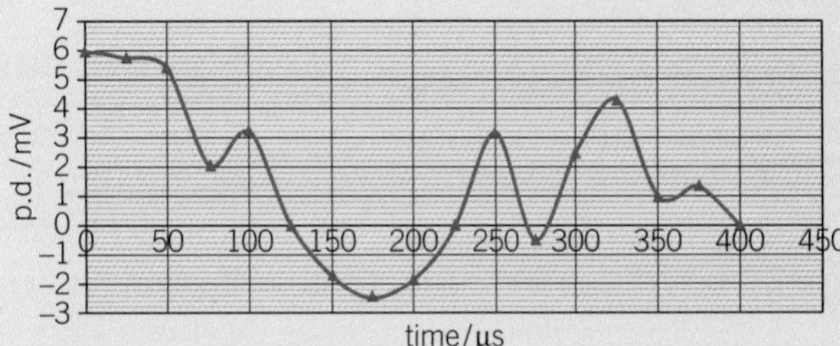


▲ Figure 2

a Show that the sample rate is 40 000 Hz. *(1 mark)*

b State an estimate for the highest frequency sound that can be accurately sampled by this system. Why might a higher frequency not be sampled accurately? *(2 marks)*

c The range of voltage variation is −3 mV to + 7 mV. There are 16 bits per sample. Calculate the resolution of the sample. *(2 marks)*

d The system is used to store a music track lasting $3\frac{1}{2}$ minutes. Calculate the number of Mbytes of memory that will be required. *(3 marks)*

e A student suggests that a digital recording can never be as accurate as an analogue recording. Suggest an argument against this point of view. State and explain an advantage of digitally recording music. *(3 marks)*

# SENSING

## 3.1 Current, p.d., and electrical power

Specification references: 3.1.2a(i), 3.1.2a(ii), 3.1.2a(v), 3.1.2a(ix), 3.1.2b(i), 3.1.2c(i), 3.1.2c(ii)

### Learning outcomes

Describe, explain, and apply:

- → the concept of current as a flow of charged particles
- → the need for charge and current to be conserved in any circuit loop
- → the equation $I = \frac{\Delta Q}{\Delta t}$
- → the term potential difference (p.d.) as the energy change per unit charge moved between two circuit points
- → the equation $V = \frac{W}{Q}$
- → the dissipation of power in electric circuits and the equations $P = IV = I^2R$ and $W = VIt$.

▲ **Figure 1** *The aurora borealis*

▲ **Figure 2** *The ion drive of the SMART-1 probe was driven by a stream of xenon ions*

Electrically charged particles are around us in abundance. You may have heard of them moving at high speed in the Large Hadron Collider at CERN, in the aurora borealis (Northern Lights), and in ion drives for propelling space probes. Ions are electrically charged atoms or molecules in which the number of electrons is different from the number of protons. They form the structure in crystals of ionic compounds, such as salt, and in metals. It is the presence of positive ions and the movement of negative electrons around them that gives metals their electrical properties.

### Electric current

A flow of charged particles produces an electric current – this could be electrons in a wire (Figure 3), or positive ions leaving a space probe's ion drive – each of which has the same electrical charge $q$, measured in **coulombs** (C). As they flow, they pass through the section of the stream shown coloured in Figure 3.

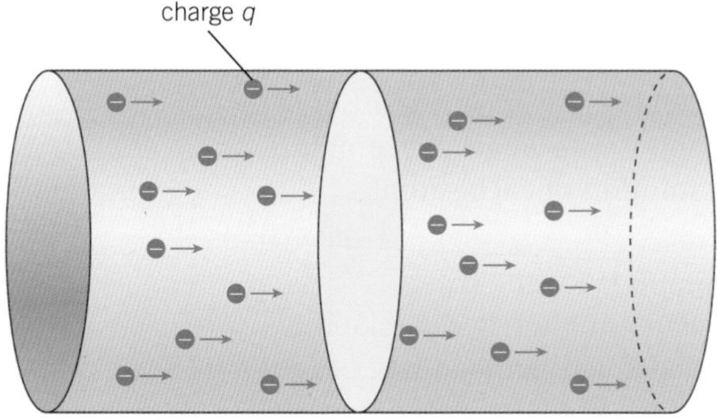


▲ **Figure 3** *Flow of charges*

If the number of charged particles passing through the pale blue section in a time $\Delta t$ is $N$, the total charge $\Delta Q$ flowing across the section is $Nq$ and the **current** $I$, defined as the rate of flow of charge, is given by

$$I = \frac{\Delta Q}{\Delta t}$$

and is measured in **amperes**, or amps (A).

The symbol $Q$ originally stood for 'quantity of electrical charge'. The French physicist André-Marie Ampère originally gave the name 'intensity of current', together with the symbol $I$, to what we now call current. The SI unit of current is named after him (but without the accent).

### Worked example: Electrons in a wire

In Figure 3, a 1.0 m length of copper wire contains $1.7 \times 10^{22}$ free electrons, each of charge $1.6 \times 10^{-19}$ C. These move along the wire with a mean speed $v = 7.4 \times 10^{-4}\,\text{m s}^{-1}$. Calculate the current $I$ in the wire.

**Step 1:** Find the time $\Delta t$ for all $1.7 \times 10^{22}$ electrons in the 1.0 m wire to move past the end point.

$$\text{speed} = \frac{\text{distance}}{\text{time}} \text{ so time } \Delta t = \frac{\text{distance}}{\text{speed}} = \frac{1.0\,\text{m}}{7.4 \times 10^{-4}\,\text{m s}^{-1}}$$

$$= 1351.3...\,\text{s}$$

**Step 2:** Find the total charge of these electrons.

Total charge passing the end point in this time =

$$\Delta Q = Nq = 1.7 \times 10^{22} \times 1.6 \times 10^{-19}\,\text{C} = 2720\,\text{C}$$

**Step 3:** Calculate the current, which is the charge flowing per unit time.

$$\text{current } I = \frac{\Delta Q}{\Delta t} = \frac{2720\,\text{C}}{1351.3...} = 2.0128 = 2.0\,\text{A (2 s.f.)}$$

Note how the combination of a tiny quantity (the electronic charge) and a huge quantity (the number of free electrons) gives a current of everyday proportions.

**Study tip: $\Delta$ and $\delta$**

In this book you will meet many expressions using the Greek letter $\Delta$ (capital delta). $\Delta Q$ (read this as 'delta-Q') does not mean $\Delta \times Q$. It is mathematical shorthand for 'a change in $Q$'. Capital delta is used for any change, which could be large. In later chapters you will also meet $\delta$ (small delta): $\delta Q$ means a tiny change in $Q$.

In Figure 3, the pale blue section could have been drawn anywhere along the stream of charges. They are all moving the same way with the same mean speed, so *the current is the same at all points* along the stream of charged particles.

If some of the charges are diverted, as in the junction in Figure 4, then the current $I$ must divide in such a way that the two divided currents $I_1$ and $I_2$ add to give the original current $I$: all of the moving charges must go one way or the other.

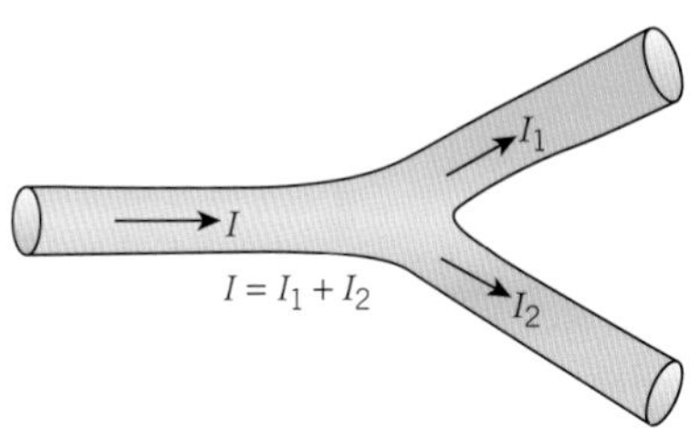


▲ **Figure 4** *Current dividing at a junction*

This rule, that the currents at a junction must add up, is called **Kirchhoff's first law**. It is a consequence of the conservation of charge – all of the charges in the stream of charges must go one way or the other.

## Potential difference

Charges will move when attracted by charges of the opposite sign or when repelled by charges of the same sign. Inside a battery cell, chemical reactions produce an electrical potential energy difference between the two terminals, resulting in a positively charged terminal and a negatively charged terminal. This can be compared with the gravitational potential energy difference down a hill, with the positive pole being the higher point and the negative the lower.

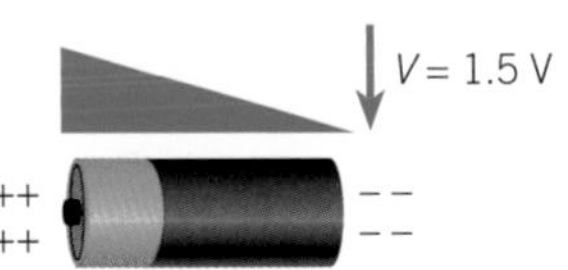


▲ **Figure 5** *Potential difference produced by a battery cell*

If the poles of the cell are joined by a conducting path, charges will flow as shown in Figure 6 (next page).

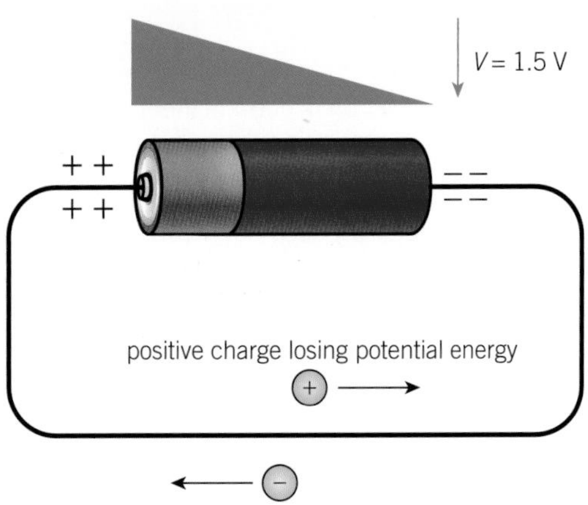


▲ **Figure 6** *Potential energy changes for charges in a circuit*

The direction in which charges flow depends on the sign of the charges: positive charges flow from the positive pole to the negative pole, and negative charges move the other way. As metals contain free electrons, which are negatively charged, the movement of negative charges is more usual in a circuit.

## Potential difference and potential energy

When charges move between two points in a circuit, their electrical potential energy changes by an amount $\Delta E$. By the principle of conservation of energy, the work done $W$ in this movement is given by $W = \Delta E$.

**Potential difference** (p.d.), or 'voltage', $V$ is the potential energy difference $\Delta E$ per unit charge moving between two points:

$$V = \frac{\Delta E}{Q} = \frac{W}{Q}$$

so the potential energy change $\Delta E = W = VQ$. For negative charges moving 'uphill' from $V = 0$ to $V = 1.5\,\text{V}$ the p.d. = +1.5 V, but as $Q$ is negative, the potential energy change, $\Delta E = VQ$, is negative – the charges have lost potential energy, just like positive charges moving 'downhill' through a p.d. of −1.5 V.

# Power dissipated in electrical circuits

A p.d. between the ends of a wire will accelerate the free electrons in it, but their movement down the wire is obstructed by their interactions with the positive ion cores of the metal atoms, so they do not gain kinetic energy. The potential energy lost does work on the wire, heating it. This 'wasted' energy is called **dissipation** and the process is often called 'Joule heating'.

$$V = \frac{W}{Q} \Rightarrow W = VQ$$

and

$$I = \frac{Q}{t} \Rightarrow Q = It$$

so

$$W = V(It) = VIt$$

**Power** $P$ is the rate at which energy is transferred, and it is measured in $\text{J s}^{-1}$, which is the **watt** (W).

$$P = \frac{W}{t} = \frac{VIt}{t} = IV$$

Figure 7 shows a component called a resistor, with p.d. $V$ between its ends and a current $I$ through it. As you will see in Topic 3.2, the ratio $\frac{V}{I}$ for any conductor is called its **resistance** $R$, measured in **ohms** ($\Omega$). The equation $R = \frac{V}{I}$ rearranges to give $V = IR$.

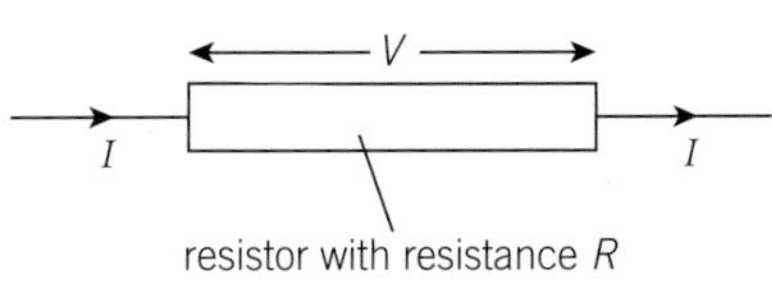


▲ **Figure 7** *The p.d. across, and current through, a resistor*

This means that there is an alternative equation to use for the dissipative heating in any component of resistance $R$ – the rate of heating or power, $P$ is given by:

$$P = IV = (IR)I = I^2R$$

## Worked example: The maximum current taken by a resistor

A resistor is labelled 10 kΩ 0.60 W. The power rating is the maximum safe power output. What is the maximum current it can take without overheating?

**Step 1:** Write down the relationship between power, resistance, and current.

$$P = I^2R \text{ so } P_{max} = I_{max}^2R$$

**Step 2:** Rearrange to give

$$I_{max}^2 = \frac{P}{R} = \frac{0.60\,\text{W}}{10 \times 10^3\,\Omega} = 6.0 \times 10^{-5}\,\text{A}^2$$

**Step 3:** Take the square root.

$$I = \sqrt{(6.0 \times 10^{-5}\,\text{A}^2)} = 0.00774...\,\text{A}$$

$$= 7.7\,\text{mA (2 s.f.)}$$

## Summary questions

**1** A charge of 10 C passes through a conductor in 2 s. The potential difference between the ends of the conductor is 12 V.

**a** Calculate the current. *(1 mark)*

**b** Calculate the power dissipated in the conductor. *(1 mark)*

**2** Copy the circuit of Figure 8, replacing $V_1$, $V_2$, and $I_1$ with their values.

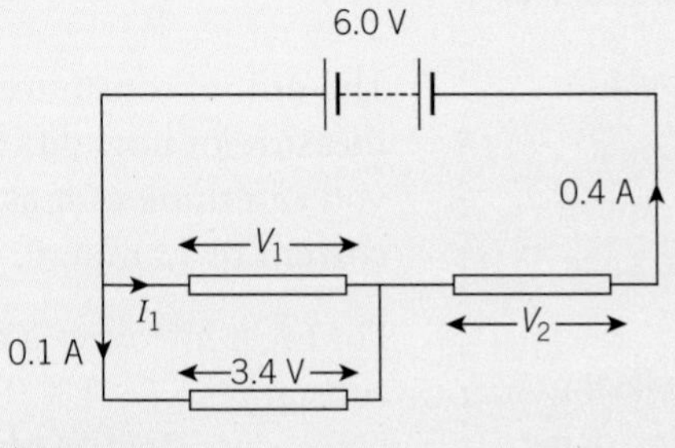


▲ Figure 8

*(3 marks)*

**3** The current in a torch bulb is 0.24 A, which is a flow of electrons, each of charge $-1.6 \times 10^{-19}$ C. Calculate the number of electrons entering (and leaving) the filament of the bulb each second. *(2 marks)*

**4** In the electrolytic cell shown in Figure 9, copper ions of charge $+3.2 \times 10^{-19}$ C move from the positive plate to the negative plate, and chloride ions of charge $-1.6 \times 10^{-19}$ C travel in the opposite direction. A current of 0.35 A flows for 2 minutes. Calculate the number of ions of each type which reach the plates.

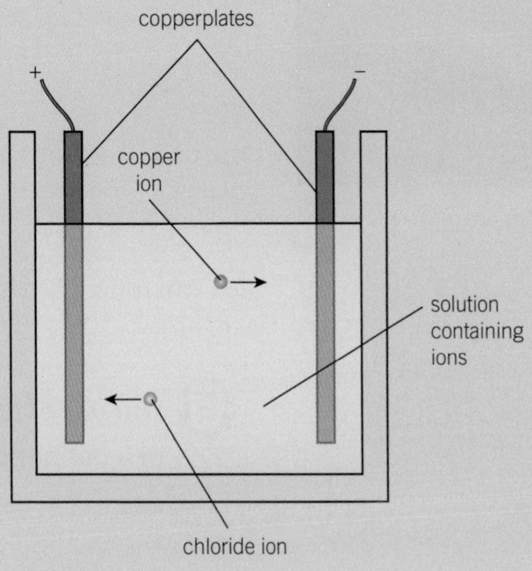


▲ Figure 9

*(3 marks)*

# 3.2 Conductors and resistors

Specification references: 3.1.2a(iii), 3.1.2a(vi), 3.1.2b(i), 3.1.2b(ii), 3.1.2b(iii), 3.1.2c(i), 3.1.2c(ii), 3.1.2d(i)

### Learning outcomes

Describe, explain, and apply:

- → the terms resistance ($R$) and conductance ($G$)
- → the equations $R = \frac{V}{I}$ and $G = \frac{I}{V}$
- → the equations $R = R_1 + R_2 + ...$ or $\frac{1}{G} = \frac{1}{G_1} + \frac{1}{G_2} + ...$ for a series combination of conductors
- → the equations $G = G_1 + G_2 + ...$ or $\frac{1}{R} = \frac{1}{R_1} + \frac{1}{R_2} + ...$ for a parallel combination of conductors
- → graphs of current against potential difference, and of resistance or conductance against temperature
- → an experiment to obtain the $I$–$V$ characteristic of an ohmic and a non-ohmic conductor.

When a car is moving at its top, constant, speed along a straight, level, road, the forwards push from the engine balances friction and drag forces from the air it is travelling through. In a similar way, when there is a constant current in a conductor, the forwards push on the moving charges from the p.d. between the ends of the conductor is balanced by the obstructing effect of the atoms in the conductor interacting with the charges. In each case we can use the word *resistance* – air resistance stops the car accelerating, and electrical resistance stops the charges accelerating.

▲ **Figure 1** *Resistance at work on a moving vehicle*

## Resistance and Ohm's law

The ratio of the p.d. $V$ across a component to the current $I$ which it produces for a component is defined as its **resistance**, $R$.

$$\text{resistance } R = \frac{V}{I}$$

The unit of resistance is V A$^{-1}$, which is the **ohm** (Ω). Resistance is a measure of how difficult it is to get current through the component – you can think of it as the number of volts of p.d. you need to get a current of 1 ampere.

For most metals at a constant temperature, the resistance of a wire is constant, so $I \propto V$ – this is **Ohm's law**. A conductor that obeys Ohm's law is described as **ohmic**.

## Conductance

Although the concept of resistance was developed first, metals are generally such good conductors it is more natural to define them by the opposite of resistance, that is, **conductance**. This is given the symbol $G$ and is defined as:

$$\text{conductance } G = \frac{I}{V}$$

The unit of conductance is A V$^{-1}$, which is the **siemen** (S). It is the inverse of resistance.

$$G = \frac{1}{R}$$

You can think of conductance as how many amperes you get from a p.d. of 1 V.

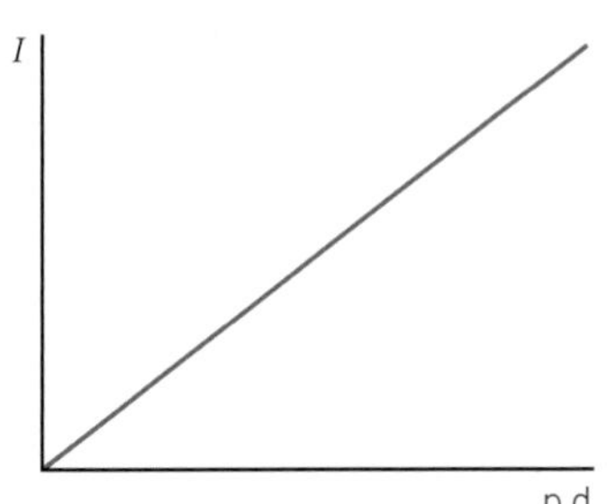


▲ **Figure 2** *Current against p.d. for an ohmic conductor*

### Worked example: Conductance and resistance of a car headlamp

A standard car headlamp bulb is rated 55 W 12 V. What are the resistance and conductance of the filament?

**Step 1:** To find the current through the bulb, write down the relationship between power, p.d., and current.

$$P = IV \text{ so current } I = \frac{P}{V} = \frac{55\,\text{W}}{12\,\text{V}} = 4.58...\,\text{A}$$

**Step 2:** Now calculate the resistance, using the definition of resistance.

$$R = \frac{V}{I} = \frac{12\,\text{V}}{4.58...\,\text{A}} = 2.61...\,\Omega = 2.6\,\Omega \text{ (2 s.f.)}$$

**Step 3:** Then calculate the conductance, using the definition of conductance.

$$G = \frac{I}{V} = \frac{4.58...\,\text{A}}{12\,\text{V}} = 0.381...\,\text{S} = 0.38\,\text{S (2 s.f.)}$$

Or you can use $G = \frac{1}{R} = \frac{1}{2.61...\,\Omega}$ to get the same answer.

## Series and parallel combinations of conductors

### Parallel circuits

Figure 3 shows two ohmic conductors **1** and **2** connected **in parallel** with a battery of voltage $V$.

The charges moving from the battery divide at one junction, pass along two different routes, and then recombine, so $I = I_1 + I_2$. The potential 'hill' down which each charge falls as it moves around the circuit is the same, whichever path it takes, so the p.d. across each conductor is $V$, assuming the connecting wires are of very high conductance (low resistance).

The conductance of the parallel combination is

$$G = \frac{I}{V} = \frac{I_1 + I_2}{V} = \frac{I_1}{V} + \frac{I_2}{V} = G_1 + G_2$$

This shows that if you have components in parallel, the total conductance is obtained by adding the conductance of each component. For more than two conductors in parallel,

$$G = G_1 + G_2 + ...$$

Because $G = \frac{1}{R}$, this expression $G = G_1 + G_2 + ...$ can be written as

$$\frac{1}{R} = \frac{1}{R_1} + \frac{1}{R_2} + ...$$

where $R$ is the resistance of the combination and $R_1$, $R_2$, etc. are the resistances of the different components.

### Study tip: Decoding circuits

Before analysing a circuit diagram, first be sure that you know the circuit symbols that you met at GCSE. Then trace circuit loops around from the positive battery terminal to the negative battery terminal, looking out for intersections where there is a choice of routes.

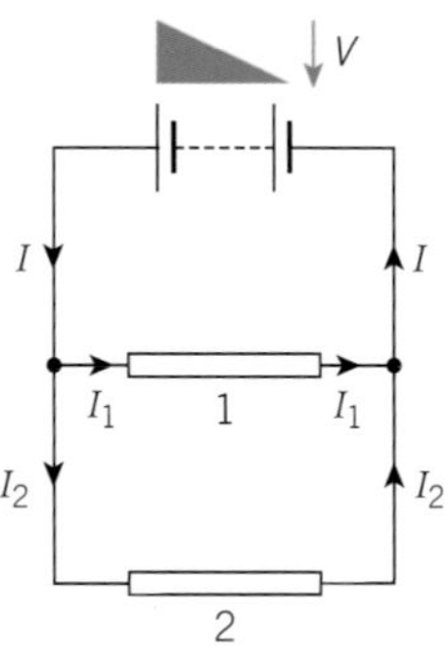


▲ **Figure 3** *Parallel conductors*

### Worked example: Adding parallel resistances

Two resistors are connected in parallel as shown in Figure 4. What is the resistance of the combination?

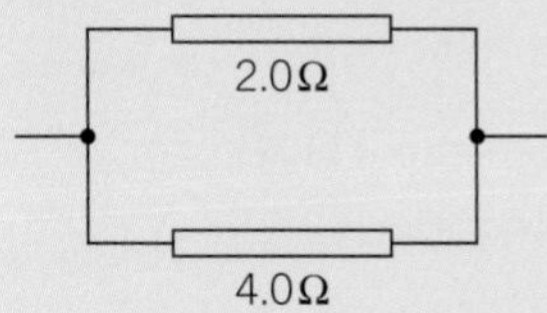


▲ **Figure 4** *Resistors in parallel*

**Step 1:** Even though resistances are given here, it is easier for a parallel circuit to work in terms of conductances. Express each resistance as a conductance.

$$\frac{1}{2.0\ \Omega} = 0.5\ \text{S and } \frac{1}{4.0\ \Omega} = 0.25\ \text{S}$$

**Step 2:** Add the conductances, because they are in parallel.

Total conductance = 0.5 S + 0.25 S = 0.75 S

**Step 3:** Find the total resistance by calculating the inverse of the conductance.

$$R = \frac{1}{G} = \frac{1}{0.75\ \text{S}} = 1.3\ \Omega\ \text{(2 s.f.)}$$

## Study tip

Note that adding resistors in parallel always results in an overall resistance that is smaller than the smallest of the values being added. This is because the overall conductance is the sum of the individual conductances, and so must be greater than the highest conductance in the combination.

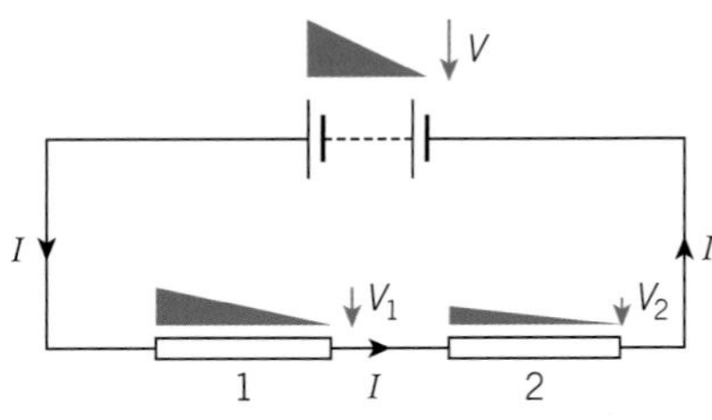


▲ **Figure 5** *Series conductors*

## Series circuits

Figure 5 shows two ohmic conductors **1** and **2** connected **in series** with a battery of voltage $V$.

The charges moving from the battery have no choice in where to go, so the current $I$ is the same throughout. As a charge $Q$ moves around the circuit, potential energy is dissipated as heat in the two conductors, and the total energy must be that obtained from the battery, so $VQ = V_1Q + V_2Q$. Dividing by $Q$ gives $V = V_1 + V_2$. The total p.d. is shared between the two conductors.

The resistance of the series combination is:

$$R = \frac{V}{I} = \frac{V_1 + V_2}{I} = \frac{V_1}{I} + \frac{V_2}{I} = R_1 + R_2$$

If you have components in series, the total resistance is obtained by adding the resistance of each component. For more than two conductors in series,

$$R = R_1 + R_2 + \ldots$$

Because $R = \frac{1}{G}$, $R_1 = \frac{1}{G_1}$, and $R_2 = \frac{1}{G_2}$, this expression $R = R_1 + R_2 + \ldots$ can be written as:

$$\frac{1}{G} = \frac{1}{G_1} + \frac{1}{G_2} + \ldots$$

Therefore, for combinations of components in series circuits it is easier to use resistance than conductance.

### Worked example: P.d. across series resistances

A 2.0 Ω and a 4.0 Ω resistor are connected in series to a 4.5 V battery, in a circuit like that in Figure 5. What is the p.d. across the 4 Ω resistor?

**Step 1:** You need to find the current through the 4 Ω resistor, which is the current throughout the series circuit. So first calculate the total resistance.

$$R = R_1 + R_2 = 2.0\ \Omega + 4.0\ \Omega = 6.0\ \Omega$$

**Step 2:** Find the current.

$$R = \frac{V}{I} \text{ so } I = \frac{V}{R} = \frac{4.5\,\text{V}}{6.0\,\Omega} = 0.75\,\text{A}$$

**Step 3:** Now you can find the p.d. across the 4 Ω resistor.

$$R = \frac{V}{I} \text{ so } V = IR = 0.75\,\text{A} \times 4.0\,\Omega = 3.0\,\text{V}$$

## Study tip

Note in this example that 3.0 V is two-thirds of the total voltage of 4.5 V. The voltages are in the same proportion as the resistances, that is, 4.0 Ω is two-thirds of the total resistance of 6.0 Ω. This useful fact is developed in Topic 3.5, Potential dividers.

Figure 6 gives a summary for dealing with parallel and series combinations of components.

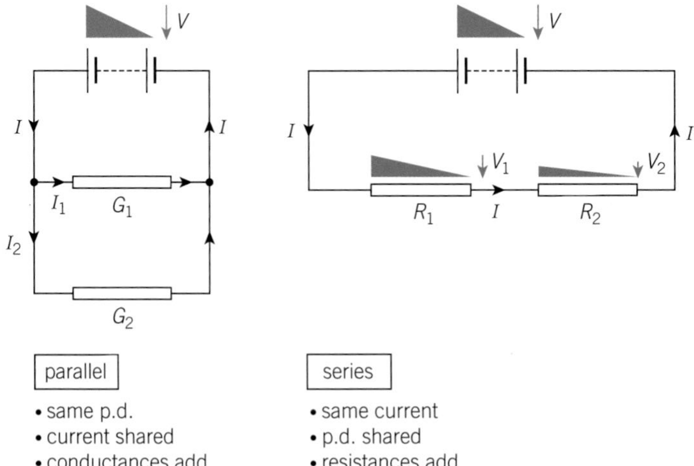


▲ **Figure 6** *Parallel and series circuits – a summary*

## More complex networks

Real circuits are often networks with combinations of both series and parallel components. To analyse these, you need to look at the separate parallel and series parts one step at a time, converting from resistance to conductance and back with $G = \frac{1}{R}$ where necessary.

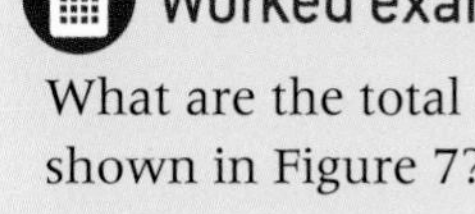

### Worked example: Step-by-step analysis of a network

What are the total conductance and resistance of the combination shown in Figure 7?

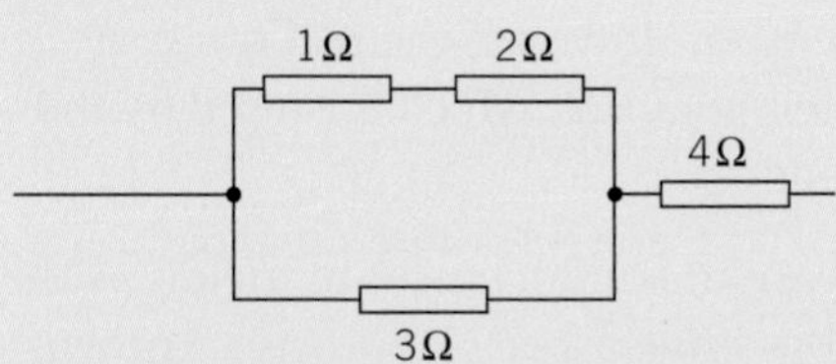


▲ **Figure 7** *The network to be analysed*

Figure 8 shows the sequence of steps in the solution.

**Step 1:** add series resistances and find the conductance of parallel components

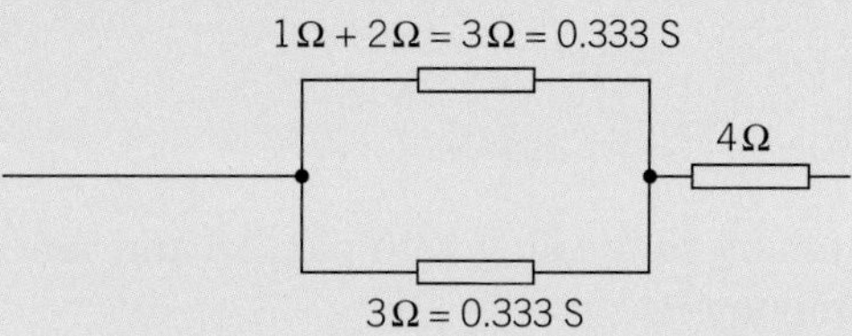


**Step 2:** add parallel conductances

0.333 S + 0.333 S = 0.666 S = 1.5Ω    4Ω

**Step 3:** add series resistances

1.5Ω + 4Ω = 5.5Ω = 0.182 S

▲ **Figure 8** *The solution*

So the circuit has an overall conductance of 0.2 S and resistance of 5 Ω.

## Resistance and temperature

Metals are ohmic conductors if the temperature can be kept constant, but in reality a current in a wire dissipates energy through Joule heating (at a rate of $P = I^2R$), which will raise the temperature. This will increase the resistance of the wire.

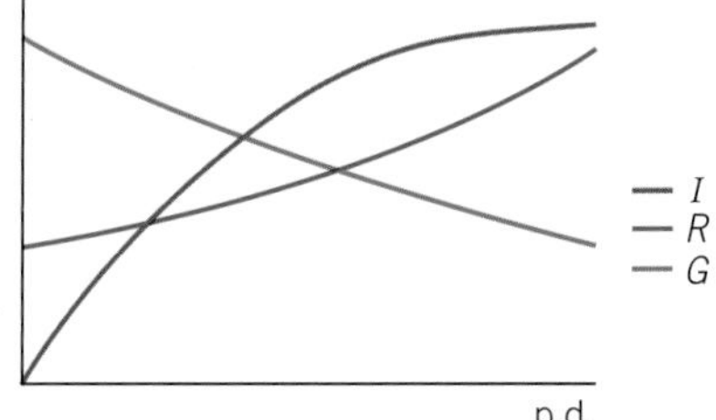


▲ **Figure 9** *Variation of current, resistance, and conductance with p.d. for a filament lamp*

As an example, if the current in a filament lamp is measured for a range of p.d.s up to the working voltage, when the filament glows white-hot it appears not to obey Ohm's law. Figure 9 shows the variation of current, resistance, and conductance for a filament lamp as the p.d. across it is increased. The $y$-axis scales have been adjusted to allow the graphs to be compared easily. Such a component, which does not obey Ohms' law under ordinary working conditions, is referred to as a **non-ohmic** conductor.

P.d. is plotted as the independent variable in Figure 9, because it is the variable you change to obtain the data. However, the real cause of variation of $R$ and $G$ is changing temperature, which is caused by the current and the $I^2R$ heating it produces.

This increase in resistance (decrease in conductance) in metals as they get hotter has both negative and positive aspects. If precise, known, and constant resistance values are needed, it can be a nuisance, so alloys such as constantan have been developed where the change is very small. On the other hand, the known variation of the resistance of platinum with temperature allows measurement of its resistance to be used as a temperature sensor.

When samples of semiconducting non-metals such as silicon are heated, the resistance does not increase, but decreases. The way that temperature changes affect resistance is discussed further in Topic 3.4, Conductance under the microscope.

### Practical 3.1.2d(i): Investigating *I–V* characteristics for ohmic and non-ohmic components

Figure 10 shows a circuit which may be used to investigate how the current through a component X, shown here as a labelled box, varies with the p.d. across it.

In this investigation, it is important to bear in mind that the component may behave differently if reversed.

It is also important to consider:

- the range of p.d.s applied to the component – changing the range will probably affect the choice of ammeter used to measure the current
- the resolution of the ammeter and the voltmeter used
- the reproducibility of the readings obtained.

The source of variable p.d. may be a variable d.c. power supply, or a battery and rheostat (variable resistor).

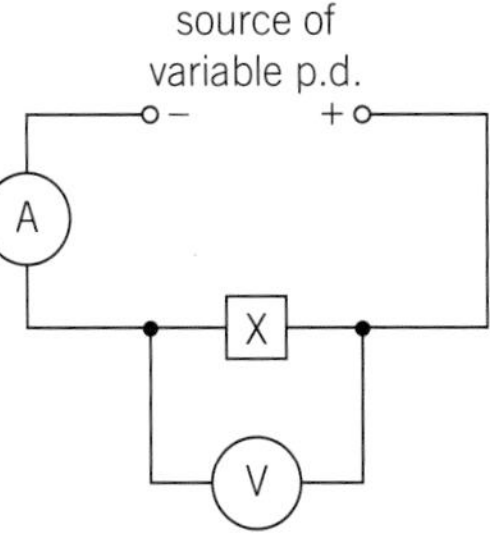


▲ **Figure 10** *Investigating an electrical component, X*

## Summary questions

**1** A conductor has a current of 250 mA when there is a p.d. of 5.0 V across it. Assuming that it is ohmic, calculate the p.d. necessary to produce a current of 1.3 A. *(2 marks)*

**2 a** Explain why the *I–V* graph in Figure 9 is not linear, but curves downwards. *(2 marks)*

**b** Explain how the shapes of the *R–V* and *G–V* graphs in Figure 9 are related to the *I–V* graph. *(2 marks)*

**3** One type of platinum resistance thermometer has a resistance of 100.00 Ω at a temperature of exactly 0 °C. The variation of the resistance of platinum with temperature is given by the equation

$$R_\theta - R_{0°C} = \alpha R_{0°C}\theta$$

where $\theta$ is the temperature in °C and $\alpha$ is a constant for platinum with value $\alpha = 0.003\,926\ °C^{-1}$. At a certain temperature $\theta_X$, the resistance, is measured to be 106.24 Ω. Calculate the temperature $\theta_X$ to 3 s.f. and explain why very accurate and precise measurement of resistance is needed in this case. *(3 marks)*

**4** The graph of Figure 11 shows how the current through a metal wire varies with temperature. The p.d. between the ends of the conductor was constant throughout. Sketch a copy of this graph, and add lines to show how the resistance and the conductance are varying with temperature. Do not do any calculations. *(2 marks)*

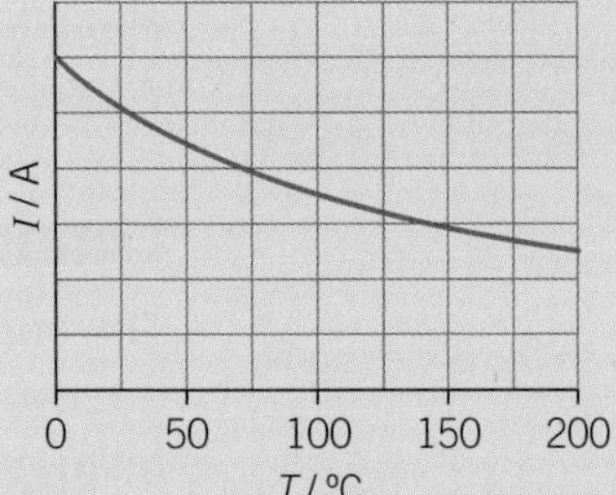


▲ **Figure 11**

**5** In the circuit shown in Figure 12, the current in the 3.0 Ω resistor is measured to be 0.45 A. Calculate the p.d. *V* of the battery. *(4 marks)*

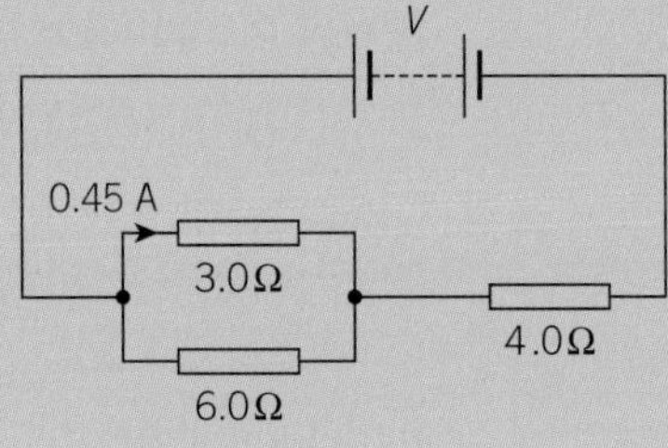


▲ **Figure 12**

# 3.3 Conductivity and resistivity

## Specification references 3.1.2b(i), 3.1.2c(iii), 3.1.2d(ii)

**Learning outcomes**

Describe, explain, and apply:

- the concepts of resistivity ($\rho$) and conductivity ($\sigma$)
- the equations $R = \frac{\rho L}{A}$ and $G = \frac{\sigma A}{L}$
- an experiment using one method of measuring resistivity or conductivity.

The resistance and the conductance of a metal wire are examples of *extensive* properties, because they depend on the length and thickness of the wire concerned, as well as the metal from which it is made. To describe and make appropriate use of the electrical properties of the metal concerned we need some measure which does not depend on the dimensions of the sample used. Such a measure is called an *intensive* or *bulk* property of the material. Other bulk properties that you will meet in the course are density, the Young modulus and refractive index.

### Dependence of resistance and conductance of a wire on its dimensions

The way in which the conductance and resistance of a sample of wire depend on the dimensions of the wire can be shown by considering identical wires in series and in parallel, as in Figure 1.

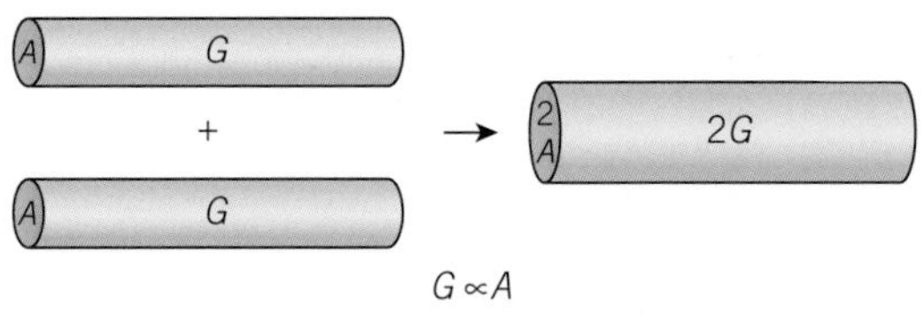


▲ **Figure 1** *Dependence of R and G on the dimensions of a wire*

Doubling the length $L$ doubles the resistance, implying that $R \propto L$ and so $G \propto \frac{1}{L}$.

In a similar way, doubling the cross-sectional area $A$ doubles the conductance, implying that $G \propto A$ and so $R \propto \frac{1}{A}$.

Combining these two relationships for $R$ and for $G$ we get:

$$R \propto \frac{L}{A} \text{ and } G \propto \frac{A}{L}$$

These two equivalent relationships apply for any given material. Each needs a constant of proportionality, or *bulk constant*, true for a particular material, to allow calculations for and comparisons between different materials.

### Conductivity and resistivity

The bulk properties for describing electrical conduction are the constants that must replace the proportionalities in the relationships above.

$$G \propto \frac{A}{L} \Rightarrow G = \frac{\sigma A}{L}$$

where $\sigma$ (sigma) is the electrical **conductivity** whose unit is $\text{S m}^{-1}$.

$$R \propto \frac{L}{A} \Rightarrow R = \frac{\rho L}{A}$$

where $\rho$ (rho) is the electrical **resistivity** whose unit is $\Omega\text{m}$.

$$\text{Just as } G = \frac{1}{R}, \text{ so } \sigma = \frac{1}{\rho}.$$

To get a feel for the meaning of resistivity and conductivity, think of the cube of material shown in Figure 2.

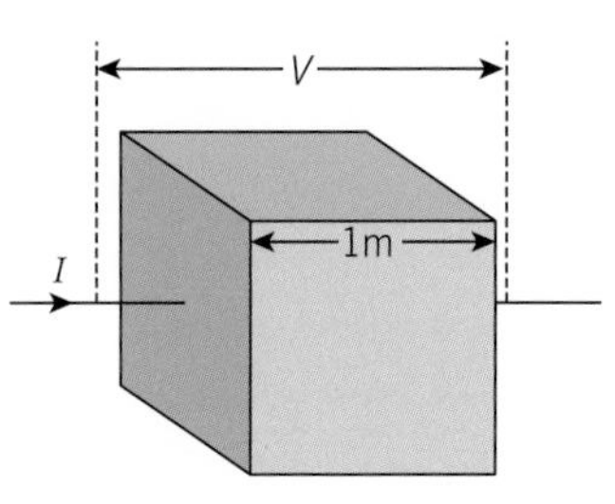


▲ **Figure 2** *Visualising conductivity and resistivity*

For this huge conductor

$$G = \frac{\sigma A}{L} = \frac{\sigma \times 1\,\text{m}^2}{1\,\text{m}}$$

so that conductivity $\sigma$, in $\text{S}\,\text{m}^{-1}$, is numerically the same as the conductance $G$ in S, and the resistivity $\rho$, in $\Omega\,\text{m}$, is numerically the same as the resistance $R$ in $\Omega$.

If the conductor is a metal, the conductance $G$ across a 1 m cube, being the current in A produced by a p.d. of 1 V, is huge, and its resistance $R$ is correspondingly tiny. Some values are shown in Table 1.

▼ **Table 1** *Values of conductivity and resistivity for some metals*

| Metal | Conductivity, $\sigma$ / $\text{S}\,\text{m}^{-1}$ | Resistivity, $\rho$ / $\Omega\,\text{m}$ |
|---|---|---|
| copper | $6.41 \times 10^{7}$ | $1.56 \times 10^{-8}$ |
| iron | $1.12 \times 10^{7}$ | $8.90 \times 10^{-8}$ |
| silver | $6.62 \times 10^{7}$ | $1.51 \times 10^{-8}$ |

## Worked example: Resistance of copper conductors

A 6 V 0.2 A bulb is connected to its battery via copper wires of diameter 0.5 mm and length 0.5 m. Do these wires add significantly to the resistance of the circuit? Use the data in Table 1.

**Step 1:** Find the resistance of the bulb.

$$R = \frac{V}{I} = \frac{6\,\text{V}}{0.2\,\text{A}} = 30\,\Omega$$

**Step 2:** You need to find the resistance of the wires. You can obtain the resistivity of copper from Table 1. From the given dimensions of the wires, calculate the cross-section area.

$$A = \pi r^2 = \pi \times (0.25 \times 10^{-3}\,\text{m})^2 = 1.96... \times 10^{-7}\,\text{m}^2$$

**Step 3:** Now calculate the resistance of the copper wires:

$$R = \rho \frac{L}{A} = \frac{1.56 \times 10^{-8}\,\Omega\,\text{m} \times 0.5\,\text{m}}{1.96... \times 10^{-7}\,\text{m}^2} = 0.040\,\Omega \text{ (2 s.f.)}$$

This is negligible compared with the resistance of the bulb, $30\,\Omega$.

## Conductors, insulators, and semiconductors

Conductivity and resistivity are measures for the electrical properties of materials that we can use to classify the materials as conductors, insulators or **semiconductors**. As Figure 3 (next page) shows, the range of values is immense. If a linear scale were used, with a resistivity of $10^{-8}\,\Omega\,\text{m}$ taking up 1 cm, then a book page over 100 000 light-years long would be needed to include polystyrene! A **logarithmic scale** must be used, with steps going up in factors of 100.

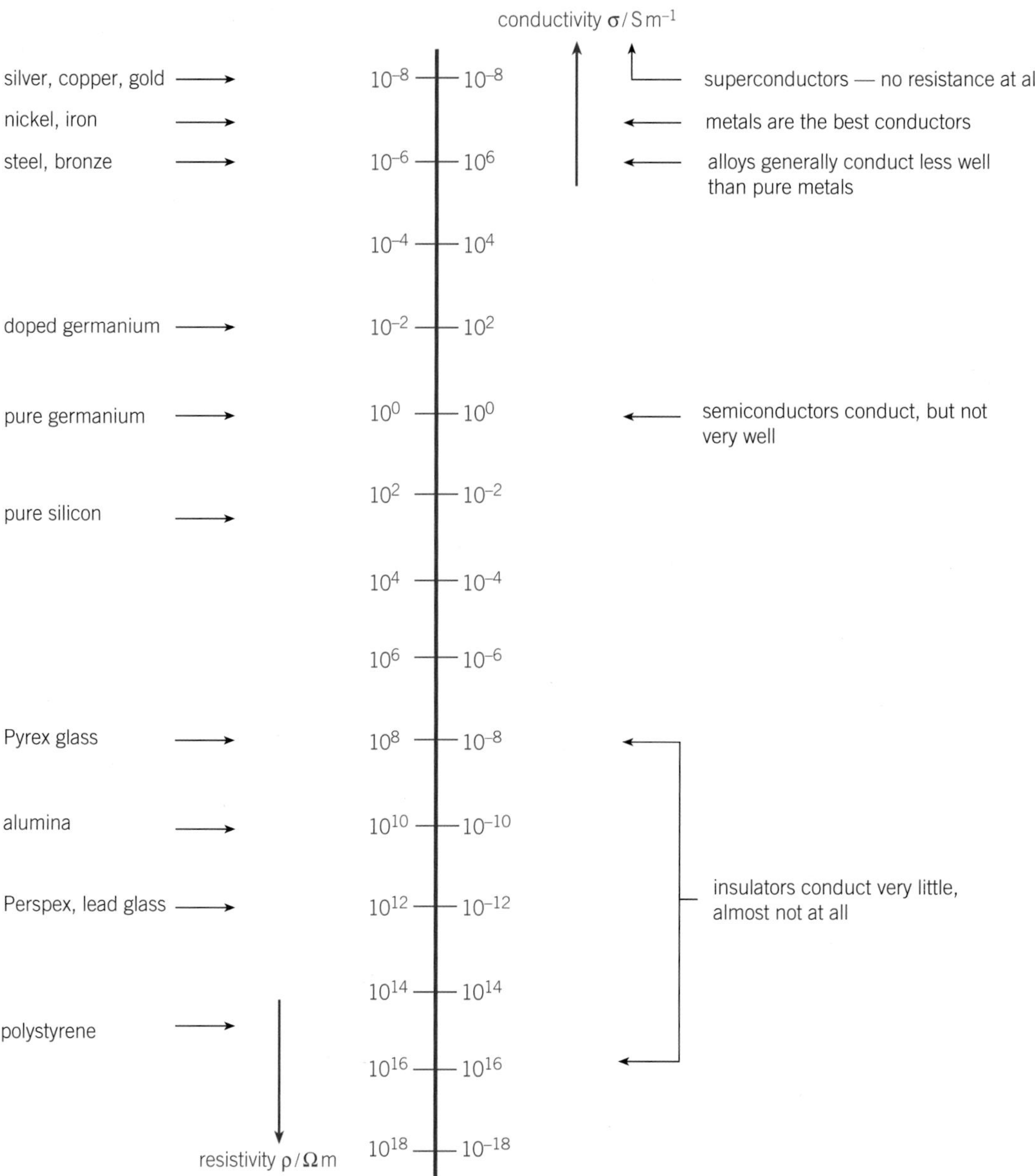


▲ **Figure 3** *Logarithmic scale of resistivity and conductivity*

## Measuring resistivity or conductivity

The particular problems in measuring these electrical quantities depend on the material you are investigating. First it is necessary to make an order of magnitude calculation of the measurements you might expect to be able to make, using approximate values taken from Figure 3.

- For an insulator, you need a tiny value of $L$, a large value of $A$, a large p.d., and a very, very sensitive ammeter to have any chance of making a measurement at all.
- For a metal, the problems are quite different. You need a large $L$ and a very small $A$, and you must be aware of complications caused by $I^2R$ heating of the wire and systematic errors such as p.d. drops in places other than across the conductor in question.

**Synoptic link**

This experiment involves measurements on a long wire. Refer to the issues involved in measurement of the Young modulus of a wire in Chapter 4, Testing materials.

## Practical 3.1.2d(ii): Determining conductivity or resistivity

The circuit used in Practical 3.1.2d(i) in Topic 3.2 can be used here, with the component **X** being a sample of the material being investigated, which could be a metal wire. It may be necessary to do a preliminary experiment to find the most suitable dimensions for this sample.
Practical considerations in this experiment include:

- appropriate ranges for the p.d. (and voltmeter) and ammeter used
- the uncertainty in the measurements, particularly of the most uncertain measurement
- how the wire is connected into the circuit, and how the p.d. between the ends of the measured length is measured.

## Extension: Semiconductors

Metals have very large conductivities ($>10^6\,\mathrm{S\,m^{-1}}$) and insulators very small ones ($<10^{-8}\,\mathrm{S\,m^{-1}}$). One class of materials has conductivities in the range 0.01–100 $\mathrm{S\,m^{-1}}$. These are **semiconductors**, of which germanium and silicon have been most important historically, although compounds such as gallium arsenide are also widely used.

The electrical properties of semiconductors are not constant, as with metals, but can be modified. Some of this modification is through *doping* a pure semiconductor with other elements which greatly affect its conductivity, and by combining thin layers of differently doped semiconductors.

Changing the voltages applied to semiconductor devices also changes their electrical behaviour, so they can actively modify signals applied to them. Since the 1950s the electronics industry has developed this behaviour and created a range of devices from transistors and amplifiers to complex computers. Each year these devices have become smaller, faster and cheaper as ever-tinier semiconductor and metal elements are 'grown' in situ on integrated circuits.

**Questions**

The integrated circuit in Figure 4 contains many tiny transistors about 22 nm in size. What is the conductance across opposite faces of a silicon cube of side length 22 nm (assume $\sigma = 1\,\mathrm{S\,m^{-1}}$)? What would the conductance be if this cube could be made half as big in each dimension?

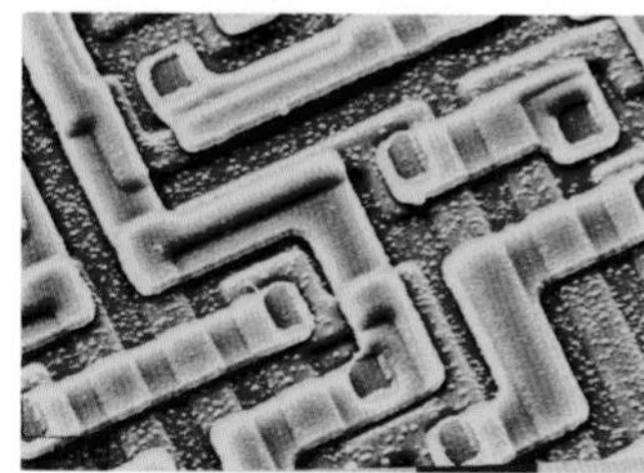

◀ **Figure 4** *Scanning electron micrograph of the surface of an integrated circuit from a computer's Arithmetic Logic Unit*

## Summary questions

1 Explain why resistivity has the unit $\Omega\,\mathrm{m}$ and conductivity has the unit $\mathrm{S\,m^{-1}}$. *(3 marks)*

2 A 230 V 50 W lamp has a filament which is a coil of tungsten wire whose resistivity, at the working temperature, is $6.5 \times 10^{-7}\,\Omega\,\mathrm{m}$. The diameter of the filament is $3.2 \times 10^{-5}$ m. Calculate the length of the filament. *(5 marks)*

3 It is planned to measure the conductivity of polythene, thought to be about $10^{-12}\,\mathrm{S\,m^{-1}}$, using a sheet of polythene 0.1 mm thick sandwiched between aluminium foil electrodes, as shown in Figure 5, with an EHT supply of 5 kV and an ammeter capable of reading currents down to 0.1 μA.
Make an appropriate calculation to predict whether this experiment could be successful. *(3 marks)*

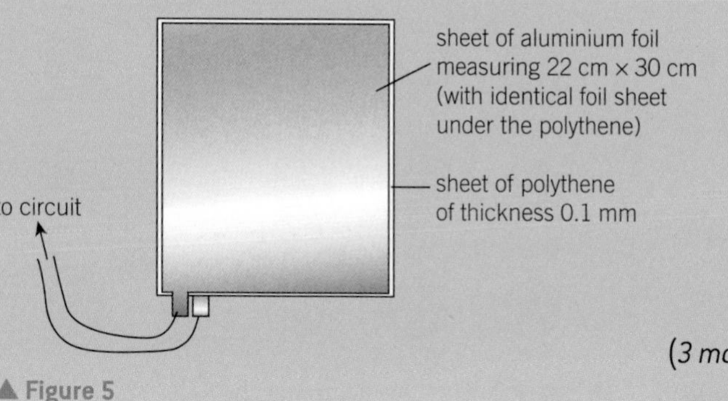


▲ **Figure 5**

# 3.4 Conduction under the microscope

Specification references 3.1.2a(viii), 3.1.2b(i), 3.1.2b(iii)

**Learning outcomes**

Describe, explain, and apply:

- simple electrical behaviour of metals and insulators in terms of the number density of charge carriers
- simple electrical behaviour of semiconductors in terms of the changing number density of mobile charge carriers on heating.

X-ray crystallography, pioneered by William and Lawrence Bragg in the early years of the 20th century, revealed the crystalline structure of a number of compounds and elements, including metals. The regular lattice structure of metallic crystals, together with the high melting point and stiffness of metals, shows that the attractive forces in metals are large.

## A conduction model for metals

In a metal, each atom contributes one or more electrons to a 'soup' of free, mobile electrons which move about in the lattice of positive ion cores. It is this delocalised cloud of electrons which provides the strong metallic bond. In Figure 1 the electrons are represented as individual charged points moving with random velocities, very much like the molecules in a gas.

When a p.d. is applied across the lattice, the electrons each experience a force pulling them to the positive side (or pushing them from the negative side). This produces a slow general drift in one direction, superimposed on the rapid random motion – imagine a swarm of bees drifting slowly in the breeze. The rate of flow of charge due to this net movement is the current produced by the p.d.

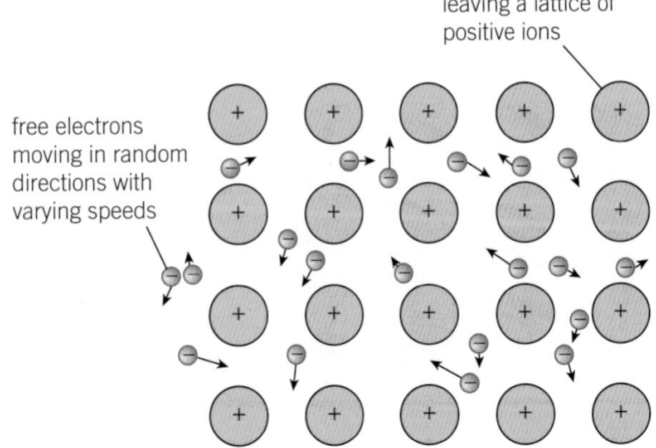


▲ **Figure 1** *A simple model of the ions and electrons in a metal*

Figure 1 shows the lattice of fixed positive ions. At room temperature, the electrons have kinetic energy and their motion is obstructed by the positive ions. The positive ions also have kinetic energy but are bound in their positions, and vibrate. These vibrations, which increase with temperature, mean that the electron paths in the lattice are obstructed more in a hotter metal than in a cooler one. This is illustrated in Figure 2, where, for clarity, only two electrons are shown in each lattice.

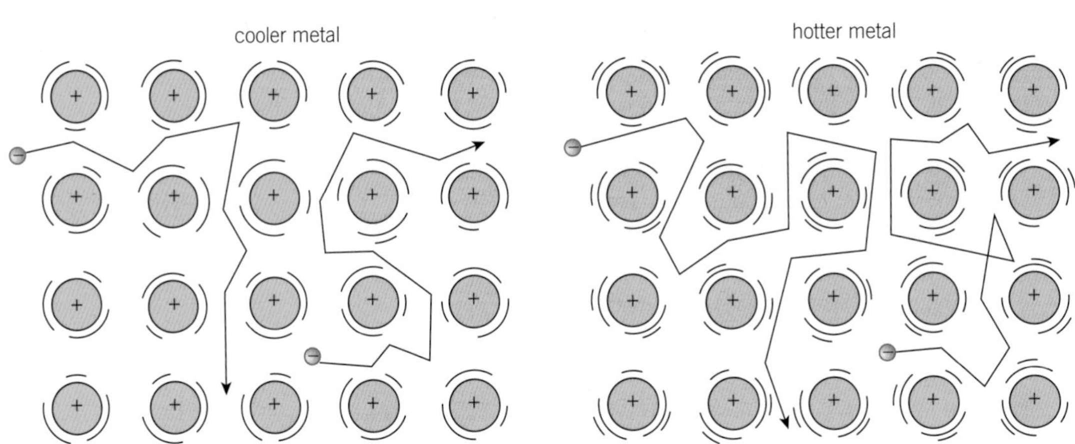


▲ **Figure 2** *Vibrating metal ions obstruct charge flow*

## Number density of charges in a metal

The simplified model in Figure 1 suggests that each atom donates one free electron to the delocalised cloud. If the **number density** $n$ of these electrons (the number per $m^3$) is known, the mean **drift velocity** of the electrons can be calculated.

### Worked example: Drift velocity of electrons

$1\,m^3$ of copper contains $8.5 \times 10^{28}$ atoms, so the number density of free electrons $n = 8.5 \times 10^{28}\,m^{-3}$. What is the drift velocity of electrons in a copper wire of cross-sectional area $3.0 \times 10^{-6}\,m^2$ carrying a current of 10 A?

**Step 1:** Think about a length of copper wire of cross-sectional area $A$, as shown in Figure 3. You need to find the length $L$ for which all the electrons in that length of wire will drift through the coloured end of wire in 1 second.

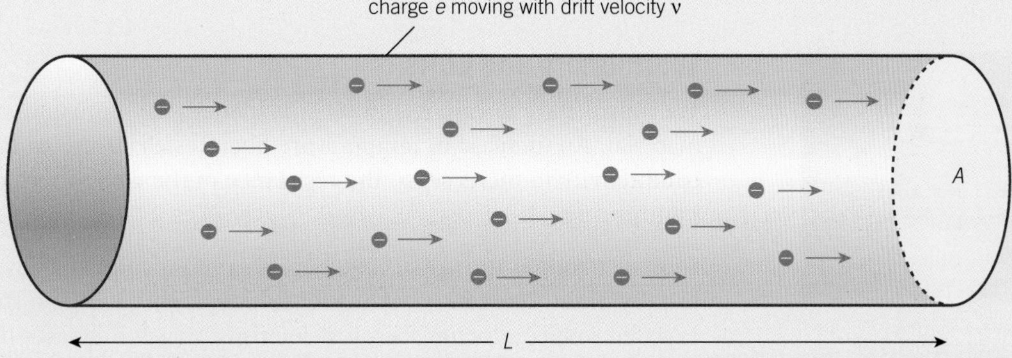


▲ **Figure 3** *Drift velocity of electrons*

**Step 2:** Find the number of electrons that drift through the coloured cross-section in 1 second.

Current $I = \frac{\Delta Q}{\Delta t} = 10\,A = \frac{\Delta Q}{1\,s}$, so $\Delta Q = 10\,C$.

Each electron has charge $e = 1.6 \times 10^{-19}\,C$, so the number of electrons is given by

$$N = \frac{\Delta Q}{e} = \frac{10\,C}{1.6 \times 10^{-19}\,C} = 6.25 \times 10^{19}$$

**Step 3:** You are given the number density of electrons, so now you can calculate the volume of the length $L$. As $1\,m^3$ contains $8.5 \times 10^{28}$ free electrons, the volume of the wire $V$ is given by

$$\frac{V}{1\,m^3} = \frac{6.25 \times 10^{19}}{8.5 \times 10^{28}}$$

giving $V = 7.4 \times 10^{-10}\,m^3$

**Step 4:** Volume $V$ is equal to $LA = L \times 3.0 \times 10^{-6}\,m^2$.

$$\text{So } L = \frac{7.4 \times 10^{-10}\,m^3}{3.0 \times 10^{-6}\,m^2} = 2.5 \times 10^{-4}\,m$$

As all the charges in the length $L$ (including those at the very back) move through the coloured end in 1 second, their drift velocity $= 2.5 \times 10^{-4}\,m\,s^{-1}$.

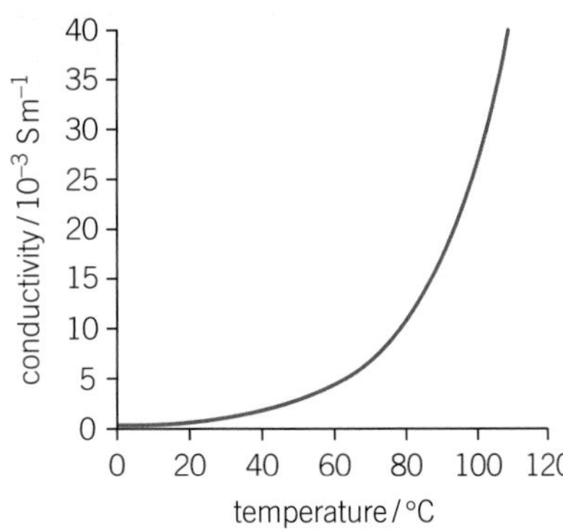


▲ **Figure 4** *The relationship between temperature and conductivity for a semiconductor*

## Insulators and semiconductors

Electrical **insulators** are materials that do not have mobile charges. There may be ions and electrons within them, but these are not free to move. An example is glass, which is a good insulator at normal temperatures, but will conduct if heated until it begins to soften, because the sodium ions in the glass can then move freely.

**Semiconductors** do have free charges, but very few – about 1 in $10^{12}$ atoms are ionised. If Figure 1 were to be drawn for a semiconductor, in order to show one mobile charge carrier the diagram would need to be 200 000 times larger in each dimension!

When semiconductors are heated, the proportion of atoms that are ionised increases greatly. This results in the rapid increase in number density of charge carriers, and therefore conductivity, as shown in Figure 4.

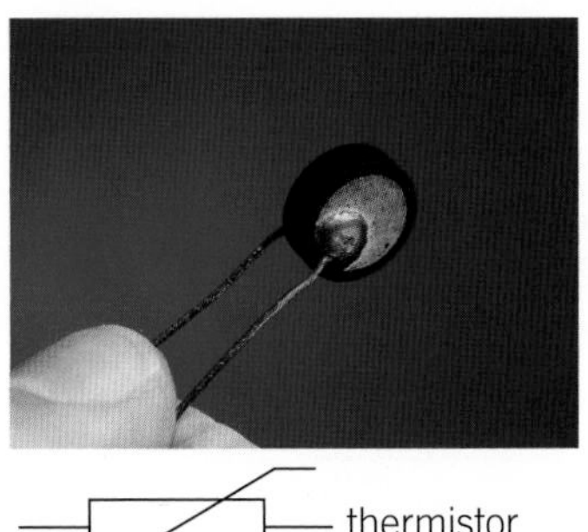


▲ **Figure 5** *Thermistor*

### Thermistors and LDRs

Thermistors are cheap, useful sensors made from semiconducting material. An increase in temperature liberates electrons, causing the conductivity to increase, so the resistance of the component drops significantly.

Light-dependent resistors (LDRs) are also semiconductor devices, usually constructed from cadmium sulfide. Light falling on the exposed semiconductor liberates electrons, which increases the conductivity much like a temperature increase does in thermistors, resulting in a drop in resistance of the LDR.

Both thermistors and LDRs can be incorporated as sensors in potential divider circuits, as you will see in Topic 3.5.

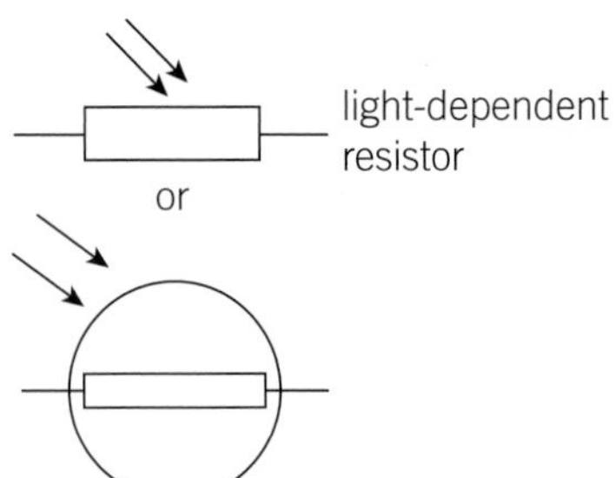


▲ **Figure 6** *Two ways of showing a light-dependent resistor (LDR)*

## Summary questions

1 Silver has an electrical conductivity nearly six times that of iron. With reference to Figure 1, suggest differences between the microscopic structure of the two which may account for this. *(2 marks)*

2 Two conductors **A** and **B** of identical length and cross-sectional area are connected in series and carry a constant current of 0.5 A. The number density *n* of free electrons in **A** is ten times that in **B**. Without calculations, explain carefully how the drift velocities of the electrons in the two conductors compare. *(2 marks)*

3 Use the approach in the worked example to show that the drift velocity *v* of electrons in a conductor is given by

$$v = \frac{I}{nAe}$$

where *n*, *A* and *e* have the same meaning as in that example. *(2 marks)*

# 3.5 Potential dividers

## Specification references 3.1.2a(vii), 3.1.2b(i), 3.1.2c(iv), 3.1.2d(iii), 3.1.2d(iv)

A prematurely born baby needs careful monitoring. The baby's temperature, pulse rate, and breathing need to be recorded, and the conditions in the incubator need careful control. These are all done with small electronic sensors, some attached to the baby and some in the hood of the incubator. For most of these sensors, a potential divider provides the output p.d. responsible for alerting medical staff, or for switching equipment on and off.

### Learning outcomes

Describe, explain, and apply:

- → the action of a potential divider and the equation: $\frac{V_1}{V_2} = \frac{R_1}{R_2}$
- → the equation: $V_{out} = \frac{R_2}{R_1 + R_2} \times V_{in}$ for a potential divider circuit
- → an experiment using a potential divider circuit
- → an experiment involving the calibration of a sensor.

## Sharing voltages

In Topic 3.2,Conductors and resistors, you saw that series resistors share the applied p.d. Consider the circuit in Figure 2.

From the definition of resistance,

$$R = \frac{V}{I} \text{ so } I = \frac{V}{R}$$

The current $I$ is the same throughout a series circuit, so for the two resistors $R_1$ and $R_2$

$$I = \frac{V_1}{R_1} = \frac{V_2}{R_2} \text{ so } \frac{V_1}{V_2} = \frac{R_1}{R_2}$$

In a similar way,

$$I = \frac{V}{R_1 + R_2}$$

giving

$$\frac{V_1}{V} = \frac{R_1}{R_1 + R_2} \text{ and } \frac{V_2}{V} = \frac{R_2}{R_1 + R_2}$$

This type of circuit is called a **potential divider**, because the applied p.d. is divided according to the relative values of $V_1$ and $V_2$.

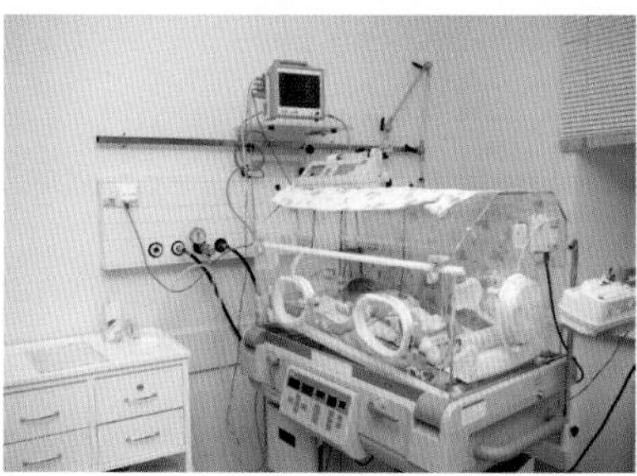

▲ **Figure 1** *Newborn premature baby in an incubator*

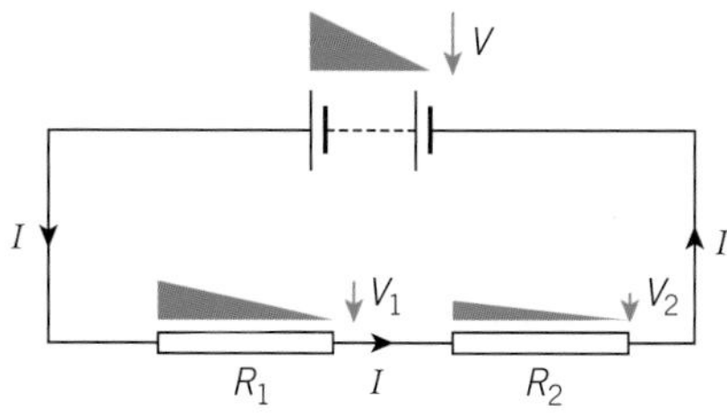


series

- same current
- p.d. shared
- resistances add

▲ **Figure 2** *The potential divider circuit*

### Worked example: The potential divider equation

Find the p.d. across each resistor in the circuit of Figure 3.

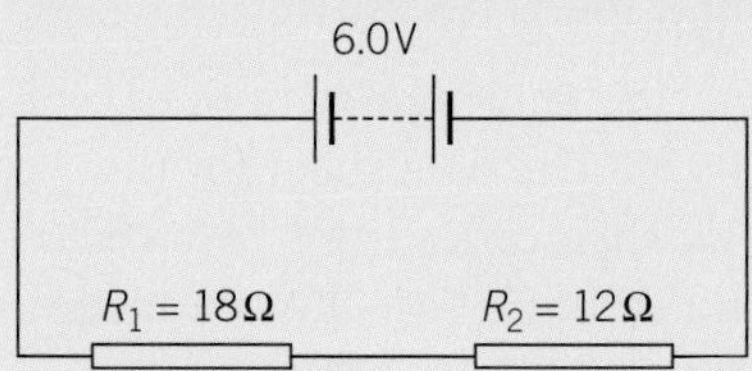


▲ **Figure 3**

**Step 1:** Find one p.d. in terms of the other, from the ratio of their resistances.

$$\frac{V_1}{V_2} = \frac{R_1}{R_2} = \frac{18\,\Omega}{12\,\Omega} = 1.5$$

Therefore, $V_1 = 1.5V_2$

➔

**Step 2:** Calculate one of the p.d.s, knowing the supply p.d.

$V_1 + V_2 = 6.0\,V \Rightarrow 1.5V_2 + V_2 = 2.5V_2 = 6.0\,V \Rightarrow V_2 = \frac{6.0\,V}{2.5} = 2.4\,V$

**Step 3:** Calculate the other p.d.

$V_1 = 6.0\,V - V_2 = 6.0\,V - 2.4\,V = 3.6\,V$

Or you could use $V_1 = 1.5V_2 = 1.5 \times 2.4\,V = 3.6\,V$.

## Changing the value of an output voltage

For the potential divider considered in Figure 2,

$$\frac{V_2}{V} = \frac{R_2}{R_1 + R_2}$$

So we can write:

$$V_2 = \frac{R_2}{R_1 + R_2} V$$

The p.d. $V_2$ across $R_2$ is the same fraction of the total voltage $V$ that the resistance $R_2$ is of the total resistance. This is useful in **sensor circuits** where one resistor (say $R_1$) is a fixed resistor and the other ($R_2$) an intrinsically variable one, such as a thermistor (the resistance of which depends on temperature) or an LDR (the resistance of which depends on light intensity). In that case, $V_2$ is the **output voltage** of the sensing circuit, generally denoted by $V_{out}$, and the battery voltage $V$ is the **input voltage**, $V_{in}$, as shown in Figure 4.

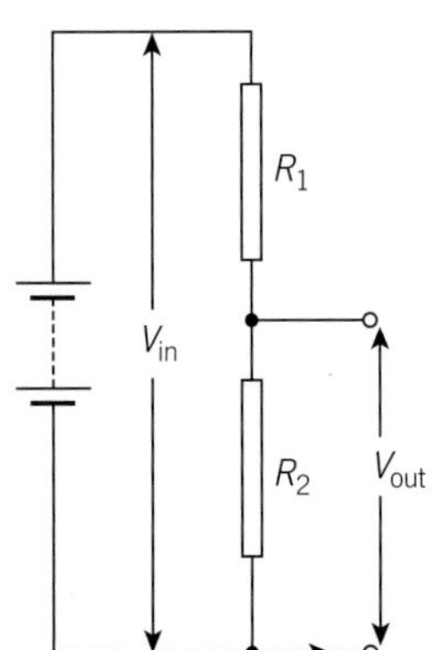


▲ **Figure 4** *Potential divider as used in a sensing circuit*

### Choice of the fixed resistor in a sensor circuit

Consider the temperature-sensing circuit in Figure 5. A thermistor is used as $R_1$ in this case, because its resistance, and hence the p.d. across it, falls as the temperature increases. The output voltage across $R_2$ will therefore increase with increasing temperature. Careful choice of the value of $R_2$ will give a large range of output voltages over the temperature range to be used.

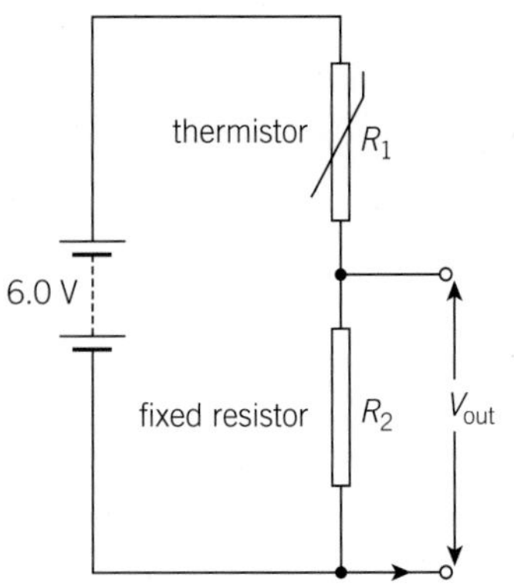


▲ **Figure 5** *Temperature-sensing circuit*

### Worked example: Choosing a fixed resistor

A thermistor is chosen for a temperature-sensing circuit that needs to work between 0 °C and 100 °C. The data sheet for the thermistor chosen states that its resistance varies from 16 300 Ω at 0 °C to 340 Ω at 100 °C. What would be the best choice for the fixed resistor? Choose from 20 000 Ω, 2000 Ω, or 200 Ω.

**Step 1:** You need to calculate the expected output voltages at 0 °C and 100 °C for each possible fixed resistor. Starting with the 20 000 Ω resistor.

At 0 °C,

$$V_{out} = \frac{R_2}{R_1 + R_2} V = \frac{20\,000\,\Omega}{16\,300\,\Omega + 20\,000\,\Omega} \times 6.0\,V = 3.3\,V$$

At 100 °C,

$$V_{out} = \frac{20\,000\ \Omega}{340\ \Omega + 20\,000\ \Omega} \times 6.0\,V = 5.9\,V$$

**Steps 2 and 3:** Do the same calculations for the other two resistors. Table 1 shows the set of possible output voltages.

▼ Table 1

| $R_2$ / Ω | 20 000 | 2000 | 200 |
|---|---|---|---|
| $V_{out}$ at 0 °C / V | 3.3 | 0.66 | 0.07 |
| $V_{out}$ at 100 °C / V | 5.9 | 5.1 | 2.2 |

The intermediate value (2000 Ω) gives the best (greatest) range of output voltages.

Note that the resistance of a thermistor does not vary in a linear way with temperature, so the output voltage will not vary uniformly over the temperature range.

## Practical 3.1.2d(iii): Using a potential divider including a sensor

Many sensor circuits use an intrinsically variable resistor, such as a thermistor or an LDR, to switch an electronic switch that operates at a fixed voltage. For some electronic switches, this is a value of 0.7 V, and for others it is half the voltage of the battery in the circuit.

Before setting up the circuit, it may be necessary to calibrate the potential divider (see Practical 3.1.2d(iv)).

When designing a sensor circuit, you need to decide whether you want the output p.d. from the sensor circuit to go from low to high as the switching point is reached, or the other way round. This will affect whether the output p.d. is across the variable component or across the fixed resistor. Figure 6 shows an example. The input to the electronic switch is the output $V_{out}$ from the sensor circuit. The switch will switch *on* the alarm when $V_{out}$ from the sensor circuit *drops*. This means that the potential divider in the sensor circuit must be set up so that $V_{out}$ is normally high, but falls below the switching value when the environment reaches the condition required for the alarm to sound.

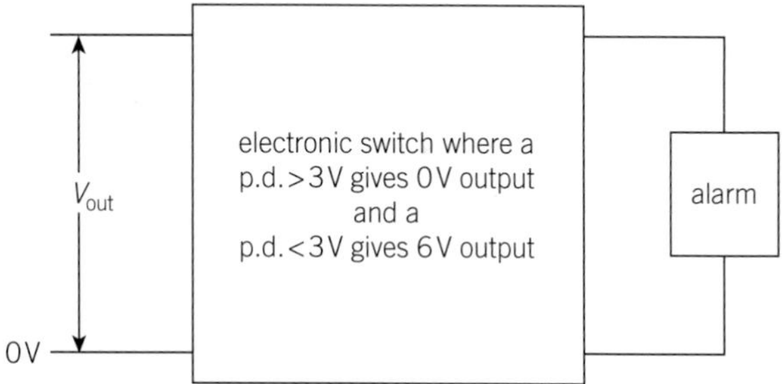


▲ **Figure 6** *Using an electronic switch*

A suitable approximate value of the fixed resistor can be obtained from knowledge of the resistance variation of the fixed resistor (see the worked example above) and from the calibration graph (see Practical 3.1.2d (iv)). It may be necessary to adjust the fixed resistor so that switching occurs at the right point. This would involve using a potentiometer (variable resistor) as the 'fixed' resistor in the potential divider.

## Practical 3.1.2d(iv): Calibrating a sensor

A simple sensor measuring the change in an environmental variable (such as temperature or light intensity) will need an intrinsically variable resistor (such as a thermistor or an LDR) as well as an appropriate fixed resistor chosen to give an output p.d. with good variation. It will be necessary also to have some way of measuring that environmental variable, so that the output p.d. can be tabulated alongside the value of that variable over an appropriate range. A suitable calibration graph will then permit the output p.d. reading to be converted to the required values of the environmental variable.

## Summary questions

**1** Calculate the p.d. across the 100 Ω resistor in Figure 7. *(2 marks)*

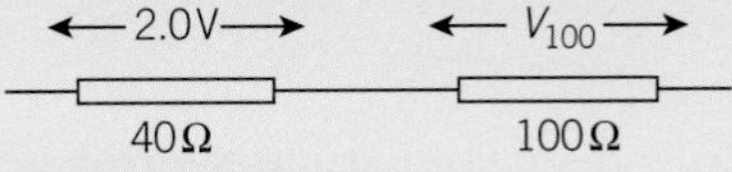


▲ **Figure 7**

**2** The resistance of an LDR varies between 5.4 kΩ in bright sunlight and 1.0 MΩ in darkness. Draw a sensor circuit to work with a 3.0 V battery, including a fixed resistor of suitable value, which will give an output voltage decreasing as the light gets brighter. *(3 marks)*

**3** Two 10 000 Ω resistors **A** and **B** are connected in series to a 6.0 V battery. A voltmeter, also of resistance 10 000 Ω, is connected across each resistor in turn, and then across the battery, giving the results in Table 2.

▼ **Table 2**

| p.d. across A / V | p.d. across B / V | p.d. across battery / V |
|---|---|---|
| 2.0 | 2.0 | 6.0 |

Explain these results. *(5 marks)*

# 3.6 E.m.f. and internal resistance

Specification references: 3.1.2a(iv), 3.1.2b(i), 3.1.2c(i), 3.1.2d(v)

The temperature sensor in Figure 1 is a thermocouple. The probe inserted into the cooking food has different metals in contact, which produces a voltage by the thermoelectric effect (discovered by the Estonian physicist Thomas Seebeck in 1820). This effect is just one of the phenomena by which a p.d. can be generated by a physical or chemical change. In this topic we will consider battery cells, but the principles developed can be applied to any source of potential difference. Other sources include generators and photocells.

## Electrical sources and e.m.f.

An electrical supply, or source, can be connected into a circuit to provide a p.d. which will make charges move. In this process, energy is given to the charges, and this is dissipated in the circuit. The potential difference (p.d.) $V$ between the external terminals of the source is the potential energy difference per coulomb as the charges move 'downhill', releasing energy as they do so, as shown in Figure 2.

The question must arise, what pushes the charges uphill to start with? Within battery cells, chemical processes liberate electrons at the negative terminal (cathode), and consume them at the positive terminal (anode), effectively moving electrons from the anode to the cathode and providing the energy to 'lift them' up the potential hill.

The **e.m.f.**, symbol $\mathcal{E}$, is the energy that the source gives to the charges for every coulomb of charge flowing from the source. The abbreviation e.m.f. stands for 'electro-motive force', but it is not strictly a force, so it is better just to use the abbreviation.

Applying the principle of conservation of energy to the circuit in Figure 2, you can see that the energy supplied by the battery is equal to the energy dissipated by the circuit resistance. This is true for every coulomb passing, so as p.d. = energy transfer per coulomb, we can say that, for any complete circuit, e.m.f. = sum of all the p.d.s across the resistances of the circuit – this is known as **Kirchhoff's second law**.

## Internal resistance

The e.m.f. produced by a battery results in a current through the entire circuit, including the body of the battery itself, and the chemicals in the battery provide resistance to the flow of charges. The battery has **internal resistance**, given the symbol $r$, so this means there is a p.d. drop inside the battery itself. Figure 2 has an 'ideal' battery with no internal resistance. For a real battery, a better representation is Figure 3 (next page). The dotted outline around the battery helps to identify where the battery terminals are.

### Learning outcomes

Describe, explain, and apply:

- → the meaning of e.m.f.
- → the effect of internal resistance on circuits connected to a source of e.m.f.
- → the equation $V = \mathcal{E} - Ir$
- → a $V$–$I$ graph showing the effect of internal resistance.

▲ **Figure 1** *A thermocouple in use in the kitchen*

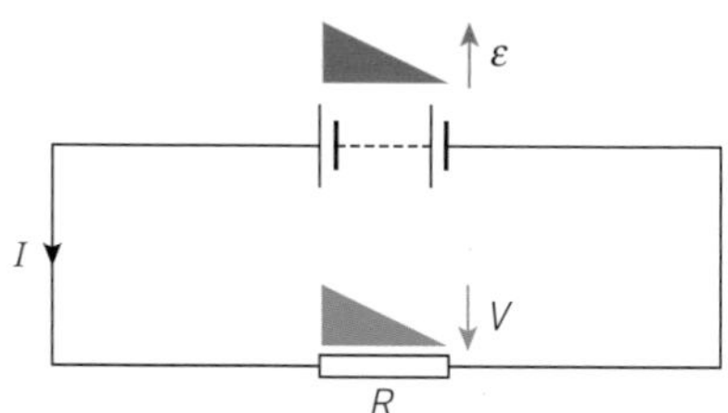


▲ **Figure 2** *The p.d. (or V) and e.m.f. ε for a battery in a circuit*

### Study tip

Sometimes you will see $E$ for e.m.f., rather than $\mathcal{E}$. In your physics course you will meet a number of variables given the symbol $E$, and they mean quite different things, so be sure you know which is meant, and what you mean when you use it.

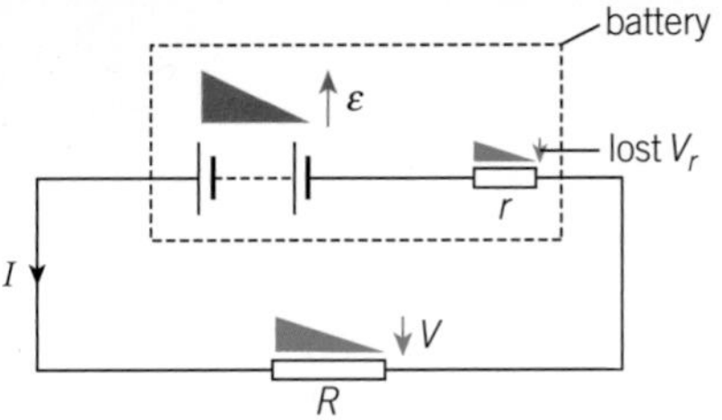


▲ **Figure 3** *A source of e.m.f. with internal resistance r*

In the circuit of Figure 3, the e.m.f. provides the total p.d. across both the external and internal resistances.

$$\mathcal{E} = V + V_r$$

where $V_r$ is the p.d. drop, or 'lost volts', within the battery itself. $V$ is the p.d. across the resistor $R$, and is also the p.d. across the battery terminals – the **terminal p.d.** – provided the resistance of the connecting cables is negligible.

The current $I$ is the same throughout the circuit. Rearranging

$$r = \frac{V_r}{I} \text{ gives } V_r = Ir$$

We can use this and the equation above for $\mathcal{E}$ to give

$$\mathcal{E} = V + Ir$$

This equation can be rearranged into

$$V = (-r)I + \mathcal{E}$$

As both $\mathcal{E}$ and $r$ are constants of the battery, this is in the form of the straight line equation $y = mx + c$ (gradient $m$ and $y$-axis intercept $c$), as shown in Figure 4. The gradient is $-r$.

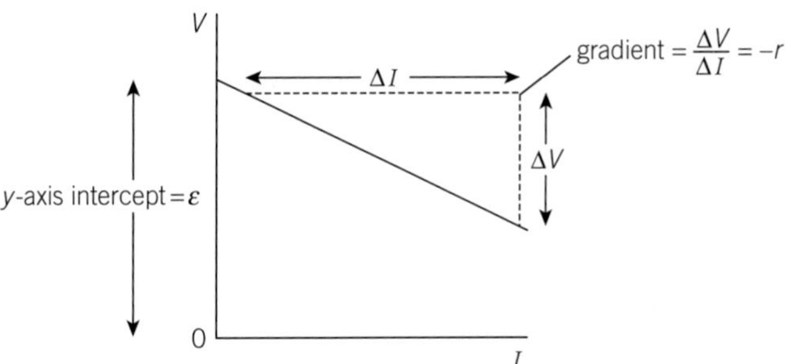


▲ **Figure 4** *Variation of terminal p.d. with current*

### Worked example: Calculating internal resistance

A battery of e.m.f. 6.0 V has a terminal p.d. of 5.6 V when delivering a current of 0.5 A. Calculate the internal resistance of the battery.

**Step 1:** Substitute into the equation $\mathcal{E} = V + Ir$

$$\Rightarrow 6.0\,\text{V} = 5.6\,\text{V} + 0.5\,\text{A} \times r$$

**Step 2:** Rearrange to find $r$.

$$r = \frac{6.0\text{ V} - 5.6\text{ V}}{0.5\text{ A}} = \frac{0.4\text{ V}}{0.5\text{ A}} = 0.8\,\Omega$$

In any real circuit, the terminal p.d. $V$ is less than the e.m.f. $\mathcal{E}$, but this difference is not constant. As the **load** resistance $R$ (the resistance of the circuit attached to the battery terminals) decreases, the current $I$ will increase. This happens because the total circuit resistance, $R + r$, has been reduced. This increase in $I$ will increase the 'lost volts' $Ir$ and so reduce the terminal p.d.

### Worked example: Calculating e.m.f. and internal resistance

A battery has a terminal p.d. of 5.6 V when delivering a current of 0.5 A, but this drops to 5.2 V when delivering a current of 1.0 A. Calculate the e.m.f. and the internal resistance of the battery.

**Step 1:** Substitute the data into the equation.

For the 0.5 A current, $\mathcal{E} = V + Ir$ becomes
$\mathcal{E} = 5.6\,\text{V} + 0.5\,\text{A} \times r$ Equation 1

For the 1.0 A current, $\mathcal{E} = V + Ir$ becomes
$\mathcal{E} = 5.2\,\text{V} + 1.0\,\text{A} \times r$ Equation 2

**Step 2:** Subtract Equation 2 from Equation 1 to give:

$$(\mathcal{E} - \mathcal{E}) = (5.6\,\text{V} - 5.2\,\text{V}) + (0.5\,\text{A} \times r - 1.0\,\text{A} \times r)$$

$$0 = 0.4\,\text{V} - (0.5\,\text{A} \times r)$$

$$r = \frac{0.4\,\text{V}}{0.5\,\text{A}} = 0.8\,\Omega$$

**Step 3:** Substitute this value of $r$ into Equation 1 or 2. Using Equation 1:

$$\mathcal{E} = 5.6\,\text{V} + 0.5\,\text{A} \times 0.8\,\Omega = 5.6\,\text{V} + 0.4\,\text{V} = 6.0\,\text{V}$$

## Measuring internal resistance

Some battery cells, such as those in lead–acid car batteries, have such a low internal resistance that it is difficult to measure. However, some cells have a significant internal resistance, and for these cells the internal resistance $r$ can be found.

### Practical 3.1.2d(v): Determining the internal resistance of a cell

If the circuit of Figure 3 is set up using a source of e.m.f. with high internal resistance (one example might be a piece of fruit or vegetable into which two dissimilar metals have been stuck), with a variable resistor in place of the load resistor $R$, it is possible to obtain a range of values of $I$ and corresponding values of $V$, the p.d. across the load resistor. $V$ and $I$ are variables, and $\mathcal{E}$ and $r$ are constants for this battery, so a suitable graph can be drawn and the internal resistance found.

Appropriate voltmeter and ammeter ranges need to be chosen. The most suitable range of values of $R$ may need to be found in a preliminary experiment.

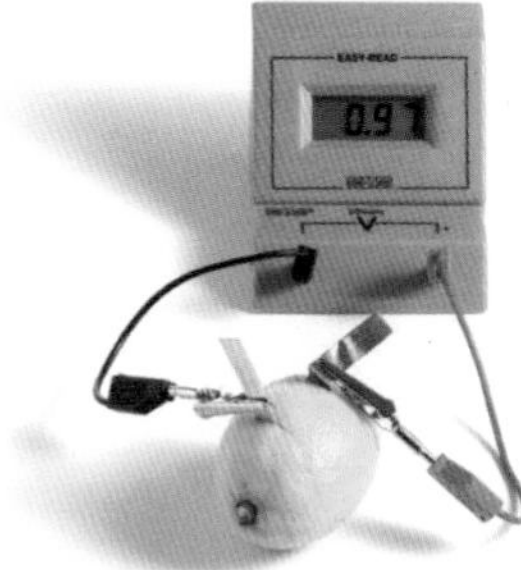


▲ **Figure 5** *Two different metals in a conduction fluid form a battery — the fluid here is lemon juice*

## Summary questions

**1** Explain why the p.d. across the internal resistance of a cell is often referred to as 'lost volts'. *(1 mark)*

**2** In an experiment to measure p.d. across the terminals of a battery for different values of current delivered, the data in Table 1 was obtained.

▼ Table 1

| Current / A | 0.5 | 1.0 | 1.5 | 2.0 | 3.0 | 4.0 |
|---|---|---|---|---|---|---|
| P.d. / V | 3.4 | 3.25 | 3.0 | 2.75 | 2.35 | 2.0 |

Plot an appropriate graph to find the e.m.f. and the internal resistance of the battery used. *(5 marks)*

**3** All electrical appliances in a car are connected in parallel to the terminals of a 12 V battery. A car is parked with its lights on. The driver starts the engine, using the powerful starter motor. The car lights become dim while the starter motor is operating. Explain this observation in terms of internal resistance. *(3 marks)*

# Module 3.1 Summary

## Imaging

### Imaging with a lens

- power of a lens = $\frac{1}{f}$ measured in dioptres (D)
- lens equation $\frac{1}{v} = \frac{1}{u} + \frac{1}{f}$ (Cartesian convention)
- linear magnification = $\frac{v}{u} = \frac{\text{object height}}{\text{image height}}$
- practical task — determination of the focal length of a converging lens

### Storing and manipulating the image

- bits, bytes and pixels
- amount of information = number of pixels × bits per pixel
- $b$ bits provide $2^b$ alternatives
- number of bits $b$ needed for $N$ alternative arrangements = $\log_2 N$
- resolution as the smallest detail that can be distinguished in an image

### Polarisation of electromagnetic waves

- $v = f\lambda$
- practical task — observing polarising effects using light and microwaves

## Signalling

### Digitising a signal

- sampling, quantisation levels and quantisation error
- resolution = $\frac{\text{potential difference range of signal}}{\text{number of quantisation levels}}$
- noise in a signal: maximum useful number of levels = $\frac{V_{total}}{V_{noise}}$

### Sampling sounds and sending a signal

- aliasing
- minimum sampling rate > 2 × maximum frequency of signal
- bit rate = samples per second × bits per sample

## Sensing

### Current, p.d., and electrical power

- $I = \Delta Q/\Delta t$, $P = IV = I^2R$, $W = VIt$
- Kirchhoff's first law
- p.d. as energy per unit charge, $V = \frac{W}{Q}$

### Conductors and resistors

- $R = V/I$, $G = I/V$
- calculating current and p.d. in series and parallel circuits
- practical task — $I$-$V$ characteristics of ohmic and non-ohmic components

### Conductivity and resistivity

- the equations $R = \frac{\rho L}{A}$ and $G = \frac{\sigma A}{L}$
- practical task — determining the resistivity or conductivity of a metal
- explaining electrical behaviour of conductors and insulators

### Potential dividers

- the action of a potential divider: $V_{out} = \frac{R_2}{R_1 + R_2} \times V_{in}$ and $\frac{V_1}{V_2} = \frac{R_1}{R_2}$
- practical task — potential divider
- practical task — calibrating a sensor

### E.m.f. and internal resistance

- $V = \varepsilon - Ir$
- Kirchhoff's second law
- practical task — determining the internal resistance of a cell

# Physics in perspective

## Strain gauges

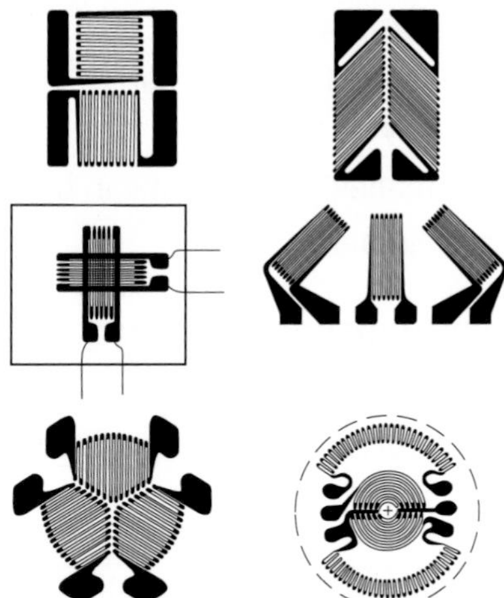

▲ **Figure 1** *These gauges are made to detect strains in several directions*

Buildings in many parts of the world can crack or even collapse if the ground beneath them moves. It is important to monitor movement to provide advance warning of a possible collapse. This was once done by cementing microscope slides to key structural parts of buildings at risk, and inspecting them regularly to check if any had cracked. Strain gauges now perform this task, with the advantage that they can detect much smaller movements and produce a continuous recording of the movement, which can be logged by digital systems.

### How does a strain gauge work?

The resistance of a metal increases when it is stretched, simply because it becomes longer and thinner. A strain gauge is a zigzag strip of metal foil glued to the surface of the component whose strain is to be monitored. The foil is stretched if the component is stretched. Stretching along the length of the strips in the foil is detected as an increase in resistance. The zigzag arrangement simply increases the length of conductor being stretched, which increases the change in resistance. Such strain gauges are cheap and convenient to use.

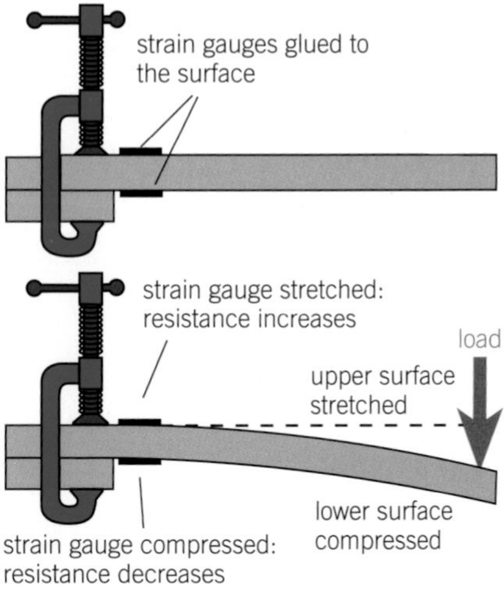


▲ **Figure 2** *The resistances of both strain gauges change when the beam bends*

Suppose the bending of a beam of material has to be measured. Such measurements are needed in investigating the strengths of materials. A clever trick is to use two strain gauges, so that as the beam bends, one is stretched and the other is compressed. See Figure 2. Connected in a potential divider, the two gauges give twice the output that one would give alone.

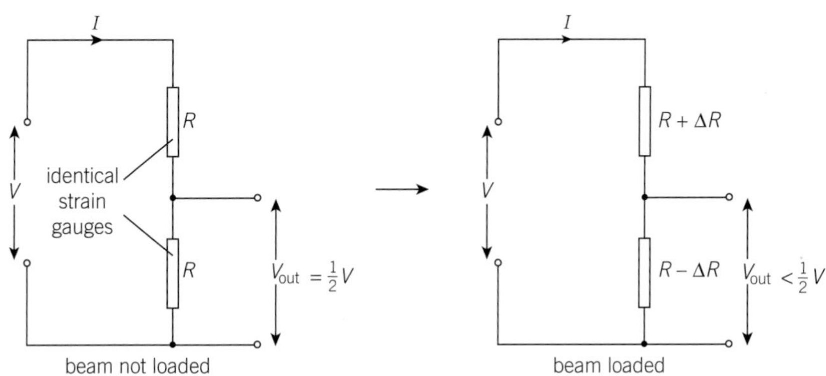


▲ **Figure 3** *Strain gauges in a potential divider*

With this setup there is a bonus — it is not so sensitive to changes in temperature. The resistance of a metal increases with temperature, so by using just one metal strain gauge in a potential divider, a rise in temperature might be registered as a strain. With two gauges, if the temperature goes up, the resistance of both gauges increases and so the output of the potential divider does not change.

This is an example of something that is important in scientific measurement. Either the apparatus is designed to eliminate the systematic effect of outside factors, or the effect is measured independently and accounted for.

A further enhancement of the circuit can be made by using not two but four identical strain gauges. Two extra strain gauges are kept unstrained and connected in parallel with the two fastened to the beam. This makes what is called a bridge circuit, shown in Figure 4.

This is done because the voltage at point **B** will vary from 0.50 times the battery voltage to about 0.49 times the battery voltage at the lowest, which is a small change.

If a meter is connected between points **A** and **B**, however, the change becomes much easier to detect. When the beam is not loaded, both **A** and **B** will be at the same p.d., so the p.d. between them will be zero. As the strain builds up, the p.d. between **A** and **B** gives the change in the p.d. at **B**, not its actual value. This can be read with a more sensitive meter.

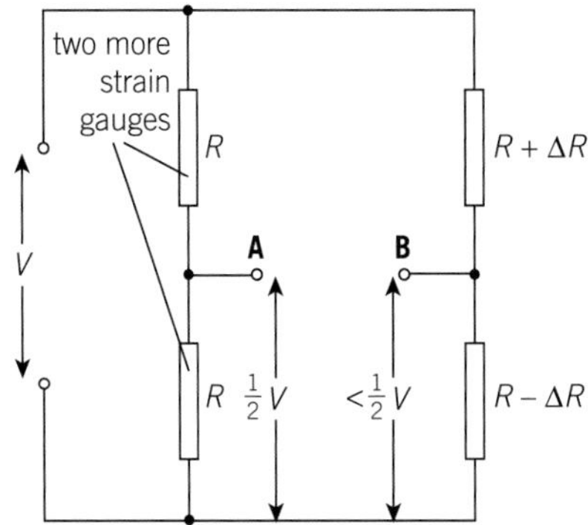


▲ **Figure 4** *Strain gauges in a bridge circuit*

## Summary questions

**1** State and explain the advantages of using strain gauges, instead of cemented microscope slides, to check for subsidence in ancient monuments such as cathedrals.

**2** One type of strain gauge is shown in Figure 5. It consists of 14 strands, each 1.0 cm long, of foil 3.5 μm thick. The width of each strip is 55 μm. The wider strips of foil holding these strands have much lower resistance, and can be ignored.
The strain gauge has a resistance of 360 Ω when unstrained. Calculate the resistivity of the metal used.

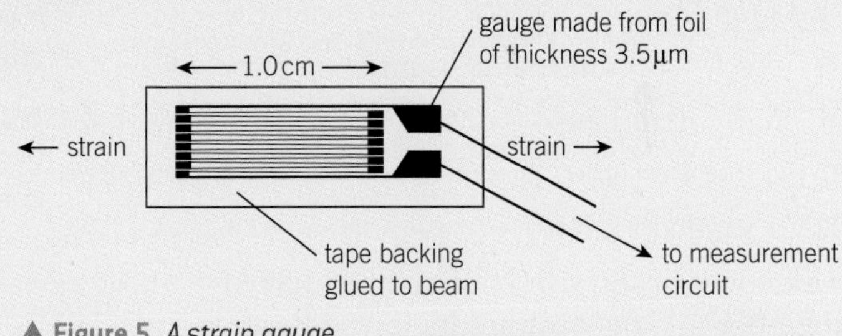


▲ **Figure 5** *A strain gauge*

**3** When the strain gauge in Figure 5 is stretched, the fourteen 1.0 cm lengths get longer, but the volume of metal in them stays the same. Show that a 1% increase in length means that there is a 1% decrease in cross-sectional area, and a 2% increase in resistance.

**4** The maximum strain that the gauge in Figure 5 can measure is 4% (that is, a length increase of 4%). Show that the change in resistance of a 360 Ω strain gauge under 4% strain is nearly 30 Ω.

**5** If the bent beam in Figure 2 produces 4% strain in each gauge, use the circuit in Figure 3 to find the output voltage $V_{out}$ from the potential divider. The circuit voltage $V = 6.0$ V.

**6** If the bridge circuit in Figure 4 had been used with the two more 360 Ω strain gauges and a 6.0 V battery, find the p.d. between **A** and **B**.

**7** Strains are usually much less than 4%. Use the answers to 5 and 6 to explain why the bridge circuit in Figure 4 is often used in practice.

**8** When strain gauges are used in bridge circuits, the p.d. between **A** and **B** is usually connected to a strain gauge amplifier rather than just measured directly with a voltmeter. Suggest a reason for this.

## Practice questions

1 Which combination of units is equivalent to J?

a As

b AV

c VC

d $Ws^{-1}$

2 Figure 1 shows an *R*-*V* graph for an electric circuit component.

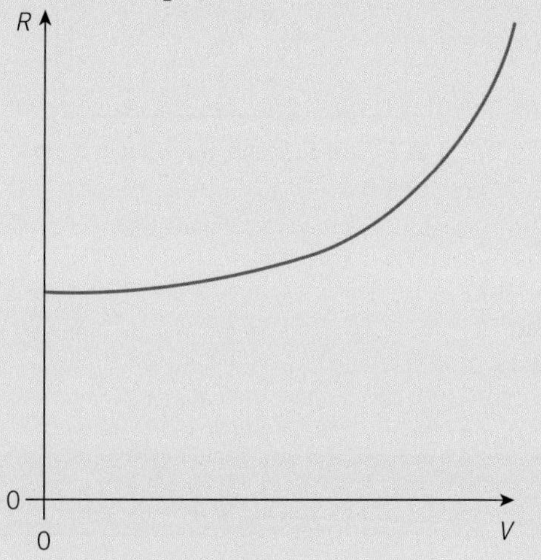


▲ Figure 1

Which electric circuit component would give this graph?

A a cell with internal resistance

B a filament lamp

C a fixed resistor

D a semiconductor

3 Three identical resistors can be connected together in a network in the four different ways shown in Figure 2

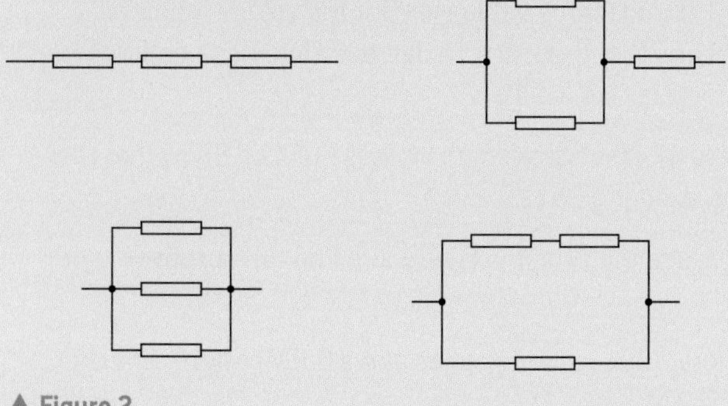

▲ Figure 2

Which of the following resistance values can be made with one of these networks of three 12 Ω resistors?

1 4 Ω

2 8 Ω

3 12 Ω

A 1, 2, and 3 are correct

B Only 1 and 2 are correct

C Only 2 and 3 are correct

D Only 1 is correct

4 An X-ray tube contains of a stream of electrons which strike a metal target. The current in the tube is 350 mA. Calculate the number of electrons striking the metal target each second.

electron charge, $e = 1.6 \times 10^{-19}$ C (*2 marks*)

5 A battery of e.m.f. 9.0 V is connected to a 5.0 Ω resistor. The p.d. across the terminals of the battery is measured to be 8.6 V. Calculate the internal resistance of the battery. (*3 marks*)

6 A nickel wire of diameter 0.10 mm and length 0.85 m has a current of 0.16 A when there is a p.d. of 1.2 V between its ends. Calculate the conductivity of nickel. (*4 marks*)

7 A potential divider containing a light-dependent resistor (LDR) and a thermistor is connected as shown in Figure 3.

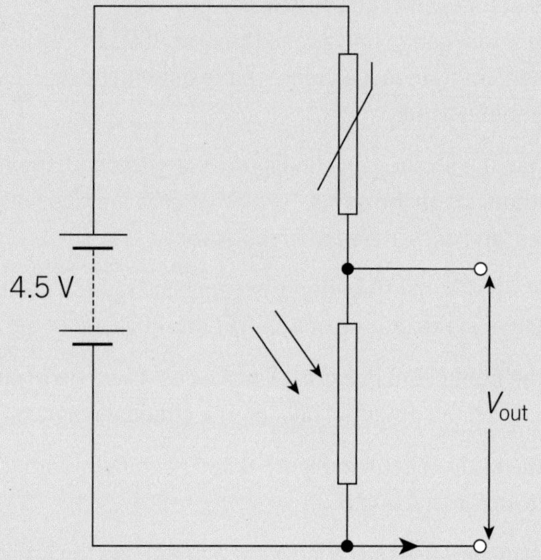


▲ Figure 3

The resistance of the thermistor and LDR under different conditions are given in Table 1.

▼ Table 1

| component | thermistor | | LDR | |
|---|---|---|---|---|
| condition | 0 °C | 25 °C | dark | bright sunlight |
| resistance / kΩ | 33 | 10 | 1000 | 5.4 |

**a** The sensors are put a light-proof box at 25 °C. Calculate the p.d. $V_{out}$. *(3 marks)*

**b** The circuit is now laid on the laboratory bench, and $V_{out}$ = 2.4 V. A cloud suddenly moves across the Sun. Without calculation, explain what will happen to $V_{out}$. *(3 marks)*

**8** In the USA, the mains p.d. is 120 V, and domestic sockets provide a maximum current of 15 A.

**a** Assuming that a US electric kettle draws a current of 10 A, calculate

(i) the power of the kettle, *(1 mark)*

(ii) the resistance of its heating element. *(1 mark)*

**b** Heating a litre of cold water to the temperature needed to make tea requires 370 kJ of energy. Calculate the time this takes using a mains connection in the USA. *(2 marks)*

**c** The kettle is brought from the USA to the UK, where the mains p.d. is 230 V. Assuming that the kettle will still work, calculate the time it takes to boil a litre of cold water. State and justify any assumptions that you make. *(4 marks)*

**d** In the UK, domestic sockets are limited to 13 A. Suggest why the kettle might work better if left in the USA. *(2 marks)*

**9** A temperature sensor circuit is to be constructed using a thermistor in a potential divider. The resistance of the thermistor is 33 kΩ at 0 °C and 680 Ω at 100 °C.

**a** Draw a circuit diagram for the temperature sensor, including suggested values of battery e.m.f. and resistance of the fixed resistor. You should arrange to have a p.d. reading which increases with increasing temperature and which covers as large a voltage range as possible. *(3 marks)*

**b** Describe how you would calibrate your temperature sensor circuit over the range 0 °C to 100 °C. *(3 marks)*

**c** The resistance of the thermistors varies as shown in Table 2.

▼ Table 2

| Temperature / °C | 0 | 20 | 40 | 60 | 80 | 100 |
|---|---|---|---|---|---|---|
| Resistance / Ω | 33 000 | 12 000 | 5300 | 2500 | 1300 | 680 |

By considering the changes in output of your circuit for 0 °C to 20 °C, and from 80 °C to 100 °C, explain how this resistance variation will affect the sensitivity of your sensor circuit at different temperatures. *(4 marks)*

# MODULE 3.2

## Mechanical properties of materials

### Chapters in this Module



### Introduction

Materials have played a crucial role in the development of human society. From the early work with stone, bronze and iron in the ancient world through to twenty-first century materials such as graphene and aerogel, the manufacture of new materials has been a driver of technological change.

In this module you will learn about the mechanical properties of materials and how these properties relate to the microscopic structure of the material. Understanding this relationship helps us predict the properties of materials and design new forms of materials for specific purposes.

Testing Materials introduces the important concepts of stress, strain and the Young Modulus — ideas which are of great importance in engineering. You will learn how to determine the Young Modulus of a material and how to classify materials by their properties. We will focus on the properties of metals, ceramics and polymers, which are all materials with a wide range of uses in the modern world. The chapter introduces many technical terms that help describe the properties of these materials.

You will also consider energy changes when a material is squashed or stretched. Understanding how energy is stored or dissipated in materials when they are deformed has many applications in automotive design, including crumple zones and car suspension systems.

Inside Materials considers the internal structure of materials. You will learn why ceramics, metals and polymers show particular properties. You will review direct evidence for the sizes of the particles and develop a model that explains the different properties of metals, ceramics and polymers. The structure of metals is looked at in detail to explain why alloying changes the property of a metal. An understanding of alloys is important in many areas of design from artificial hip joints to jet aircraft and space vehicles.

## Knowledge and understanding checklist

**From your Key Stage 4 study you should be able to do the following. Work through each point, using your Key Stage 4 notes and the support available on Kerboodle.**

- ☐ Describe the difference between elastic and plastic deformation.
- ☐ Calculate work done in stretching a spring.
- ☐ Describe the relationship between force and extension for a spring, describing the difference between linear and non-linear relationship between force and extension. Calculate the spring constant (force constant) in linear cases.
- ☐ Recall the typical size (order of magnitude) of atoms and small molecules.

## Maths skills checklist

**All physicists use maths. In this unit you will need to use many different maths skills, including the following examples. You can find support for these skills on Kerboodle and through MyMaths.**

- ☐ **Change the subject of an equation, including nonlinear equations**, to solve mathematical problems about the energy stored in a spring and the Young Modulus.
- ☐ **Use an appropriate number of significant figures** when answering problems involving different numbers of significant figures and approximations. For example, when calculating the cross-sectional area of a wire.
- ☐ **Plot two variables from experimental or other data and use $y = mx + c$**, such as when studying Hooke's law.
- ☐ **Calculate the gradient from a graph**, for example in experiments to determine the Young Modulus of materials.
- ☐ **Understand the possible physical significance of the area between a curve and the *x*-axis, and be able to calculate it or estimate it by graphical methods**, such as when using a graph to calculate the energy stored in a stretched spring.
- ☐ **Interpret logarithmic plots**, such as when looking at materials selection charts that compare the properties of a range of materials.

# 4 TESTING MATERIALS

## 4.1 Describing materials

Specification references: 3.2a(iii), 3.2b(i)

**Learning outcomes**

Describe, explain, and apply:

- → behaviour of classes of materials (limited to metals, ceramics, and polymers)
- → the terms: stiff, ductile, hard, brittle, tough.

Why do adults choose plates made from china but use other materials for baby bowls? Throughout history, materials have been chosen and developed that improve the properties of the finished object. For example, iron axes are superior to stone axes, and cars with lightweight body shells can accelerate more quickly than heavier vehicles. The development of new materials continues – aerogel and graphene were unknown until a few years ago, but are now beginning to find uses in areas such as firefighter suits, composite materials, and photovoltaic cells.

▲ **Figure 1** *Graphene may improve efficiency of photovoltaic cells*

### Classes of materials

We are going to consider the mechanical properties of three classes of materials – ceramics, metals, and polymers.

*Ceramics* are **hard**, meaning that they are difficult to scratch. They are **brittle**, meaning that they will shatter into jagged pieces – think of a broken plate. Ceramics are also **stiff** – they are difficult to stretch or bend. Examples of ceramics include china and other pottery, as well as more modern 'engineering ceramics' such as alumina and silicon carbide.

*Metals* have a wide range of mechanical properties. Pure metals tend to be soft. Lead can be bent into shape by hand – gold and copper can be easily hammered into shape. Metals that can be shaped easily are called **malleable**, whilst those that can be drawn into wires are called **ductile**. Metal alloys, such as steel, are usually harder than the pure metals used to make them.

*Polymers* include familiar synthetic materials such as polythene and Perspex, as well as natural materials such as leather and cotton. *Glassy polymers* have properties similar to glass and often replace glass in spectacle lenses. These materials are brittle. *Semi-crystalline polymers* are **tough** – they can undergo considerable deformation and can absorb more energy before breaking (compared to more brittle polymers).

### Choosing materials

The choice of materials to use in a product depends to a large extent on the mechanical properties of the material, but other factors such as cost and the look of the finished article are also considered. The properties of each class of material in the hip joint in Figure 2 are used to produce an implant that is effective, long-lasting, and comfortable.

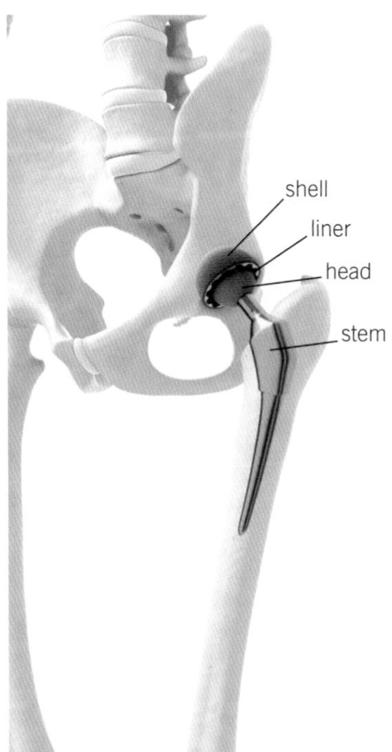


▲ **Figure 2** *The hip joint pictured uses a porous metal shell, a polythene liner, a ceramic head, and a metal stem*

## Summary questions

1 **a** Classify each of the following materials as metal, ceramic, or polymer: rubber, acrylic, steel, china. *(1 mark)*
**b** Explain your answer. *(3 marks)*

2 Use objects around you to name two brittle materials and two tough materials. *(2 marks)*

3 Identify some of the mechanical properties required for a scalpel used in surgery. Give reasons for your choice. *(3 marks)*

4 **a** Suggest desirable mechanical properties for the outer casing of a mobile phone. *(2 marks)*
**b** Compare the properties identified in **a** with the properties required for a mobile phone cover or sleeve. *(2 marks)*
**c** Suggest, using your answers to **a** and **b**, the choice of materials for the casing and the sleeve, giving the classes of materials that these belong to. *(4 marks)*

# 4.2 Stretching wires and springs

## Specification references: 3.2a(i), 3.2b(i), 3.2b(ii), 3.2c(i), 3.2d(i)

### Learning outcomes

Describe, explain, and apply:

- → simple mechanical behaviour: elastic and plastic deformation, and fracture
- → the terms: plastic, elastic, compression
- → force–extension graphs up to fracture
- → the equation for Hooke's law: $F = kx$
- → the equation for energy stored in an elastic material: energy $= \frac{1}{2}kx^2$
- → energy as the area under a force–extension graph for elastic materials
- → an experiment to plot force–extension characteristics for arrangements of springs, rubber bands, and polythene strips.

▲ **Figure 1** *The suspension spring on this vehicle is clearly visible once the wheel has been removed*

For centuries armies relied heavily on storing energy in springy things. The bowmen at Agincourt had yew bows that needed great strength to pull back – crossbows stored even more energy by having a thicker bow that had to be wound back mechanically. A more everyday example of energy stored in a spring is the suspension of a car, designed to absorb the energy of shocks to the wheels from bumps in the ground. In this topic we will look at how materials stretch and how they store energy.

## Hooke's law

Hang a weight on the end of a spring and the spring will stretch. In 1678 the English experimenter Robert Hooke found that, for small extensions, the force $F$ is proportional to extension $x$. This relationship is now known as **Hooke's law**.

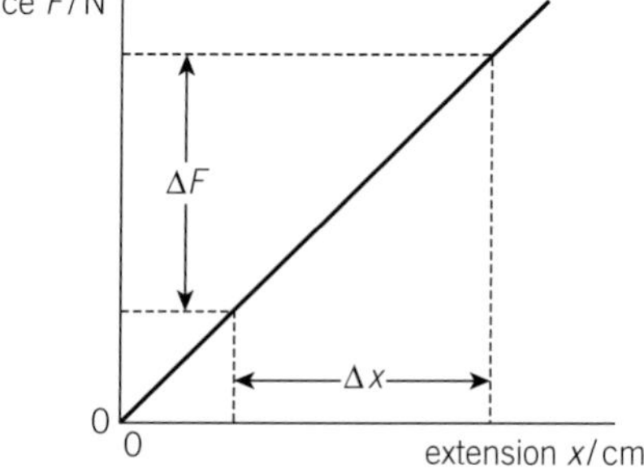

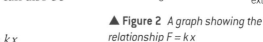

▲ **Figure 2** *A graph showing the relationship $F = kx$*

The relationship can also be given as

$$F = kx$$

where $k$ is a constant of proportionality called the spring (or force) constant.

### Worked example: Finding *k*

Figure 3 shows results from an experiment to measure the extension of a spring as weights are added to it. Use the graph to find a value for the spring constant, $k$.

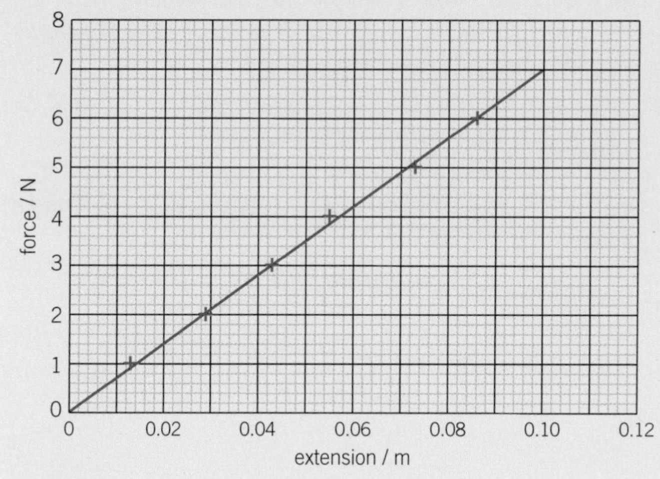


▲ **Figure 3**

**Step 1:** Identify the equation required.

$$F = kx$$

**Step 2:** Rearrange the equation to find $k$.

$$k = \frac{F}{x}$$

This is the gradient of a force–extension graph.

**Step 3:** Find the gradient of the graph.

Taking the points (0,0) and (0.1,7).

$$\text{gradient} = \frac{7.0\,\text{N} - 0.0\,\text{N}}{0.10\,\text{m} - 0.00\,\text{m}} = 70\,\text{N}\,\text{m}^{-1}$$

Every spring or wire has its own value for the spring constant. This value gives a measure of how stiff the specimen is. A large spring constant means the specimen is difficult to stretch. The value of $k$ depends on the material of the wire, its length, and its cross-sectional area. It is important to remember that the spring constant is a value for a specimen rather than for a material, for example, a thin wire will have a much smaller $k$ than that of a thick wire made from the same metal.

## Practical 3.2d(i): Plotting a force–extension graph for a rubber band

The extension of a rubber band can be investigated using the experiment shown in Figure 4.

A piece of stiff wire can be wrapped around the bottom of the weight hanger to provide a pointer. The unstretched length of the band is measured, and then measurements of the length of the band are taken after adding each mass. The extension is found by subtracting the original length from the length when the band is stretched.

The measurements can be repeated as masses are taken off the band and a graph plotted of force (weight) against extension can be plotted.

This experiment can be repeated for a number of similar materials, for example, arrangements of springs and polythene strips.

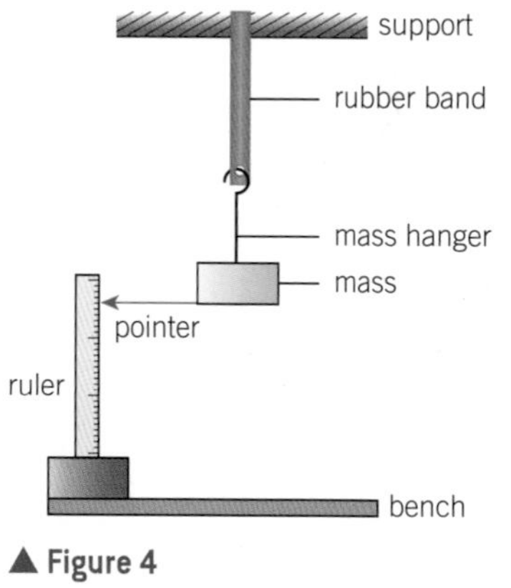


▲ Figure 4

## Elastic and plastic deformation

Hooke's law can also be applied to **compression**, that is, squashing the springs rather than stretching them, but it does have its limitations. Extend a spring too far and it will become easier to stretch and will not return to its original length when you let go. The graph in Figure 5 (next page) illustrates how force varies with extension for larger extensions.

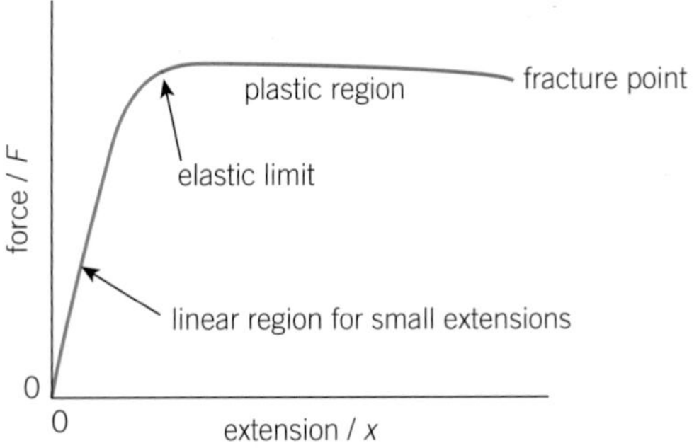


▲ **Figure 5** *The variation of force with extension up to fracture*

Over the first part of the graph, the wire (or spring) stretches or deforms **elastically**. A wire that has deformed elastically will return to its original length when the load is removed. When the extension exceeds the **elastic limit** the wire deforms **plastically**. A wire that has deformed plastically will not return to its original length when the load is removed. The graph is linear for nearly all of the elastic region, curving very slightly near the elastic limit. The plastic region of the graph is non-linear. The wire **fractures** (breaks) at the fracture point.

It is not only wires and springs that follow Hooke's law. For small extensions and compressions all materials will show elastic behaviour, but it is much easier to observe this with a spring than it is with a wire, a glass fibre, or a brick.

## Energy stored in a spring

▲ **Figure 6** *The more you stretch a catapult the more energy it stores*

When you stretch a catapult the first little bit is easy – you don't have to pull hard, and you store little energy in the elastic. Let go, and the stone drops feebly from it. The small force from your hand did little work in extending it. Pull harder, store more energy for the extra stretch, and the stone flies off impressively when you let go.

The force $F$ required to stretch a spring by $x$ is $kx$ if the force is proportional to extension. However, the energy stored in an elastically stretched spring is not simply force × displacement, $Fx$. This is because the force grows steadily larger as the spring is stretched. As the force starts at zero and rises linearly to finish at $F$ the average force is $\frac{F}{2}$.

As a result, the energy stored in an elastically stretched spring $E = \frac{1}{2}Fx$

As $F = kx$, this can also be written as $E = \frac{1}{2}(kx)x = \frac{1}{2}kx^2$

**Synoptic link**

You will meet the relationship between distance and force in Topic 9.3, Conservation of energy

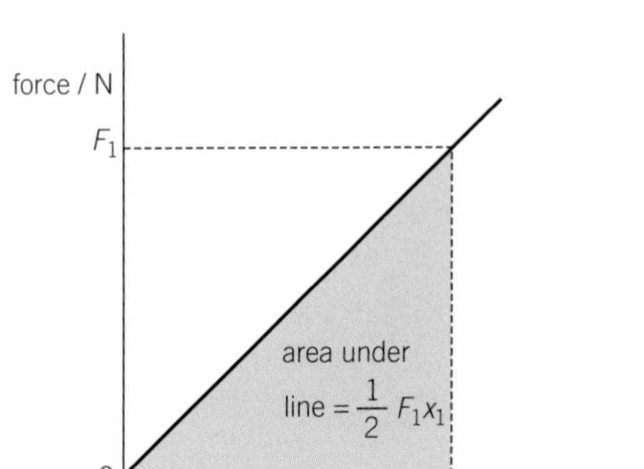


◀ **Figure 7** *Energy stored in a stretched spring* $= \frac{1}{2}kx^2$

Figure 7 shows elastic deformation. When a body deforms elastically the energy stored is equal to the energy transferred stretching the spring. The energy transferred stretching a material can be found from the area under the line.

### Worked example: Finding the energy stored in a spring

A spring has a spring constant $k$ of $70\,\mathrm{N\,m^{-1}}$. It is extended elastically by $0.055\,\mathrm{m}$. Calculate the energy stored in the spring.

**Step 1:** Select the appropriate equation.

$$E = \frac{1}{2}kx^2$$

**Step 2:** Substitute and evaluate.

$$E = \frac{1}{2} \times 70\,\mathrm{N\,m^{-1}} \times 0.055^2\,\mathrm{m^2} = 0.11\,\mathrm{J}\ (2\text{ s.f.})$$

## Summary questions

1 Sketch a force–extension graph showing a wire that deforms elastically followed by a region of plastic deformation up to fracture. Label the sections of the graph showing elastic deformation, plastic deformation, and the fracture point. *(4 marks)*

2 A spring is compressed from a length of 5.2 cm to 4.8 cm when a weight of 22.0 N is placed on it. Calculate the spring constant of the spring. Remember to include the units in your answer. *(3 marks)*

3 **a** Study the graph in Figure 8, which shows force against extension for a spring. State the features of the graph which show that the spring is behaving elastically. *(2 marks)*

**b** Calculate the spring constant of the spring in Figure 8. Give your answer in $\mathrm{N\,m^{-1}}$. *(2 marks)*

**c** Calculate the energy stored in the spring when it has extended by 0.070 m (7.0 cm). *(2 marks)*

4 For Figure 8, a student suggests that the spring would extend by 42.0 cm when it supported a load of 35.0 N. Deduce how the student must have reached this conclusion, state the assumptions the student has made, and suggest reasons why this may be incorrect. *(4 marks)*

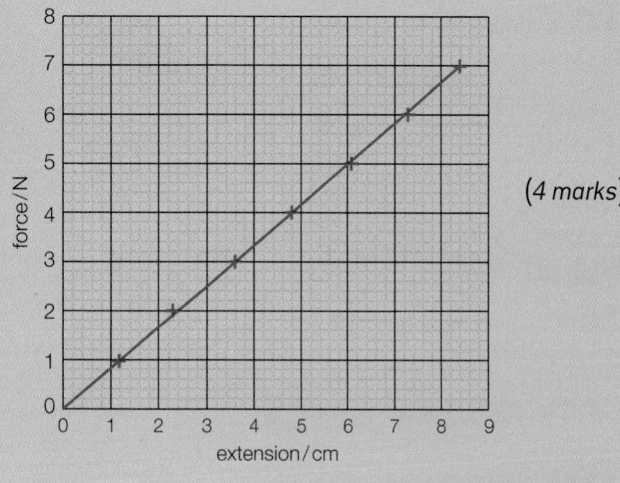


▲ Figure 8

# 4.3 Stress, strain, and the Young modulus

Specification references: 3.2a(iv), 3.2b(i), 3.2b(ii), 3.2c(ii), 3.2d(ii)

### Learning outcomes

Describe, explain, and apply:

- → one method of measuring Young modulus and fracture stress
- → the terms: stress, strain, Young modulus, fracture stress, yield stress
- → a stress–strain graph to fracture
- → the equation for Young modulus: $E = \frac{\text{stress}}{\text{strain}}$
- → an experiment to determine Young modulus for a metal such as copper or steel wire.

▲ **Figure 1** *Is this web really stronger than this girder?*

### Study tip

Always consider units and powers of ten carefully when calculating stress and strain. Try not to round your calculations until the final answer.

Which is stronger, a steel girder or a thread from a spider's web? The answer seems obvious – the girder must be stronger. However, you may have read that spider silk is stronger than steel, so which statement is correct?

## Stress

When you compare the strength of a girder to that of a spider thread you are comparing specimens, not materials. It will take a bigger force to break or fracture the girder than the thread. But the girder has a far greater cross-sectional area than the thread. To make a fair comparison we use the concept of **stress**, which is force per unit area.

**Fracture stress is the stress at which a material breaks. Stress is found by determining the cross-sectional area of the specimen under investigation, measuring the force on the specimen and dividing the force by the cross-sectional area.**

It is also important to know the **yield stress** of materials. This is the stress at which a material begins to deform plastically and become permanently deformed. Imagine you are designing a steel bridge – the breaking stress is an important factor but so is the yield stress as this is the value at which the materials will begin to bend or buckle.

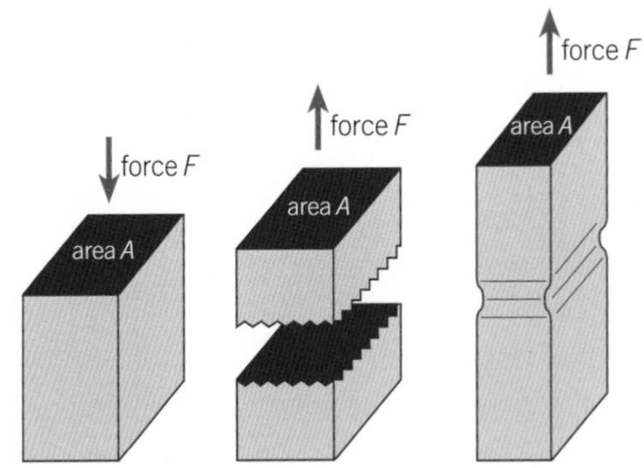

force F
area A
force F
area A
force F
area A
stress is force per unit area
breaking stress is the stress which breaks a material
yield stress is the stress which causes a material to yield

common units of stress:

MN m$^{-2}$ (meganewton per square metre)
or MPa (megapascal)

The units of stress are the same as the units of pressure, pascal. One pascal (1 Pa) is 1 newton per square metre.

Useful rule of thumb – a mass of 1 kg weighs about 10 N on the Earth's surface.

◀ **Figure 2** *Force and stress*

Stress is force per unit area.

$$\text{stress } \sigma\ (\text{N m}^{-2}) = \frac{\text{force } F(\text{N})}{\text{cross-sectional area } A(\text{m}^2)}$$

## Worked example: Calculating stress

A steel wire of diameter 0.84 mm fractures when under a tension (stretching force) of 152 N. Calculate the fracture stress.

**Step 1:** Select the appropriate equation for fracture stress.

$$\text{fracture stress } \sigma = \frac{F}{A}$$

**Step 2:** Calculate the cross-sectional area of the wire in m$^2$.

$$\text{Cross-sectional area} = \pi r^2$$

$$\text{radius of wire} = 0.42\text{ mm} = 4.2 \times 10^{-4}\text{ m}$$

$$\text{cross-sectional area of wire} = \pi \times (4.2 \times 10^{-4})^2 = 5.541... \times 10^{-7}\text{ m}^2$$

**Step 3:** Substitute values back into the equation to evaluate $\sigma$.

$$\text{fracture stress } \sigma = \frac{F}{A}$$

$$\frac{152\text{ N}}{5.541... \times 10^{-7}\text{ m}^2} = 2.7 \times 10^8\text{ N m}^{-2}\ (2\text{ s.f.})$$

### Hint

Remember that the final answer is rounded to the same number of significant figures as the least precise value in the data.

## Strain

A long wire will stretch more than a shorter wire of the same cross-sectional area if the same load is hung from both. Therefore, to make a fair comparison, we can calculate the **strain**. Strain is the fractional increase in length. Unlike extension, strain does not depend on the original length of the specimen.

$$\text{Strain } \varepsilon = \frac{\text{extension}}{\text{original length}} = \frac{x}{L}$$

Strain is a ratio of two lengths and is often given as a percentage, for example, a strain of 5%.

How much does a material stretch? For a given force $F$ a long piece of material will stretch more than a short one. Doubling the length $L$ doubles the extension $x$

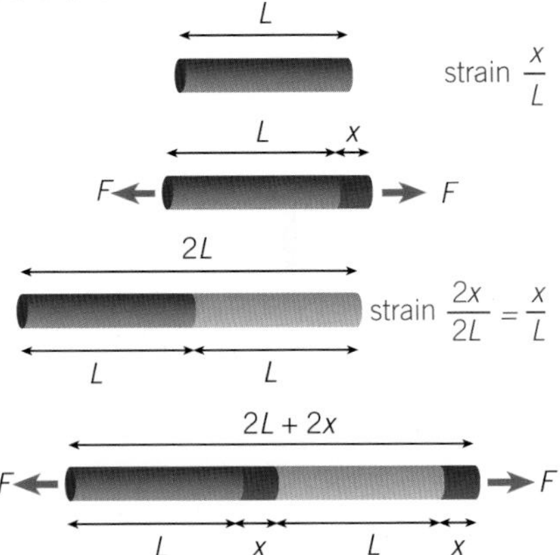


extension $x$ depends on original length
strain = fractional increase in length and does not depend on the original length

▲ **Figure 3** *Stress produces strain*

### Study tip

Remember, stress is dependent on *cross-sectional area*, whereas strain is dependent on *length*.

## Worked example: Using strain

A rubber cord of original length 3.2 m is extended until it is under a strain of 18%. Calculate the total length of the extended cord.

**Step 1:** Select and rearrange the appropriate equation.

$$\text{strain} = \frac{\text{extension}}{\text{original length}}$$

$$\therefore \text{extension} = \text{strain} \times \text{original length}$$

**Step 2:** Convert percentage strain into a decimal figure.

$$18\% = \frac{18}{100} = 0.18$$

### Hint

You must remember to convert percentage strain into the ratio of $\frac{\text{extension}}{\text{original length}}, \frac{x}{L}$, before using the value in a calculation.

**Step 3:** Substitute values and evaluate.

$$\text{extension} = 0.18 \times 3.2\,\text{m} = 0.576\,\text{m}$$

**Step 4:** Total length = original length + extension

$$= 3.2\,\text{m} + 0.576\,\text{m} = 3.8\,\text{m}\ (2\ \text{s.f.})$$

## Stress–strain graphs

The stress–strain graph of a brittle material, such as glass, shows very little plastic deformation. The graph is linear for all its length.

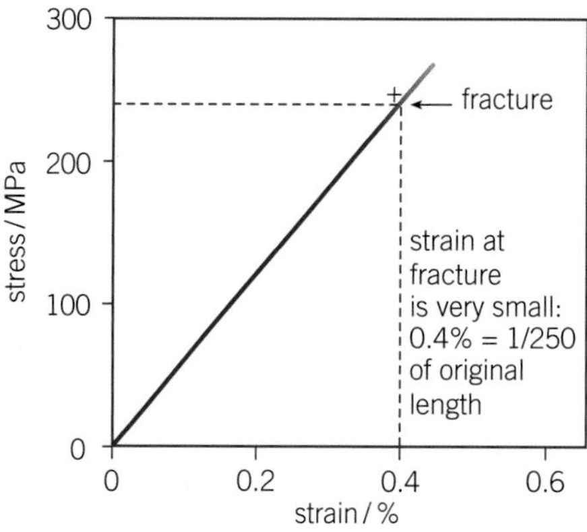


▲ **Figure 4** *A stress–strain graph for glass*

Mild steel is a tough material and undergoes considerable plastic deformation before fracture. This is why tough materials have rounded edges on fracturing rather than the sharp, jagged edges of brittle materials. A rod of mild steel may 'neck' under tension. This means that part of it becomes narrower than the rest.

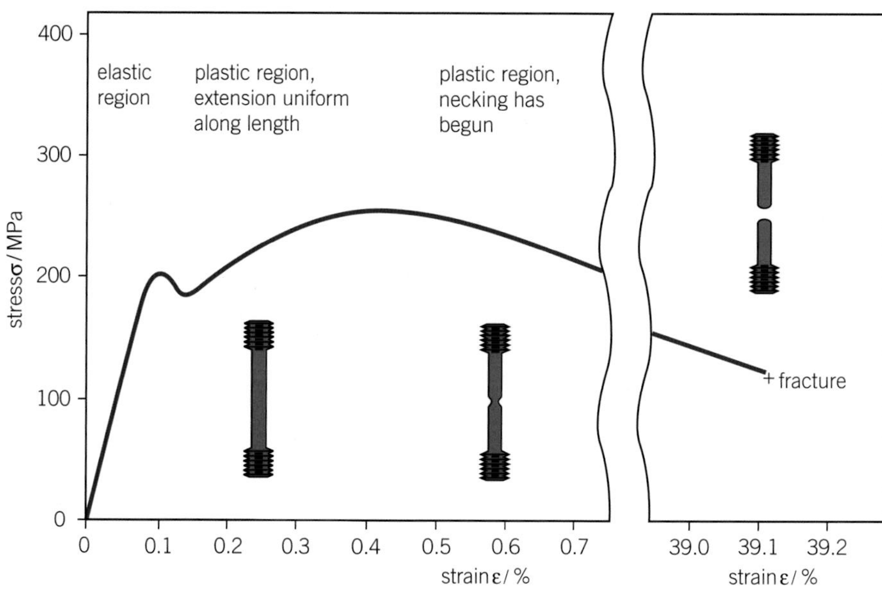


▲ **Figure 5** *A stress–strain graph for mild steel with an extensive plastic region*

## Young modulus

The **Young modulus** $E$ gives a measure of the stiffness of a material rather than the stiffness of a particular specimen.

$$E = \frac{stress}{strain} = \frac{\sigma}{\varepsilon} = \frac{F/A}{x/L} = \frac{FL}{xA}$$

The units for Young modulus are $\mathrm{N\,m^{-2}}$ or Pa. These are the same as the units of stress. This is because strain is a ratio of two lengths and so has no units – strain is dimensionless. Since $F$ and $L$ are large values, divided by the product of the much smaller $x$ and $A$, Young modulus is often a very large number.

Many materials stretch in a uniform way. Increase the stretching force in equal steps, and the extension increases in equal steps too, in proportion. That is, the strain is proportional to the stress producing it. This is the same as Hooke's law – the stretching of a spring is proportional to the stretching force you apply.

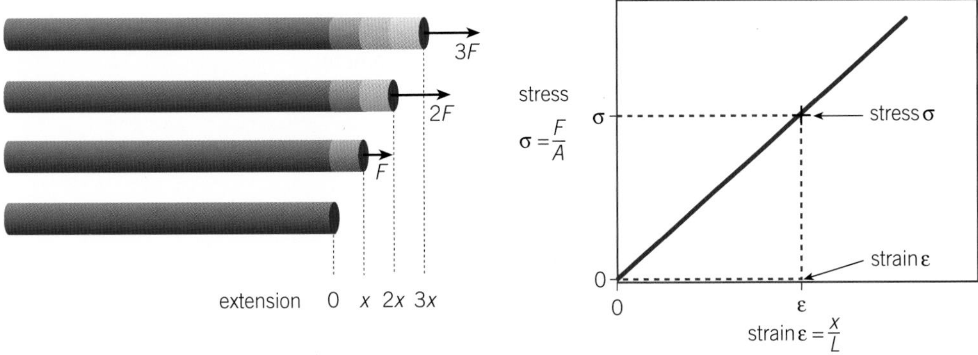


strain ∝ stress .................... graph is straight line

ratio $\frac{\text{stress}}{\text{strain}}$ is constant

$$\text{Young modulus} = \frac{\text{stress}}{\text{strain}}$$

$$E = \frac{\sigma}{\varepsilon}$$

▲ **Figure 6** *The Young modulus*

### Worked example: The Young modulus

Data about a wire that is under tension and has extended elastically is shown below.

cross-sectional area = $1.7 \times 10^{-6}\,\mathrm{m^2}$

original length = 2.5 m

force on wire = 220 N

extension = $2.0 \times 10^{-3}\,\mathrm{m}$

Calculate the Young modulus of the wire.

**Step 1:** Select the appropriate equation

$$E = \frac{\sigma}{\varepsilon} = \frac{F/A}{x/L}$$

**Step 2:** Substitute values and evaluate.

$$E = \frac{220\,\mathrm{N}/(1.7 \times 10^{-6}\,\mathrm{m^2})}{(2.0 \times 10^{-3}\,\mathrm{m})/2.5\,\mathrm{m}} = 1.6 \times 10^{11}\,\mathrm{N\,m^{-2}}\ \text{(or Pa) (2 s.f.)}$$

## Practical 3.2 d(ii): Determining the Young modulus and fracture stress of a metal

A wire is put under tension and its extension is measured as the force is increased to find its Young modulus. The gradient of a linear stress–strain graph will give you the value of the Young modulus.

Although the method is simple, doing this well is a challenge. The golden rule for any experiment is to carry out a trial experiment first, or make a rough estimate, to find out what to expect.

In the case of steel, a rough estimate or a trial experiment quickly tells you two things:

- for any reasonable length of wire you will have to measure an extension of only a few millimetres
- to stretch the wire with any reasonable force you will have to use a thin wire, a fraction of a millimetre in diameter.

A typical experimental set-up is shown in Figure 7.

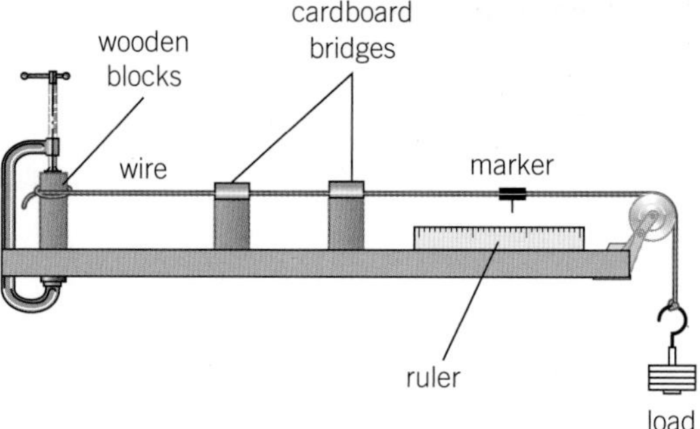


▲ **Figure 7** *Experimental set-up for determining the Young Modulus of a wire*

The diameter and the original length of the wire are measured.

The extension of the wire is measured as more weights are added to the wire, increasing the force. This can be continued until the wire breaks, allowing you to calculate breaking stress and strain.

Stress and strain values can be calculated from the data and a graph of stress ($y$-axis) against strain ($x$-axis) drawn. The gradient of the linear region is the best estimate for the Young modulus of the wire. The uncertainty can be estimated by choosing the largest source of uncertainty as a guide.

## Summary questions

**1** **a** Explain the difference between fracture stress and yield stress. *(2 marks)*

**b** A concrete block of cross-sectional dimensions of 10 cm × 10 cm fractures when a force of 1 MN is applied. Calculate the fracture stress of the concrete in $\text{MN m}^{-2}$ and in $\text{N mm}^{-2}$. *(3 marks)*

**2** **a** A spring stretches from an original length of 5.1 cm to a new length of 5.7 cm when a force of 1 N is applied. Calculate the strain of the spring. Give your answer as a decimal and as a percentage. *(3 marks)*

**b** What might you expect the length of the spring to be if a 2 N force were applied? Explain your answer, stating any assumptions you make. *(2 marks)*

**3** A rubber cord, 5 m long and 50 mm thick, hangs from the top of a zoo cage. A chimp weighing 300 N hangs on the end of the cord. Calculate the extension of the cord. The Young modulus of the rubber is 100 MPa. *(3 marks)*

**4** In measuring the Young modulus of copper wire you obtain the following values:

length of wire = $1.5 \pm 0.005$ m

diameter of wire = $0.33 \pm 0.02$ mm

extension = $3.5 \pm 0.02$ mm

force = $11.5 \pm 0.2$ N.

**a** Show that the largest percentage uncertainty is ± 6%. *(4 marks)*

**b** Calculate the Young modulus of copper. *(2 marks)*

**c** Assuming that the uncertainty in the final value is also ± 6%, calculate the range of possible values of the Young modulus from this result. *(3 marks)*

**d** The experiment can be improved if uncertainty in the cross-sectional area can be minimised. This can be done by measuring the mass of the wire and using known density values for the material. Explain how this method can be used to determine the cross-sectional area of the wire and why it may be an improvement over determining the cross-sectional area from the diameter of the wire. *(4 marks)*

**e** The experiment is repeated to determine the fracture stress of the copper. Sketch the stress–strain graph you would expect from copper, a ductile metal, up to fracture. *(2 marks)*

### Hint

percentage uncertainty = $\dfrac{\textit{uncertainty in measurement}}{\textit{measured value}} \times 100\%$

# 4.4 Choosing materials

## Specification references: 3.2b(i), 3.2b(iii)

### Learning outcomes

Describe, explain, and apply:

→ the term: tension

→ tables and diagrams comparing materials by properties.

### Study tip

It is useful to have a 'ball-park' figure in mind for typical strengths and Young modulus values.

- The Young modulus of steel is of the order 100 GPa, whilst that of polymers is of the order 1 GPa.
- The tensile strength of steel is of the order 1000 MPa, whilst that of polymers is of the order 100 MPa.

Knowing the fracture stress, yield stress, and Young modulus of different materials is very important to architects, engineers, and designers. It allows them to predict the behaviour of materials under stress, and calculate the requirements of a particular construction project.

Choosing the best material for the job is important in all areas of life. The surgeon's scalpel is made from a particular steel that is strong, hard, and non-corrosive. The clothes you are wearing will be made from natural or synthetic polymers which stretch to fit around your body, showing that the fibres have a degree of elasticity. Bamboo has a low density and is chosen as a material for scaffolding in many countries.

▲ **Figure 1** *Polymers are used extensively in clothing*

▲ **Figure 2** *Bamboo as scaffolding*

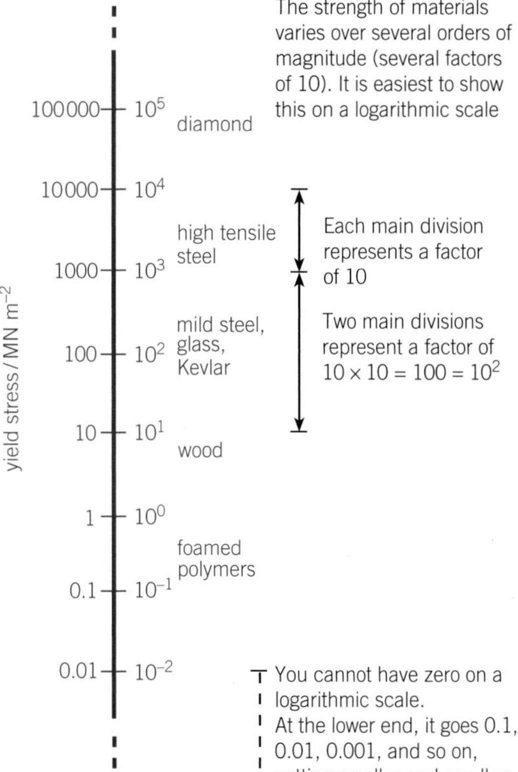


▲ **Figure 3** *Logarithmic scale of stress*

A material is in **tension** when a force is acting in a direction to stretch the material. The force in this case is described as a tensile force. On the other hand, a compressive force tends to compress or squash the material. Some materials can be strong in compression but weak in tension. For example, Portland cement concrete has a compressive strength of around 30 MPa and a tensile strength of about 3 MPa.

## Selection charts

The choice of a material for a particular use does not only depend on the properties that we have been considering so far. Other physical properties such as **density** (where density = $\frac{\text{mass}}{\text{volume}}$) and toughness are also important, as are non-physical properties such as cost and availability. Material selection charts allow quick comparisons to be made between different classes of materials.

As numerical values of stress and strain vary greatly between different classes of materials, scientists use logarithmic scales on charts comparing different materials to display the large range of values.

The chart in Figure 4 compares Young modulus with density of different materials. It shows that polymers have a wide range of stiffness values and a narrower range of densities.

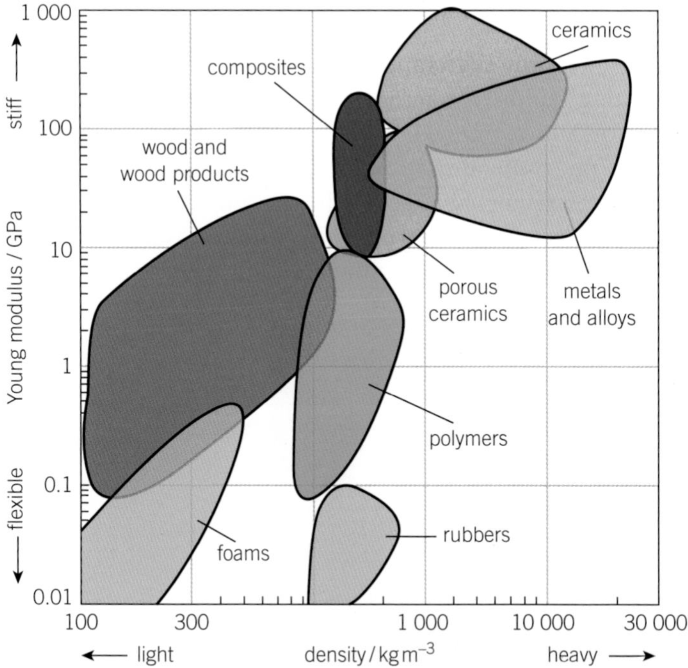


▲ **Figure 4** *Young modulus and density values for different classes of materials*

Polymers also show a wide range of toughness. You will have noticed that some 'plastic' rulers can be bent into all kinds of shapes but that others will shatter – this is an example of the variation in toughness of polymers.

There are also huge variations between strength and toughness for different classes of materials.

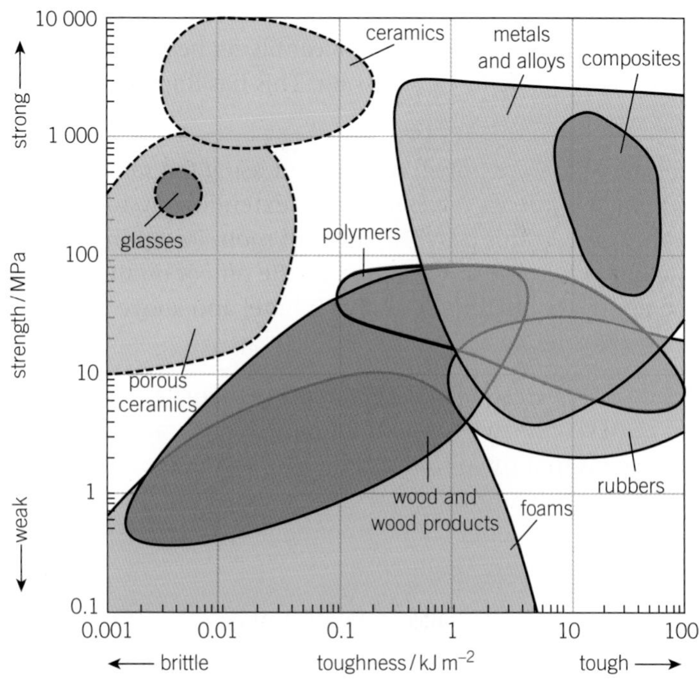


◀ **Figure 5** *Strength and toughness values for different classes of materials*

## Synoptic link

You will find a section about logarithms in the Maths Appendix.

## Summary questions

1 Using the chart of Young modulus and density in Figure 4, show that the stiffest metal has a Young modulus about 30 times that of the least stiff. *(2 marks)*

2 Using the chart of strength and toughness in Figure 5, calculate the ratio of the strength of the strongest metal to that of the weakest. *(2 marks)*

3 A lift has a weight of 5000 N. It carries up to eight passengers.
   a Taking the average weight of an adult as 650 N, calculate the tension in the steel cable when the lift is fully loaded. *(2 marks)*
   b Calculate the minimum diameter of steel cable needed to support the lift and passengers. Assume the fracture stress of the steel is 1000 MPa. *(2 marks)*
   c Explain why a much thicker cable will be used in practice. *(2 marks)*

# Physics in perspective

## How to build a skyscraper

The Shard is an 87-storey skyscraper topped with a steel and glass spire, transforming the skyline across the Thames from the Tower of London. At a height of 306 m it was the tallest building in the European Union on completion in 2013. Building such a tall tower in a confined space was a challenge for the designers and engineers working on the project. The construction schedule was also demanding – one floor was added to the building each week.

▲ **Figure 1** *The Shard and tower bridge at night*

▲ **Figure 2** *The Shard during construction, showing the core, some of the floors of the building, and the glass cladding*

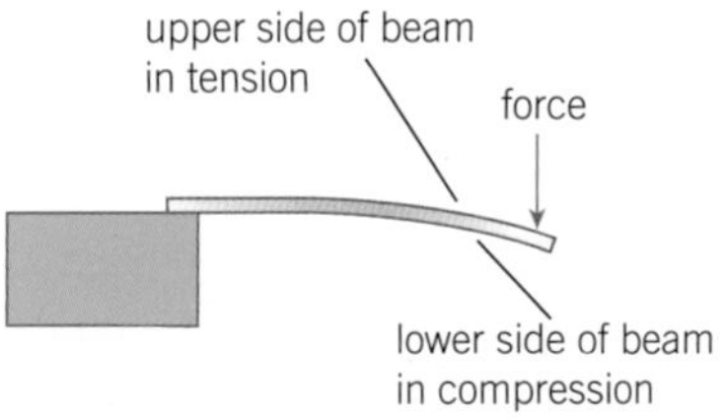


▲ **Figure 3** *A simple cantilever*

The building is formed around a core reinforced-concrete tower rising from the foundations. The core accounts for much of the 100 000 tonnes of concrete used in the building. Beneath the core are piles — columns of steel and concrete extending 53 metres down into the ground.

The floors, which you can see in Figure 2, extend from the core tower rather like branches of a tree extending from a trunk. This is called a cantilever arrangement. The upper of a cantilever beam will be in tension and the lower side in compression. This has important design implications.

The floors of the Shard are not all constructed using the same materials. The first 40 floors use steel beams extending up to 15 m from the core. These beams provide plenty of room for wiring, air conditioning and other services required by the offices on these levels. Above these floors the building is used as a hotel and above the hotel are residential apartments.

These floors do not need so many services in the floors and ceilings but do need to have effective soundproofing. The pyramid-like shape of the Shard means that the distance from the core to the outer glass wall has reduced to 9 m at this level. Post-tensioned concrete is used for these floors as it has better soundproofing properties than steel beams and can span the 9 m required.

The building is topped with the glass-covered pinnacle that gives the Shard its name. It is certainly at the sharp end of design and construction.

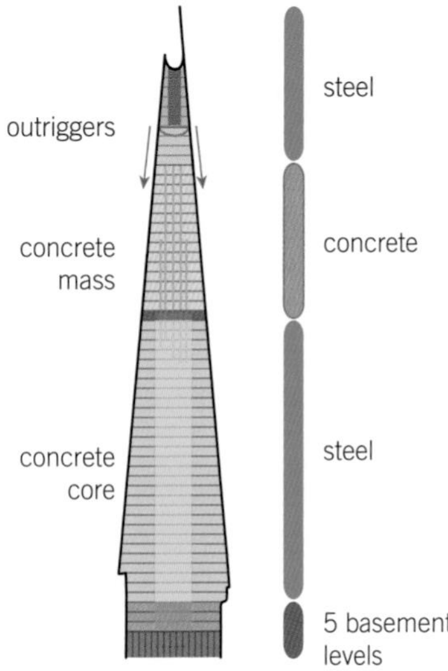


▲ **Figure 4** *The main materials used to construct the shard*

## Summary questions

1 In most structures, the stresses and strains within components are designed to remain within the elastic region of the material used. They tend not to stray too close to the elastic limit, or the plastic region which lies beyond.

   a Explain why this is so.

   b Give an example of a structure where plastic deformation of the material is used to advantage.

2 Assume all 100 000 tonnes of concrete used in the Shard went into the core tower.

   a Calculate the smallest cross-sectional area of tower that could support this weight of concrete. The compressive strength of the concrete is 60 MPa.

   b Use your answer from **(a)** to estimate the maximum height of the tower. The density of concrete is 2400 kg $m^{-3}$.

   c Compare your value to the height of the Shard and suggest reasons for any difference between the two values.

3 The graphs in Figure 5 show the stress against strain % for the concrete and steel used for reinforcing the core tower and concrete floors of the Shard, up to their yield points.

   a Explain why the graph for steel is symmetrical in tension and compression but the graph for concrete isn't.

   b For concrete, calculate the ratio $\frac{\text{strength in compression}}{\text{strength in tension}}$.

   c Calculate the Young modulus for the steel.

   d Calculate the ratio $\frac{\text{Young modulus for steel}}{\text{Young modulus for concrete}}$

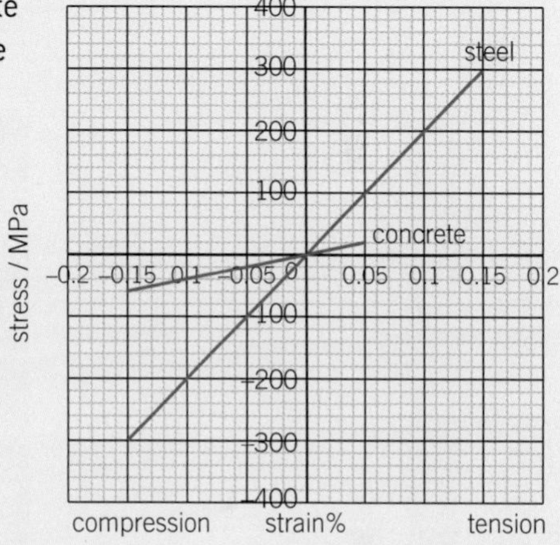


▲ Figure 5

4 For the Shard's core tower construction, liquid concrete was poured around a network of bound steel mesh (this is known as reinforced concrete). Suggest why this composite material is used, and why the surface of the steel used to form the mesh is designed to be rough and knobbly.

5 For the concrete composite floors post-tensioned concrete was used as shown in Figure 6.

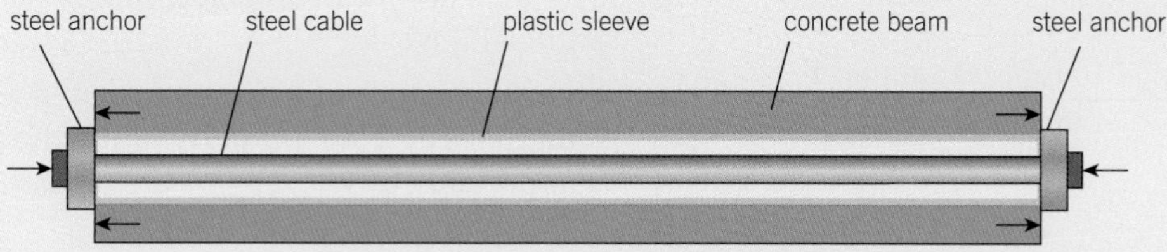


▲ Figure 6

Here steel cables are inserted through plastic sleeves set into the concrete floors. They are stretched with hydraulic levers and then secured by anchor plates, which place the floor beams into a state of compression.

   a Suggest advantages for this method of construction.

   b These concrete composite floors are used for the hotel and living accommodation floors of the Shard, whereas steel floors are used for the commercial offices. Suggest why this might be the case.

## Practice questions

1 Which of the following are units of stress?

A $N kg^{-1}$  B $N s$

C $N m$  D $N m^{-2}$

2 Which of the properties below is a measure of a material's resistance to stretching or bending?

A brittleness  B strength

C stiffness  D toughness

3 Two wires of the same material are compared. Sample 1 has diameter *d*. Sample 2 has diameter 2*d*.

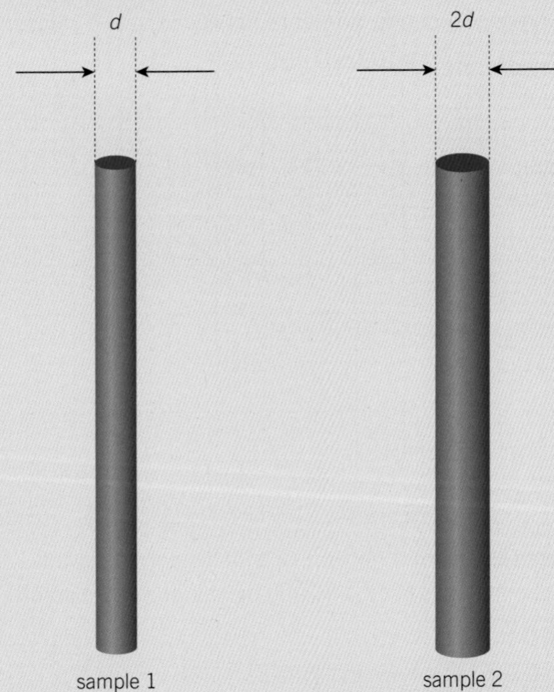


▲ Figure 1

What is the ratio $\frac{\text{Young Modulus of sample 1}}{\text{Young Modulus of sample 2}}$?

A $\frac{1}{4}$  B $\frac{1}{2}$

C $\frac{1}{1}$  D $\frac{2}{1}$

4 A weight of 5 N is hung from a steel spring which behaves elastically. The spring extends by 9 cm.

Calculate the energy stored in the spring. Give your answer in J. *(2 marks)*

5 The figure shows the stress-strain graph for a mild steel.

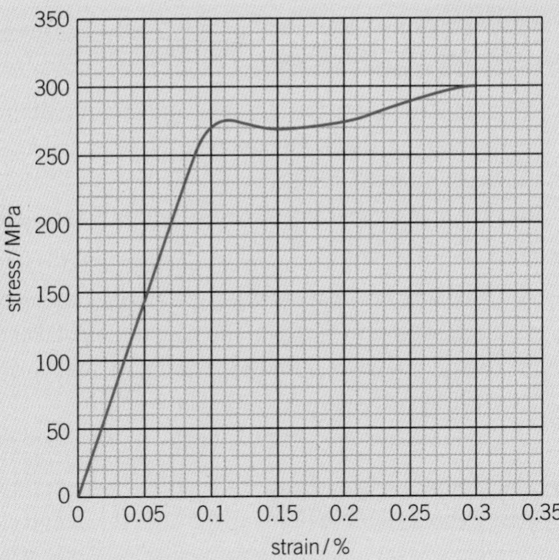


▲ Figure 2

a Calculate the Young Modulus for the linear section of the graph *(2 marks)*

b State how the material is deforming in the non-linear section of the graph. *(1 mark)*

6 Two students are discussing the properties of a cast iron frying pan. The first student correctly describes cast iron as a brittle material. The second student argues that cast iron cannot be brittle because it is strong.

a Describe the properties of brittle materials to show why the first student is correct. *(2 marks)*

b Give an example of an object that is brittle and weak. *(1 mark)*

c Give an example of a material used in a situation where the property of toughness is important. *(2 marks)*

7 A wire breaks when a force of 700 N is applied. The breaking stress of the wire is 250 MPa. Calculate the diameter of the wire. *(3 marks)*

8 Figure 3 shows the force-extension graph for a metal wire.

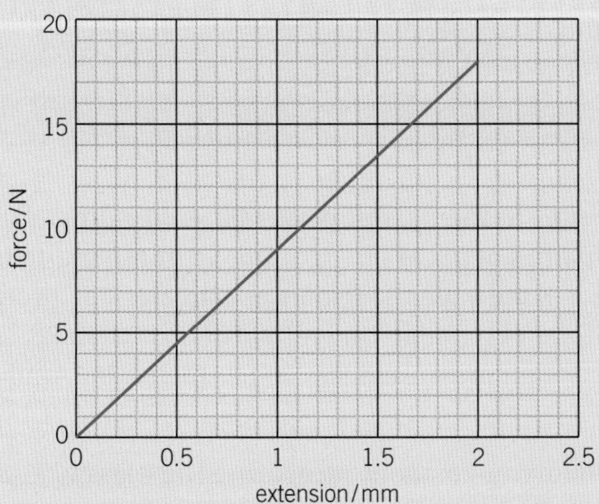


▲ Figure 3

a State how the graph shows that the force on the wire is proportional to the extension. *(2 marks)*

b The wire has a Young Modulus of $1.8 \times 10^{11}$ Pa and an original length of 1.9 m. Use data from the graph to calculate the diameter of the wire. *(4 marks)*

c The wire yields at a force of 18 N and breaks at a force of 18.5 N. At this point the extension of the wire is 2.1mm.

(i) Explain the difference between yield stress and breaking stress. *(2 marks)*

(ii) Use your answer from **(b)** to calculate the yield stress of the specimen. *(2 marks)*

(iii) Explain why the yield stress of a specimen is important in the choice of materials for construction. *(1 mark)*

9 A student is investigating the region of linear strain of a rubber band. The following measurements are taken to calculate the Young Modulus of the rubber.

force = 0.50 N ± 1%

cross-sectional area = 4.0 $mm^2$ ± 3%

extension = 0.0080 ± 0.0005 m

original length = 0.145 ± 0.001 m

a (i) Explain which measurement contributes most to the overall uncertainty in the calculated value of the Young modulus. *(3 marks)*

(ii) Suggest and explain how the uncertainty in the measurement you have chosen in **(a)**(i) can be reduced. *(2 marks)*

b (i) Using the uncertainties given in the data, show that the maximum calculated value of the Young modulus of the rubber is about $2.5 \times 10^6$ Pa. *(4 marks)*

(ii) Without taking into account the uncertainties in the data, the student calculates the Young modulus to be $2.3 \times 10^6$ Pa. Use this value and your answer to (**b**)(i) to estimate the percentage uncertainty in the final result. *(2 marks)*

(iii) The student suggests that the uncertainty in the original length can be ignored.

Comment on this suggestion. *(1 mark)*

# 5 LOOKING INSIDE MATERIALS

## 5.1 Materials under the microscope

Specification references: 3.2 a(ii)

**Learning outcomes**

Describe, explain, and apply:

- → direct evidence of the size of particles and their spacing, for example, using Rayleigh's oil drop experiment or a scanning tunnelling microscope.

How tall are you in metres to the nearest power of ten (or order of magnitude)? Since you are not 0.1 m ($10^{-1}$ m) nor 10 m ($10^{1}$ m) tall, you must be closest to 1 m tall ($10^{0}$ m). From the diameter of a proton ($10^{-15}$ m) to the size of the observable Universe ($10^{27}$ m) scientists have estimated sizes through measurements, experimentation, and careful thought.

For example, how would you go about measuring the mass of one page of this book using bathroom scales? This is relatively straightforward – measure the mass of the whole book and divide by the number of pages. This is an example of a calculated estimate. It is an estimate because you haven't taken into account the difference in mass for the covers of this book and that bathroom scales have poor resolution.

▲ **Figure 1** *The average height of these giraffes is $10^{0}$ m to the nearest order of magnitude*

**Hint**

The word 'estimate' does *not* mean guess. Making an estimate of a measurement can often involve calculations.

**Synoptic link**

Estimations are covered in the Maths Appendix.

### Estimating the size of atoms and molecules

At the end of the 19th century, Lord Rayleigh performed a simple experiment that enabled him to estimate the thickness of an oil layer floating on water. If the oil spreads out as far as possible the thickness of the layer will be equal to the length of the oil molecule.

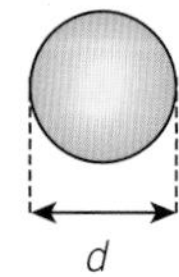


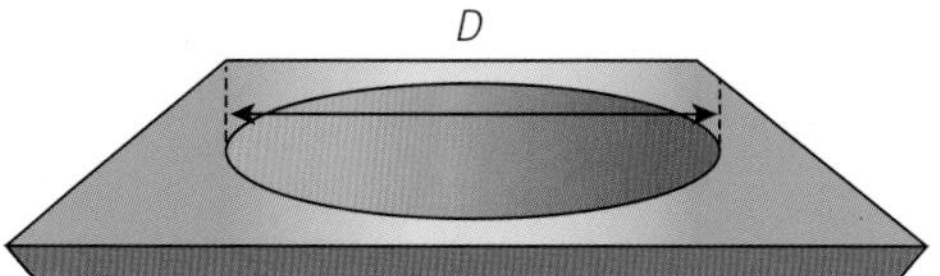


▲ **Figure 2** *Rayleigh's oil drop experiment*

The thickness of the layer, $h$, is an order of magnitude measure of the largest possible value for the length of an oil molecule.

Rayleigh carried out his experiment by:

1 measuring the diameter $d$ of the oil drop

2 calculating the radius $r$

3 placing the oil drop on still water to observe it spreading

4 measuring the diameter of the patch of oil $D$ after the oil had spread

5 calculating the radius $R$ of the oil patch.

volume of oil drop $= \frac{4}{3}\pi r^3$

volume of oil patch $= \pi R^2 h$, where $h$ = height of the oil patch

As the volume of drop = volume of patch

$$\frac{4}{3}\pi r^3 = \pi R^2 h$$

Rearranging and cancelling gives

$$h = \frac{4r^3}{3R^2}$$

## Uncertainties in estimation

There are a number of sources of uncertainty in this experiment. It is always important to recognise the largest source of uncertainty and to try to minimise it. In this case the largest percentage uncertainty comes from the difficulty of measuring the diameter of an oil drop more precisely than about ± 0.5 mm. For an oil drop of about 1 or 2 mm diameter this is a considerable factor. The value for radius of the drop is cubed in the calculation so any uncertainty in this reading will produce considerable uncertainty in the final result.

### Worked example: The length of an oil molecule

A drop of oil of diameter 1 mm spreads into a disc of approximately 36 cm in diameter. Assuming that the oil spreads into a disc that is uniformly one molecule thick, calculate the length of an oil molecule.

**Step 1:** Find the radius of the drop and the disc in metres.

$$\text{drop radius} = \frac{1 \times 10^{-3}\,\text{m}}{2} = 5 \times 10^{-4}\,\text{m}$$

$$\text{disc radius} = \frac{0.36\,\text{m}}{2} = 0.18\,\text{m}$$

**Step 2:** Equate the volume of the drop with the volume of the disc and substitute.

$$\frac{4}{3}\pi r^3 = \pi R^2 h$$

$$\frac{4}{3}\pi \times (5 \times 10^{-4}\,\text{m})^3 = \pi \times 0.18^2\,\text{m}^2 \times h$$

**Step 3:** Rearrange to make $h$ the subject of the equation and evaluate.

$$h = \frac{4 \times (5 \times 10^{-4}\,\text{m})^3}{3 \times 0.18^2\,\text{m}^2} = 5 \times 10^{-9}\,\text{m (1 s.f.)}$$

## Estimating the order of magnitude of the size of a gold atom

A similar method can be used to estimate the size of an atom in a solid by calculating the volume it occupies. For example, if we know the mass of a gold atom and the density of gold we can estimate the diameter of a single atom.

mass of gold atom $= 3.27 \times 10^{-25}$ kg

density of gold $= 19\,300\,\text{kg}\,\text{m}^{-3}$

Since density $= \frac{\text{mass}}{\text{volume}}$, we can calculate the number of atoms in $1\,m^3$ of gold.

mass of $1\,m^3$ gold = 19 300 kg

$$\text{number of atoms in } 1\,\text{m}^3 \text{ of gold} = \frac{\text{total mass of gold}}{\text{mass of one gold atom}}$$

$$= \frac{19\,300\,\text{kg}}{3.279 \times 10^{-25}\,\text{kg}}$$

$$= 5.90... \times 10^{28} \text{ atoms}$$

Since $5.90... \times 10^{28}$ gold atoms have a volume of $1\,m^3$,

volume of one atom is $\frac{1}{5.90... \times 10^{28}} = 1.69... \times 10^{-29}\,m^3$.

If we make the assumption that gold atoms are cubes (of course they are not, but it simplifies the arithmetic and this is an estimation)

length of side of the cube $= \sqrt[3]{1.69... \times 10^{-29}\,m^3} = 2.6 \times 10^{-10}\,m$ (2 s.f.)

The order of magnitude of a gold atom is therefore $10^{-10}\,m$.

**Study tip**

It is useful to remember that the diameter of an atom is of the order $10^{-10}$ m.

## Atoms – from imagination to images

Modern atomic theory developed throughout the 20th century, but it was only at the end of the century that scientists began to make images which showed the atomic structure of matter. Atomic force microscopes (AFMs) and scanning tunnelling microscopes (STMs) can be used to show individual atoms. Scanning electron microscopes (SEMs) show larger scale structures.

▲ **Figure 3** *STM image of gold atoms on a graphite surface*

The photograph in Figure 3 shows a false-colour STM image of graphite (carbon) atoms (in green) in a regular array, with atoms of gold (orange coloured) piled on the graphite surface. The length of the near side is $1 \times 10^{-9}\,m$. There are roughly ten green blobs representing individual carbon atoms along this side, showing that the length taken up by an individual carbon atom is about $1 \times 10^{-10}\,m$. This agrees with our order of magnitude estimate for the gold atom.

**How an AFM works**

Rather like an old-fashioned record player, an AFM moves a needle over a sample to detect the contours of the surface. It can detect changes on an atomic scale. A fine point is mounted on the arm and forces between the surface and the tip make the arm bend. A laser beam reflected from the arm detects the bending. One way of using this apparatus is to move the specimen to keep the force on the tip constant. The up and down movement of the specimen as it is scanned under the tip corresponds to the surface profile.

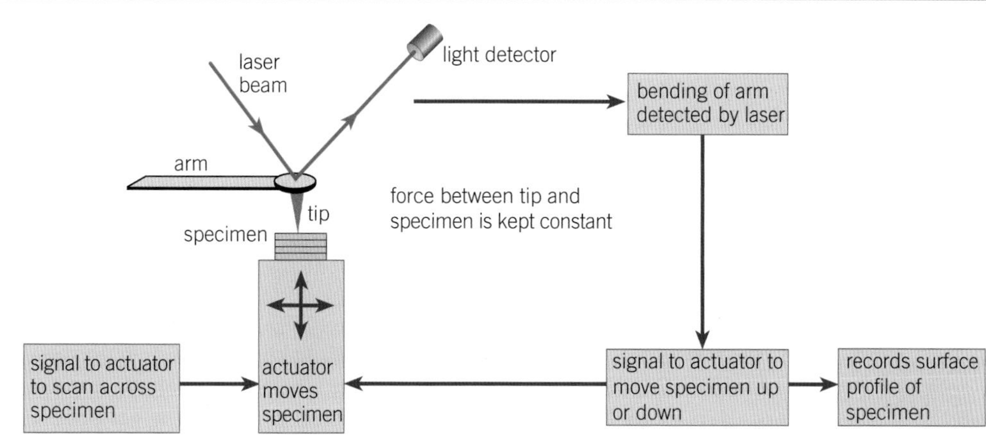


The AFM does not need the specimen's surface to be specially prepared. The specimen doesn't need to conduct electricity or be in a vacuum. The AFM is suitable for biological work.

▲ **Figure 4** *How an AFM works*

**Questions**

1. Suggest why the specimen moves rather than the tip mechanism.
2. One scientist states that an AFM image of gold on graphite shows a picture of atoms. Discuss whether or not you agree with this statement.

## Summary questions

1. You are shown an STM image of graphite (carbon) atoms in a regular array. Explain how you can use the image to make an order of magnitude calculation for the size of a carbon atom, stating any other information you need. *(3 marks)*
2. Suggest reasons why the oil drop experiment gives a ***maximum*** value for the length of an oil molecule. *(2 marks)*
3. In performing the oil drop experiment, a student measures the diameter of the drop as 0.5 mm and the diameter of the oil patch as 20 cm. Use these values to calculate an order of magnitude estimate for the height of the oil patch. *(3 marks)*
4. The calculation of the diameter of the gold atom using the mass of the atom and the density of gold can only give an order of magnitude answer. Suggest why this is the case. *(2 marks)*

# 5.2 Modelling material behaviour

## Specification references: 3.2a(iii), 3.2b(i), 3.2b(iv)

**Learning outcomes**

Describe, explain, and apply:

- → structures of metals and ceramics
- → dislocations leading to slip in metals with brittle materials not having mobile dislocations
- → the term: strong
- → images showing structure of materials.

**Synoptic link**

You will learn more about diffraction in Chapter 6, Wave behaviour.

AFMs and STMs tell us about the arrangement of atoms on the surface of a material. However, the atoms within the material may be arranged differently.

Techniques such as X-ray diffraction crystallography can tell us more about the arrangement of atoms beneath the surface of a material. The photograph in Figure 1 shows a famous image of the X-ray diffraction pattern of DNA, produced by Rosalind Franklin in 1953. This, and other images, helped establish the shape of DNA, the crossed bands suggesting its helical structure.

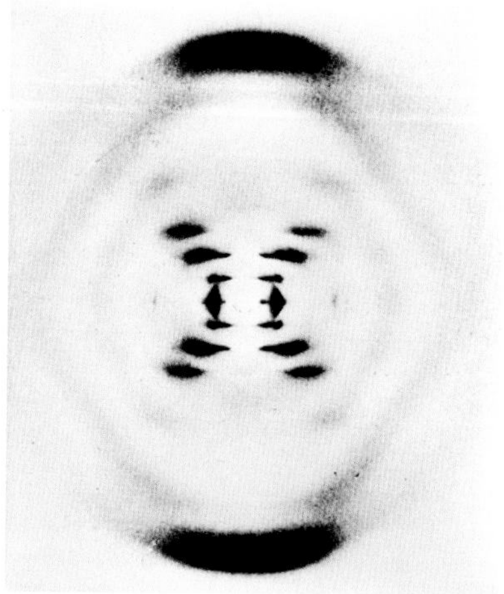

▲ **Figure 1** *X-ray diffraction image of DNA*

## Metal structure and ductility

Metals are **crystalline**. This means that the individual particles are arranged in a regular pattern over distances many times the spacing between the particles. Often the crystalline structure is not obvious from the appearance of the metal – a metal saucepan doesn't look crystalline, but techniques such as X-ray crystallography reveal the hidden structure.

Some pure metals are malleable and ductile. We explain this behaviour with the idea of **dislocations**. These are mismatches in the regular rows of atoms – missing atoms in the otherwise orderly arrangement. It is the movement of dislocations that makes metals ductile. The diagram shows how a dislocation moving through the layers of atoms allows layers to move one atom at a time. This greatly reduces the energy needed to deform the metal. Without mobile dislocations, metals such as gold could not be hammered into shape.

▲ **Figure 2** *SEM image showing crystals of tungsten*

▲ **Figure 3** *'The Mask of Agamemnon' – a beaten gold mask dating from around 1500 BC*

Ceramic materials also have dislocations within their structures, but the dislocations are not mobile and so they do not move through the material. Ceramic materials are therefore brittle.

### Changing the ductility of metals

Metal **alloys** tend to be less ductile than pure metals. Metal alloys can be formed by the addition of other metallic elements that usually have different sized atoms. These can pin down the dislocations in the metal structure, making slippages between the layers of atoms more difficult.

Atoms in gold are in a regular array: a crystal lattice. To shape the metal, one layer must be made to slide over another.

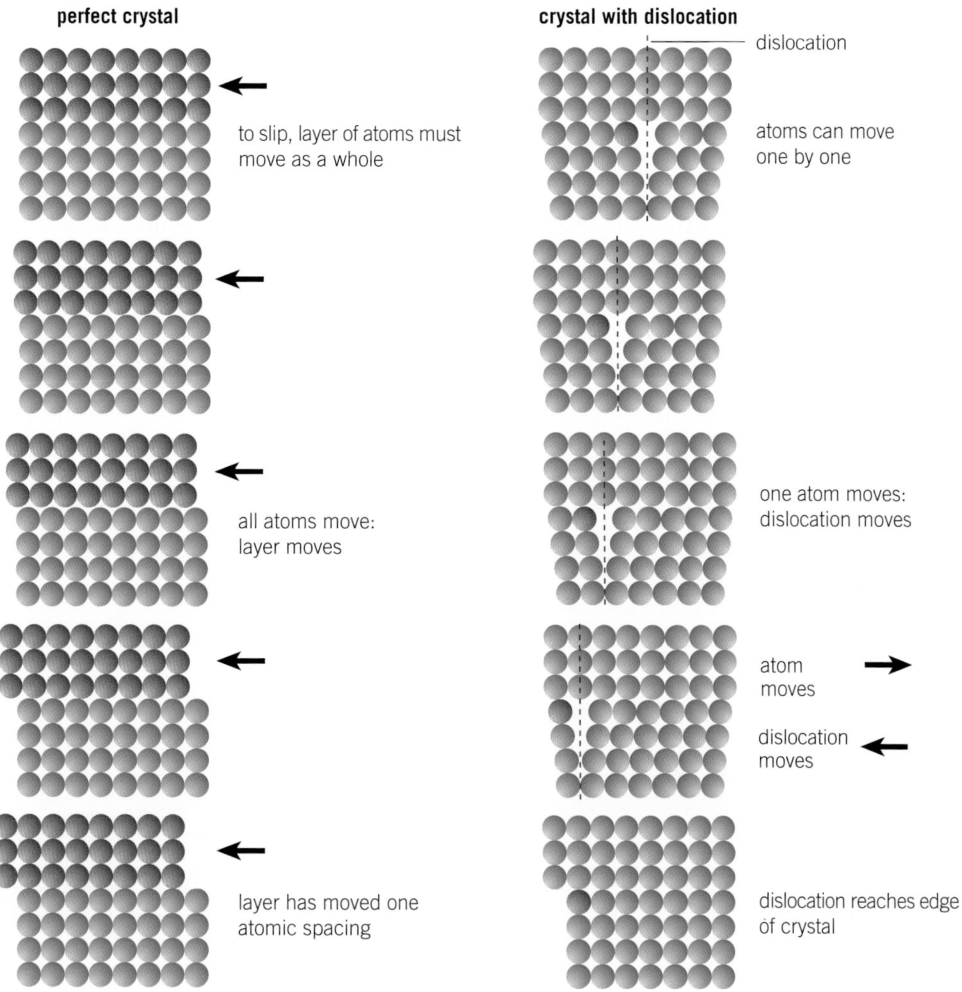


in both examples a layer has slipped by one atomic spacing

**Wrong model:**

Making all the atoms slip together needs considerable energy
This model predicts metals to be 1000 times as strong as they actually are

**Better model:**

One atom slipping at a time needs much less energy

The dislocation model predicts the strength of metals much better

▲ **Figure 4** *Slipping of particles in metals*

The proportion of alloy atoms in the lattice changes the properties of the alloy — see Figure 5.

## Amorphous and crystalline materials

Solids form when liquids cool. The internal structure of the resulting solid may be crystalline, like metals, or **amorphous** (disordered), like glass. Rapid cooling tends to trap particles in an amorphous state, resembling the disordered arrangement in a liquid.

Slow, controlled cooling of a liquid can lead to a single, pure crystal. High-purity silicon is such a crystal, used for making microchips. A single crystal can have a mass of several kilogrammes and contain $10^{26}$ atoms arranged in a near-perfect array.

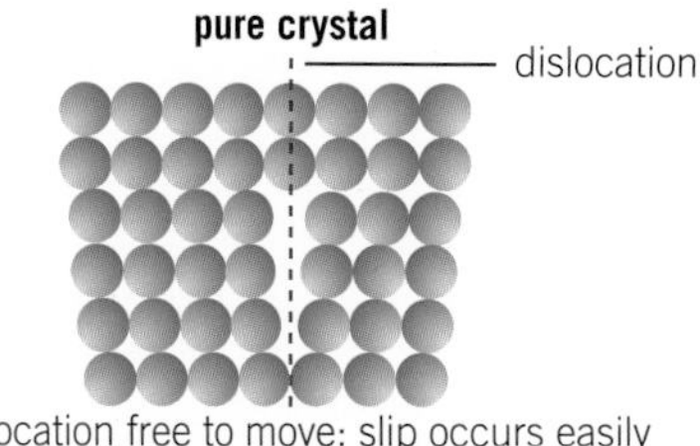


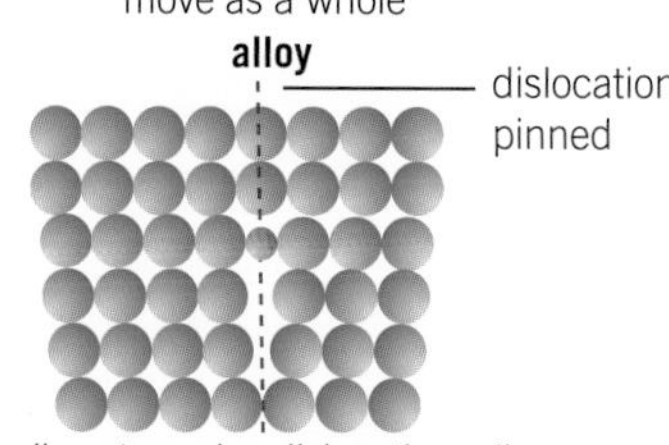


▲ **Figure 5** *Particles in metals and metal alloys*

▲ **Figure 6** *Polycrystalline structure of brass (an alloy of copper and zinc) showing the grain boundaries*

## Polycrystalline materials

Many solids are neither purely crystalline nor completely amorphous – rather, they are **polycrystalline**. A polycrystalline material consists of a number of grains all orientated differently relative to one another but with an ordered, regular structure within each individual grain.

As a liquid cools, crystals start to form at different points within it. Each crystal grows out into the remaining liquid until it runs into its neighbours. The result is a patchwork of tiny crystals or grains. The interface where these grains meet is known as the **grain boundary**.

## Stress concentration and crack propagation

The strength of materials is affected by tiny cracks and flaws in the structure of the material. The stress concentration around such cracks can be hundreds or even thousands of times the applied stress. This can lead to cracks working through a specimen until it fractures. Think about bending a glass rod that has a tiny scratch on the surface.

- The glass becomes strained elastically.
- At the tip of a crack, two neighbouring atoms are pulled apart.
- The next two atoms are pulled apart, and the next two, and so on.
- The crack moves through the material like a zip being undone, propagating through the material.

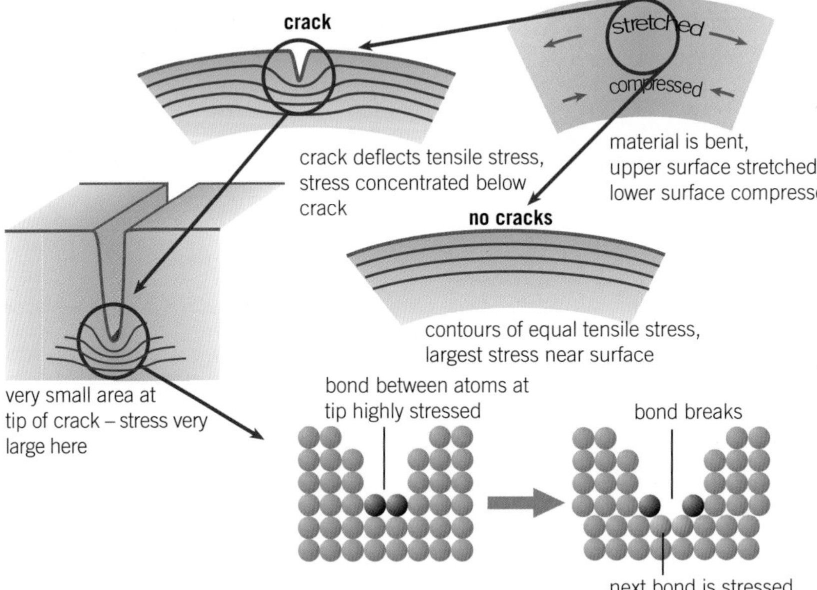


▲ **Figure 7** *Cracks propagate through materials under stress*

## Toughness in metals

Many metals are tough. *Toughness* is a measure of the energy needed to extend cracks through a material – it takes more energy to extend a crack in a tough metal than in a brittle substance. This is because they are ductile. When a stress is applied to a crack the metal deforms plastically in the region of the crack which makes the crack broader, reducing the stress around the crack.

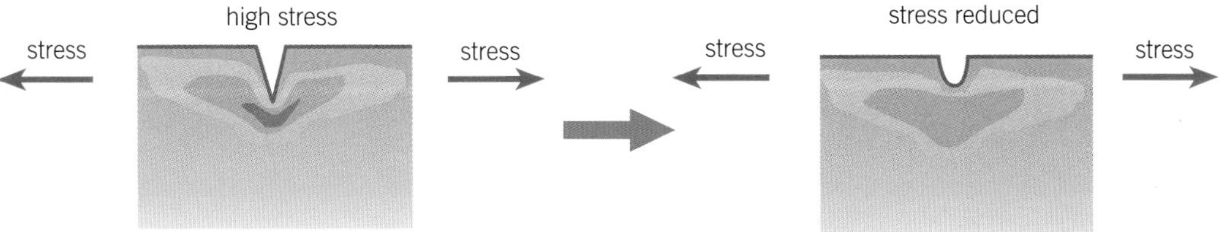


Metals resist cracking because they are ductile. Under stress, cracks are broadened and blunted – they do not propagate.

▲ **Figure 8** *Stopping cracks in metals*

It is important to remember that strength and toughness are not the same thing. A material is **strong** if it has a large breaking stress whereas a tough material is one which will not break by shattering — tough is the opposite of brittle. Glass and steel have similar breaking strengths but steel is much tougher.

## Summary questions

**1** Describe the differences between an amorphous material and a polycrystalline material. *(2 marks)*

**2** A student is heard to say: "Glass shatters because it is weak." Explain why this statement is incorrect and write a few sentences explaining the terms weak, strong and brittle to help the student understand the error. *(4 marks)*

**3** Using the concept of mobile dislocations, explain why alloying tends to make metals harder and more brittle. You should include diagrams in your answer. *(4 marks)*

**4** Explain why the stress at the tip of a crack in a brittle material can be very large. *(2 marks)*

**5** Ceramic materials can have dislocations in their structures, but these do not lead to ductile behaviour. Explain why dislocations produce ductile behaviour in metals but not in ceramics. *(2 marks)*

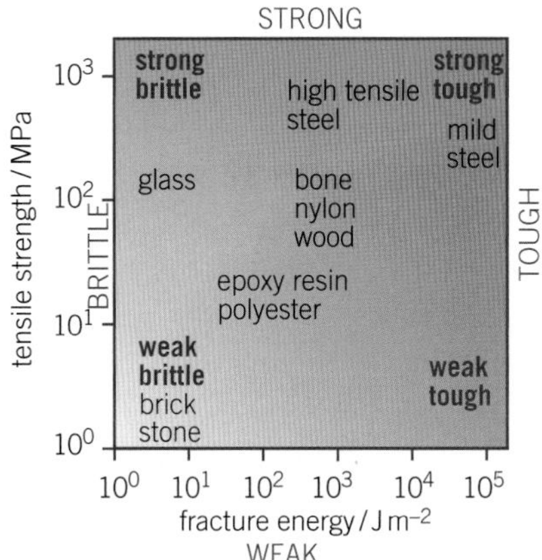


▲ **Figure 9** *Strength and toughness of materials on a logarithmic scale*

# 5.3 Microscopic structures and macroscopic properties

Specification references: 3.2a(iii)

**Learning outcomes**

Describe, explain, and apply:

- structure of metals, ceramics, and polymers
- polymer behaviour in terms of chain entanglement/ unravelling.

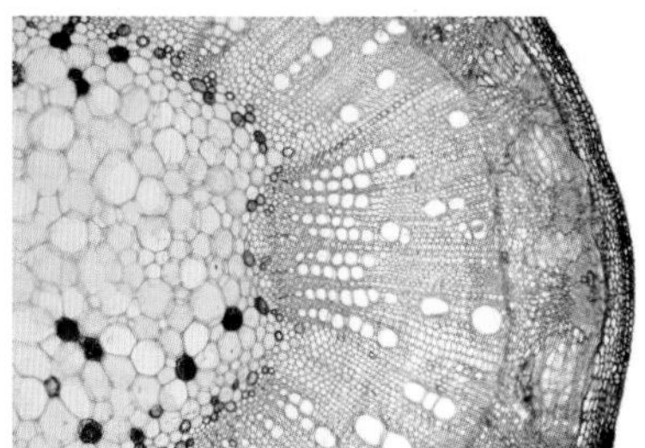

▲ **Figure 1** *The stem of a cotton plant as seen under a microscope*

**Synoptic link**

Further macroscopic properties of gases, for example, pressure, will be linked back to the microscopic behaviour of the particles of the gas in Topic 14.1, The gas laws in the A level course.

**Study tip**

Make sure that you can remember an example of a metal, a ceramic, and a polymer. You should also be able to explain their properties using microscopic structures.

Microscopic means 'not visible to the unaided eye'. When we talk about the microscopic structures of materials, we mean those structures which are too small to see. At the smallest scale, this means the bonds and patterns of particles making up the material. On the other hand, macroscopic implies something that you can see or a property that is detectable without using sophisticated equipment.

## More about ceramics and metals

Ceramics undergo brittle fracture. We have explained macroscopic properties like brittleness and toughness by considering the microscopic structure of the materials. Many of these properties depend upon the bonds between atoms in the materials.

There are three types of bonds between atoms – covalent, ionic, and metallic. However, the bonds in ceramics and ionic compounds have a further property in that they are directional. This means that the atoms are locked in place and cannot slip, making the material hard and brittle.

**Ceramics** have rigid structures

**Covalent structures,** for example, silica, diamond, and carborundum

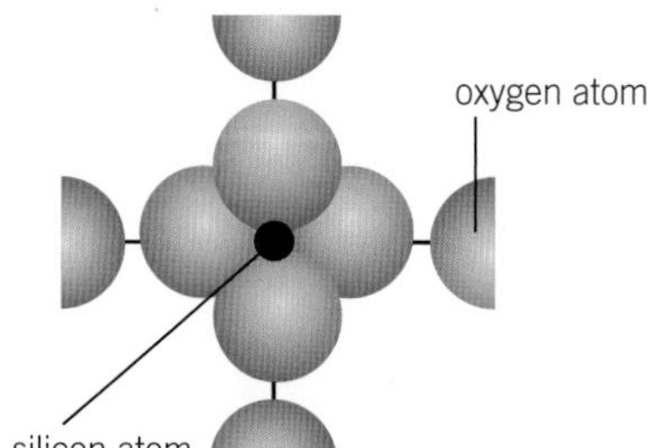


Atoms share electrons with neighbouring atoms to form covalent bonds. These bonds are directional – they lock atoms in place, like scaffolding.

The bonds are strong – silica is stiff.
The atoms cannot slip – silica is hard and brittle.

**Metals** have non-directional bonds

**Metallic structures,** for example, gold

negative electron 'glue'

gold ion

Atoms in metals are ionised. The free electrons move between the ions. The negative charge of the electrons 'glues' the ions together, but the ions can easily change places.

The bonds are strong – metals are stiff.

The ions can slip – metals are ductile and tough.

▲ **Figure 2** *Microscopic diagram of ceramics and metals*

## Stiffness and elasticity in metals and polymers

Metals behave elastically for small strains, that is, up to strains of around 0.1%. Up to this point the metal extends because the spacing between the positive ions increases. When the tensile force is removed the metal returns to its original length.

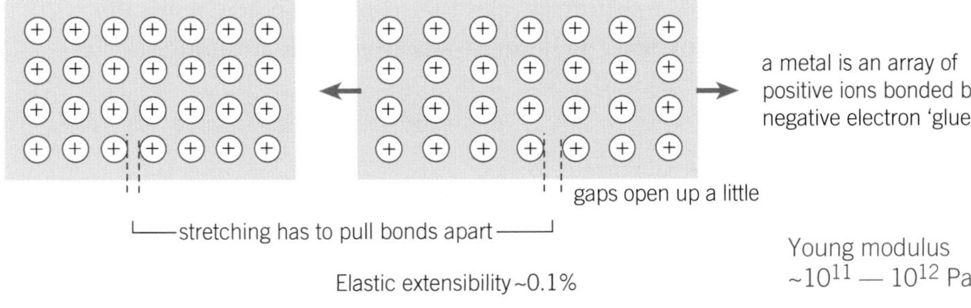


gaps open up a little
stretching has to pull bonds apart

Elastic extensibility ~0.1%

▲ **Figure 3** *Stretching a metal pulls bonds apart*

Polymers such as polythene are long-chained molecules that can extend elastically up to about 1% strains. Polythene is very 'floppy' because it is free to rotate about its bonds. The bonds are strong so although they can rotate they are difficult to break. This gives polythene the macroscopic properties of strength and flexibility.

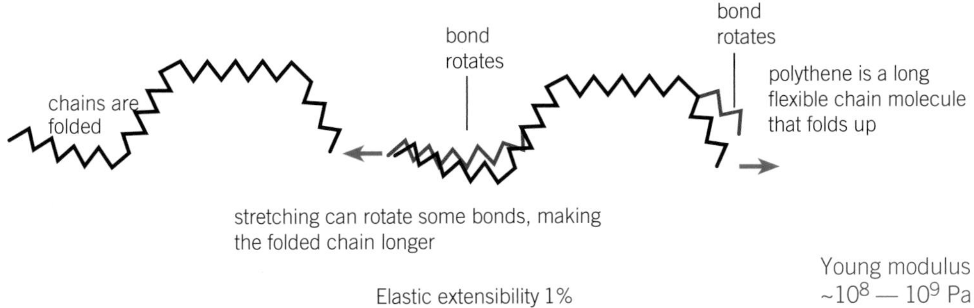


bond rotates
bond rotates
chains are folded
polythene is a long flexible chain molecule that folds up
stretching can rotate some bonds, making the folded chain longer

Elastic extensibility 1%

▲ **Figure 4** *Stretching polythene rotates bonds*

# Designing polymers

Not all polymers are flexible. Think of all the different uses for plastics you see around you. Polymers can be stiff if the rotation or unfolding of the chains of molecules is difficult. Adding cross-linkages, where polymer chains are *tied together* at regular intervals along the chains, produces a stiffer material.

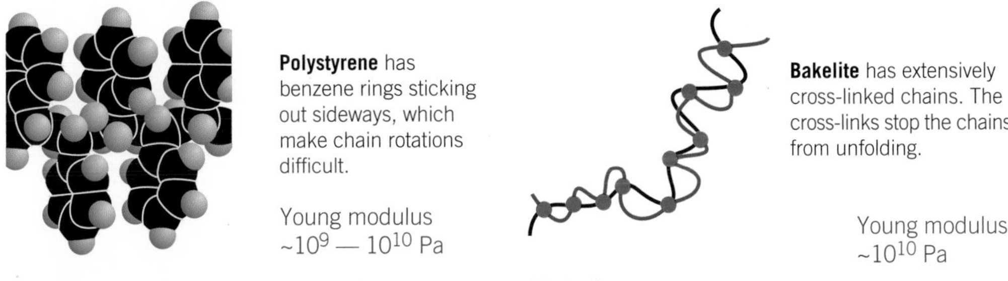

**Polystyrene** has benzene rings sticking out sideways, which make chain rotations difficult.
Young modulus ~$10^{9}$ — $10^{10}$ Pa
**Bakelite** has extensively cross-linked chains. The cross-links stop the chains from unfolding.
Young modulus ~$10^{10}$ Pa


▲ **Figure 5** *Reducing chain rotation in polystyrene and Bakelite*

Natural rubber is a runny white liquid of limited uses. In 1839 Charles Goodyear invented the process of *vulcanisation* of rubber in which natural rubber is heated with sulfur. The sulfur atoms form cross-links with the polymer chains. The more sulfur you add, the more cross-links form and the stiffer the rubber. Controlling the microscopic structure of the material allows control of its macroscopic properties. The technique of vulcanisation means that rubber can be chemically adapted for many uses including tyres and shock absorbers.

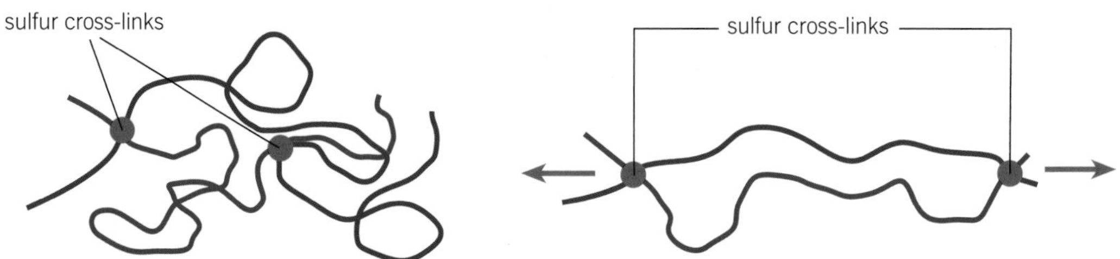
sulfur cross-links
sulfur cross-links
In unstretched rubber, chains meander randomly between sulfur cross-links.
In stretched rubber the chain bonds rotate, and chains follow straighter paths between cross-links. When let go, the chains fold up again and the rubber contracts.
Elastic extensibility > 100%

▲ **Figure 6** *Cross-linkages in rubber*

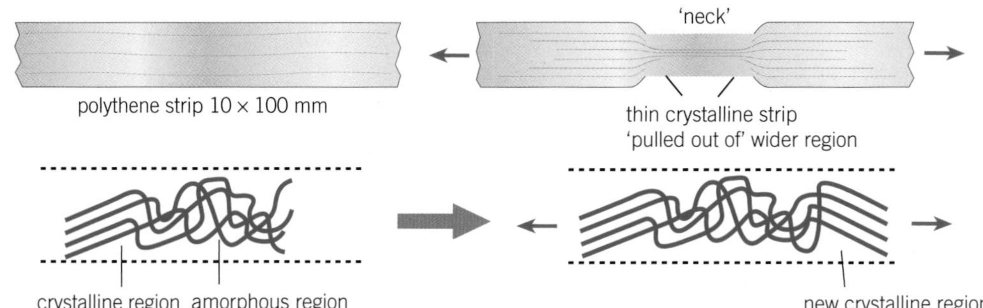
'neck'
polythene strip 10 × 100 mm
thin crystalline strip
'pulled out of' wider region
crystalline region
amorphous region
new crystalline region
Polythene is semicrystalline. Think of polythene as like cooked spaghetti. In amorphous regions the chains fold randomly. In crystalline regions the chains line up.
When stretched plastically, the chains slip past each other. More of the material has lined up chains. More of it is crystalline.
Plastic extensibility > 100%

▲ **Figure 7** *Plasticity in polymers*

## Summary questions

1 Polymer chains can become tangled. State the effect(s) this would have on the properties of the material. *(1 mark)*

2 Explain why ions in a metal can slip but atoms in a ceramic do not. *(2 marks)*

3 Suggest why metals only extend elastically up to strains of about 0.1%, but polythene is elastic up to strains of 1%. *(2 marks)*

4 Explain how cross-linking makes polymers stiffer. *(2 marks)*

# Module 3.2 Summary

## Testing Materials

### Describing materials

- describing materials using the terms: hard, brittle, stiff, tough, malleable, ductile, fracture
- the behaviour of metals, ceramics, and polymers
- choosing the right material for the job

### Stretching wires and springs

- elastic and plastic deformation and fracture
- force-extension graphs
- Hooke's law, $F = kx$
- energy stored in an elastic material, $E = \frac{1}{2}kx^2$
- energy as the area under a force-extension graph for elastic materials
- practical task: plotting force-extension characteristics for springs, rubber bands, and polythene strips

### Stress, strain, and the Young modulus

- using and understanding the terms: tension, stress, strain, fracture stress, yield stress, Young modulus
- stress-strain graphs up to fracture
- stress $\sigma = F/A$
- strain $\varepsilon = x/l$
- Young modulus = $\sigma/\varepsilon$
- practical task: determining the Young modulus and fracture stress of a metal

### Choosing materials

- using materials selection charts and other logarithmic charts to compare mechanical characteristics of materials

## Looking inside materials

### Materials under the microscope

- evidence of the size of particles and their spacing (Rayleigh's oil drop experiment, scanning tunnelling microscope images)

### Modelling material behaviour

- describe and explain the structures of metals and ceramics
- use the model of mobile dislocations allowing slip in metals with brittle materials not having mobile dislocations
- understand the meaning of strong in terms of breaking stress
- interpret images showing the structure of materials

### Microscopic structures and macroscopic properties

- describe the microscopic structure of metals, ceramics, and polymers
- explain polymer behaviour in terms of polymer chains entangling and unravelling

# Physics in perspective

## Graphite to graphene

Graphite and diamond are two forms of pure carbon. Graphite, used as the incorrectly-named pencil lead, has very different properties to those of diamond. This is due to the different arrangements and bonding of the carbon atoms in the two forms — a clear example of microscopic structure producing macroscopic properties.

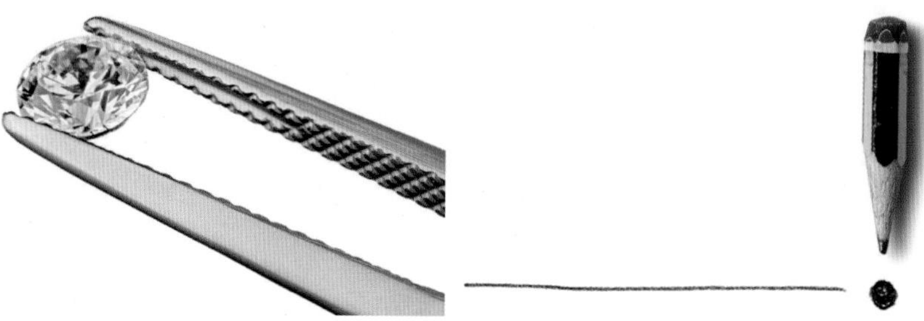

▲ **Figure 1** *The line of graphite from the pencil and the diamond are both carbon*

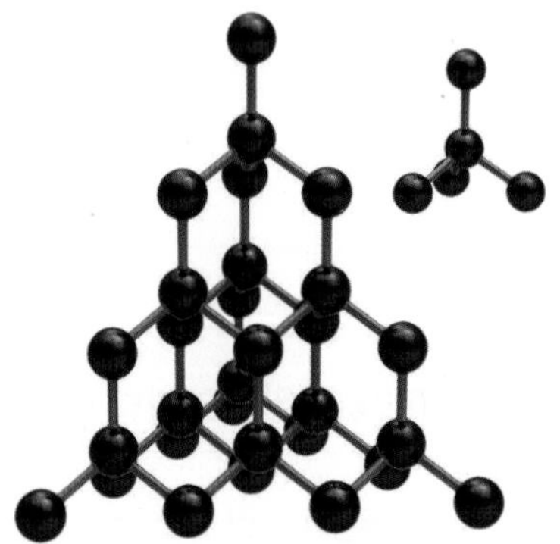

▲ **Figure 2** *The structure of diamond*

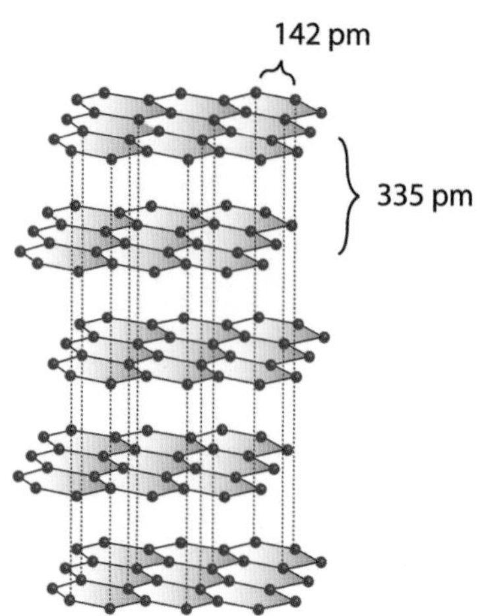


▲ **Figure 3** *The structure of graphite*

Graphite and diamond are examples of giant covalent structures. In diamond, as shown in Figure 2, each carbon atom bonds with four others in a rigid, three-dimensional framework.

In graphite, each carbon atom is bonded to three others in a two-dimensional sheet. The distance between each carbon atom in the sheet (the bond length) is $1.42 \times 10^{-10}$ m or 142 pm. Each layer is attracted to the layers above and below. The distance between the layers is about two and a half times the distance between atoms within a sheet, which explains why graphite is less dense than diamond.

The force of attraction between the layers of graphite is much weaker than the attraction between carbon atoms within a layer, which means layers can slide over each other fairly easily. However, the bonds between atoms within a layer of graphite are stronger than those between atoms in diamond. When you write with a pencil you are simply transferring layers of graphite from the pencil to the paper.

Because each carbon atom in graphite makes covalent bonds with three other atoms rather than four, there are delocalised electrons on each layer — electrons can easily move along the sheets of carbon (though not through or between them). This is why graphite has a much higher conductivity than diamond. This suggests that graphite should only conduct electricity in a direction parallel to the layers, but if you take a piece of graphite (from a pencil, say) you will find it conducts in all directions. This is because graphite is not one perfect crystal — it is made from many crystals clumped together in many different orientations, allowing the current to find a route through.

## Graphene

A perfect, single layer of graphite would only conduct electricity *along* the sheet. Such a layer was produced at Manchester University in 2004, when researchers Andre Geim and Kostya Novoselov used sticky tape to pull thin layers from a sample of graphite. This material, now known as graphene, is one atom thick and shows remarkable properties. It is the first two-dimensional material to be fabricated, an achievement recognised in the award of the 2010 Nobel Prize to Geim and Novoselov.

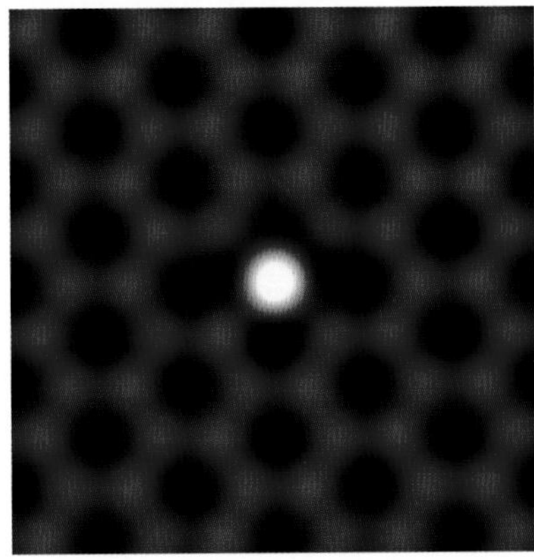

▲ **Figure 5** *SEM image of a silicon impurity atom in a graphene layer. The hexagonal arrangement of carbon atoms can be seen.*

A sheet of graphene is strong, flexible, and a very good conductor of electricity. It's also light — a single layer of graphene large enough to cover a football field would have a mass of less than 1 g. There is much research happening into ways of putting this remarkable material to use. For example, it is thought that graphene could be used in flexible touchscreens, protective paint, and even solar cells. Many claims have been made about this remarkable material and it will be fascinating to watch new applications develop in the years ahead.

Next time you draw a line with a pencil, remember that a small fraction of the graphite you transfer to the paper will be only a few layers thick, or even a single layer. You may have produced graphene without knowing.

Data: graphene layer
Breaking stress: 130 G Pa.
There is one atom in every $2.6 \times 10^{-20}\,m^2$ in a graphene layer.
Spacing between layers of graphene = 0.34 nm
Mass of $6.0 \times 10^{23}$ carbon atoms = 0.012 kg

## Summary questions

1 State why graphene is described as a two-dimensional material.

2 It has been said that a single layer of graphene spread over a football pitch would have a mass of less than 1 g. Is this true?
Area of a typical football pitch = $7140\,m^2$

3 a A graphene 'nanoplatelet' has dimensions 7 nm × 50 μm × 50 μm. It consists of layers of graphene of area 50 μm × 50 μm stacked to form a platelet of height 7 nm. Use the data given to calculate the density of the graphene.

b The density of graphite varies between $2.3 \times 10^3\,kg\,m^{-3}$ and $2.7 \times 10^3\,kg\,m^{-3}$. Compare these values to your answer from **(a)** and state what this may suggest about the structure of the two forms of carbon.

4 Many claims have been made about graphene, including that it is the world's strongest material. The breaking stress of structural steel is about $5 \times 10^8$ Pa, whilst the breaking stress of graphene is 130 GPa. Explain why the comparison of these two values might be misleading when considering possible uses for graphene.

5 The distance between neighbouring carbon atoms in a graphene layer is about 0.14 nm. Use this value to estimate the area of the SEM image in $m^2$.

6 After a talk on graphene, a student was heard to comment that "graphene is basically just graphite — I don't know what all the fuss is about." Describe and explain the similarities and differences between the two materials.

## Practice questions

1 Which of the following statements apply to the microscopic structure of metals?

**Statement 1** The bonds between metal ions are non-directional

**Statement 2** The bonds between metal ions are covalent

**Statement 3** The bonds between metal ions can rotate

**A** 1, 2 and 3 are correct

**B** Only 1 and 2 are correct

**C** Only 2 and 3 are correct

**D** Only 1 is correct

2 The density of silver is 10 490 kg m$^{-3}$.

In one mole of silver there are $6.0 \times 10^{23}$ atoms. The mass of one mole of silver is 0.108 kg.

Use this data to calculate the volume of one atom of silver. *(3 marks)*

3 Figure 1 shows the microscopic structure of a metal alloy.

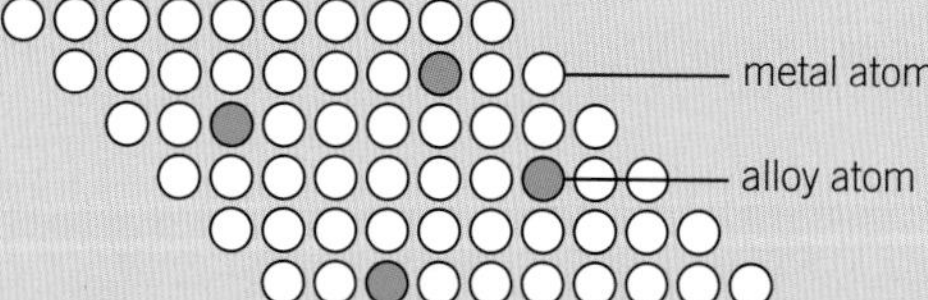


▲ Figure 1

The presence of the alloy atoms reduces the movement of *dislocations* through the metal structure. Explain what is meant by dislocation in this context, and state the effect of the presence of the alloying atoms on the mechanical properties of the material. *(2 marks)*

4 Rubber is a long chain molecule. The molecule can be modelled as a series of repeating units of length $L$ joined by bonds that are free to rotate. Rubber can undergo elastic strains greater than 100%.

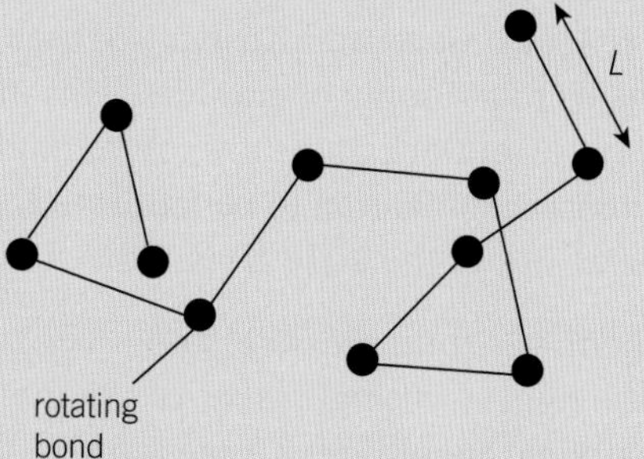


▲ Figure 2

**a** A rubber band has an unstretched length of 6 cm. How long will it be when the strain is 100%? *(1 mark)*

**b** Explain why rotating bonds allow the rubber to stretch to such a degree. *(2 marks)*

**c** A molecule with $N$ rotating bonds linked by units each of length $L$ will have an average total length of $L\sqrt{N}$.
The molecule stretches out into an approximately straight line when it is at maximum elastic strain.

Calculate the maximum strain of a rubber molecule of 8 units of length $L$. *(3 marks)*

5 The microscopic structure of an ionic crystal such as sodium chloride is an ionic lattice. The bonds between ions are strong and directional. There are no mobile dislocations in the structure.

Explain how this structure makes the material hard and brittle. *(3 marks)*

6 Polythene bags can be permanently stretched quite a lot before the material tears. Rubber bands stretch easily but relax back to their original length. Explain the behaviour of these materials in terms of the behaviour of molecules. *(3 marks)*

**7** Molecules of DNA have been stretched using 'optical tweezers'. A tensile force of $400 \times 10^{-12}$ N applied to a single DNA strand produces a strain of 20%. The cross-sectional area of a DNA strand = $2 \times 10^{-17}\,m^2$.

Calculate an estimate of the Young Modulus of DNA. *(2 marks)*

**8** This question relates macroscopic properties of copper to its microscopic structure. The density of copper is $8960\,kg\,m^{-3}$.

**a** There are $6.0 \times 10^{23}$ atoms in 0.063 kg of copper. Show that the volume occupied by one copper atom is about $1.2 \times 10^{-29}\,m^3$. *(2 marks)*

**b** The separation between copper atoms is given by

$$\text{separation} = \sqrt[3]{\text{volume occupied by a copper atom.}}$$

Calculate the separation between copper atoms. *(1 mark)*

**c** A copper wire has a cross-sectional area of $0.5\,mm^2$. The wire can be pictured as a layer of atoms laid on top of one another. Calculate the number of copper atoms in one layer of wire of cross-sectional area $0.5\,mm^2$.

Cross-sectional area occupied by one copper atom is $5.6 \times 10^{-20}\,m^2$. *(2 marks)*

**d** A force of about $5 \times 10^{-11}$ N is required to separate a pair of copper atoms. Use your answers to (c) to estimate the force required to break a copper wire of cross-sectional area $0.5\,mm^2$. *(2 marks)*

**e** Calculate the theoretical breaking stress of copper from your answers. *(2 marks)*

**f** The accepted value for the breaking stress of copper is about 70 MPa. Compare this with your calculated value in **(e)**, explaining any difference in terms of the microscopic structure of the metal. *(3 marks)*

**9** Figure 3 shows the structure of glass.

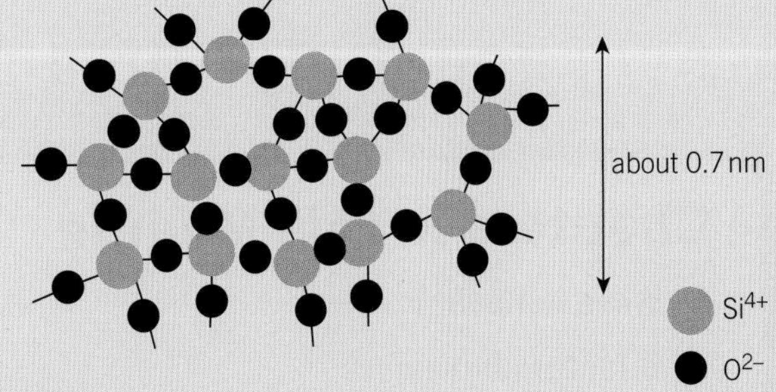


▲ **Figure 3**

The bonding is strong, stiff, and directional within groups of ions, but with random orientations between neighbouring groups. There is little short range order in the structure.

Use features of this micro-structure of glass to suggest **explanations** for the following macroscopic properties of glass.

**a** Glass fibres are strong but show no plastic deformation before fracture. *(3 marks)*

**b** A sheet of glass can be broken cleanly and accurately into two pieces, if a scratch is drawn across its surface and the glass is slightly bent. *(3 marks)*

**c** Solid glass at room temperature is a good electrical insulator, but when heated near its melting temperature it can conduct electricity. *(3 marks)*

*May 2009 G481*

**10 a** In a simple cubic arrangement the atoms sit at the corners of imaginary cubes with their curved surfaces just touching. Show that the ratio of filled space to empty space between the atoms is $\frac{\pi}{6}$, or about 0.5. This 'packing fraction' is related to the density of the material.

The volume of a sphere of radius $r$ is $\frac{4}{3}\pi r^3$. *(3 marks)*

**b** If the atoms in the simple cubic arrangement in (a) have diameter $d$, show that the largest impurity atom that can just fit in between the host atoms has a diameter of $0.4d$. *(3 marks)*

# MODULE 4.1

## Waves and quantum behaviour

### Chapters in this Module



### Introduction

The behaviour of light has intrigued scientists for centuries — does it travel instantaneously, or does it have a speed that can be measured? Does it behave as a wave or as a particle? As knowledge of the structure of the atom developed at the turn of the twentieth century, similar questions were asked about the nature of the newly-discovered electron.

In this module you will learn about the properties of waves and how light (and all electromagnetic radiation) demonstrates these properties. You will also learn that the behaviour of light cannot be adequately described by picturing it as only a wave or only a particle. You will be introduced to the phasor picture of light that uses the idea of superposition to provide a more comprehensive explanation of the behaviour of light, an explanation that can be extended to the behaviour of electrons.

Wave Behaviour introduces the concepts of wave superposition, path difference and phase difference. You will learn how to measure the wavelength of light and you will experiment with standing waves on strings and in air. The behaviour of light as it travels from one medium to another is investigated experimentally and explained using wave superposition.

Quantum Behaviour focuses on phenomena that were investigated in the early years of the twentieth century, including the photoelectric effect and electron diffraction. Photons, discrete 'packets' of electromagnetic energy, are introduced and the problems of relating light as photons to light as a wave are considered. At this point, the phasor model is introduced as an alternative way of looking at light. The chapter draws to a close by considering electrons as 'quantum objects' that can show 'wave-like' and 'particle-like' behaviour, which can also be described using phasors.

## Knowledge and understanding checklist

**From your Key Stage 4 study you should be able to do the following. Work through each point, using your Key Stage 4 notes and the support available on Kerboodle.**

- ☐ Describe wave motion in terms of amplitude, wavelength, frequency and period.
- ☐ Define wavelength and frequency, and describe and apply the relationship between these and the wave velocity.
- ☐ Describe the differences between transverse and longitudinal waves.
- ☐ Recall that light is an electromagnetic wave, and recall that electromagnetic waves are transverse and all have the same velocity in space.
- ☐ Recall that wavelength may affect refraction of waves in different substances, and that refraction is related to the difference in velocity of waves in different substances.
- ☐ Recall that atoms and nuclei can generate and absorb electromagnetic radiation.

## Maths skills checklist

**In this unit, you will need to use the following skills. You can find support for these skills on Kerboodle and through MyMaths.**

- ☐ **Use of the small angle approximation sin $\theta \approx$ tan $\theta$**, when analysing Young's double-slit experiment.
- ☐ **Understand the relationship between degrees and radians**, and know that phase differences between waves can be expressed in both units. You will also need to find sines of angles on your calculator for angles in degrees and in radians.
- ☐ **Find the slope of a linear graph**, such as in the experiment to determine the Planck constant using LEDs.

# 6 WAVE BEHAVIOUR

## 6.1 Superposition of waves

Specification reference: 4.1a(i), 4.1b(i), 4.1c(i), 4.1d(i) 4.1d(iii), 4.1d(v)

### Learning outcomes

Describe, explain, and apply:

- → the formation of standing waves
- → the terms: phase, phasor, amplitude, superposition
- → the wavelength of standing waves
- → how an oscilloscope can be used to determine frequencies of waves
- → the use of a rubber cord to demonstrate superposition
- → how the speed of sound in air can be determined using stationary waves.

You may have noticed that at the instant that the crests of two water waves pass through each other a bigger crest is formed. This is an example of **superposition**. The principle of superposition states that when two or more waves overlap (superpose), the resultant displacement at a given instant and position is equal to the sum of the individual displacements at that instant and position. This explains a wide range of phenomena.

The colours of a peacock's tail (Figure 1) are iridescent – they shimmer and change as you view it from different angles. This is superposition in action. The colours seen on the surface of a bubble provide another example (Figure 2).

▲ **Figure 1** *An iridescent peacock*

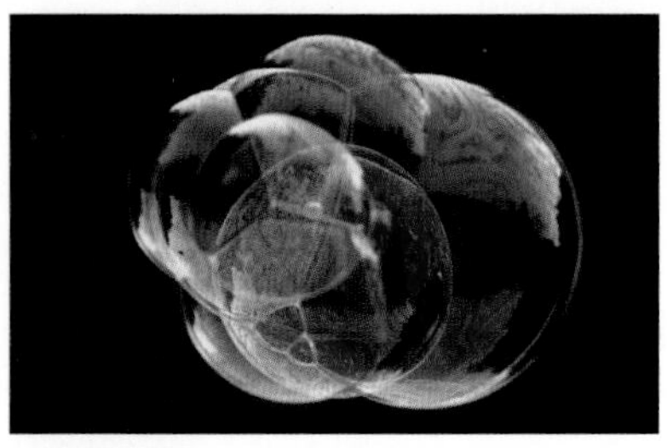

▲ **Figure 2** *Colours on a soap film*

### Representing wave motion

The diagrams in Figure 3 and Figure 4 look very similar but there is an important difference between them. Figure 3 is a displacement–displacement diagram, a snapshot of a wave at a single instant in time. The distance between any two points at the same part of the wave cycle is the wavelength, $\lambda$. The **amplitude**, $A$, of the wave is the maximum displacement from the equilibrium position.

Figure 4 shows how the displacement of a point along a wave changes over time. For example, if point P (Figure 3) is at the equilibrium position at time $t = 0$, its displacement will vary with time in the manner shown in Figure 4. The time it takes to return to the same position in the cycle moving in the same direction is the period of the wave, $T$. The time for one period of the wave is shown.

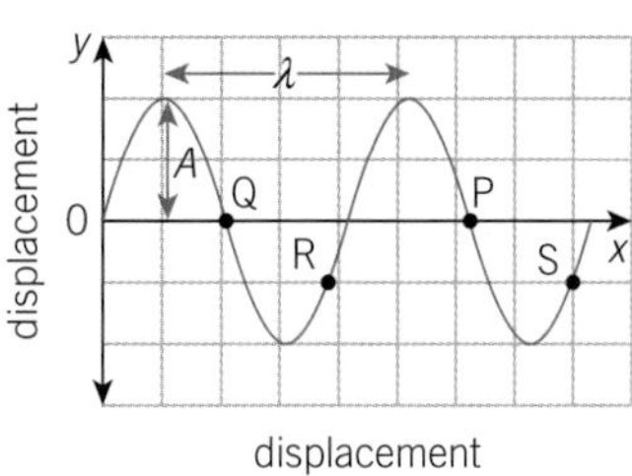


▲ **Figure 3** *A snapshot of a wave at an instant in time*

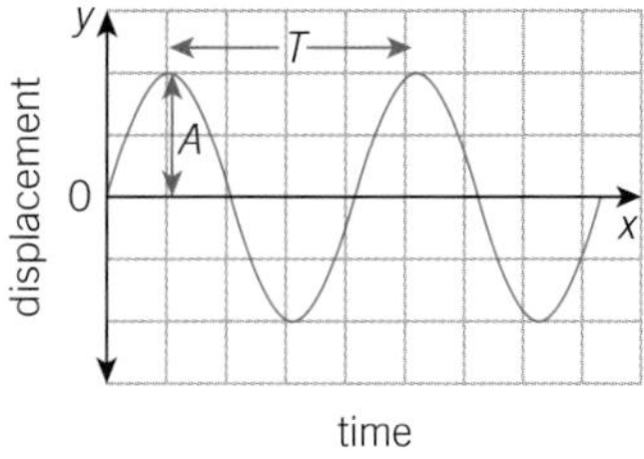


▲ **Figure 4** *The position of particle P on a wave plotted against time*

### Time period and frequency

You can use an oscilloscope to measure the time period of a wave or waveform and use this value to calculate the frequency.

## Practical 4.1 d(i): Measuring frequency using an oscilloscope

An oscilloscope shows how the potential difference across a component or supply varies with time. The $y$-value shows the potential difference across the component and the $x$-value shows the time. The sensitivity of the $y$-axis (in $\mathrm{V\,cm^{-1}}$) and the time base (in $\mathrm{ms\,cm^{-1}}$) can be controlled.

When a stable source, such as a signal generator, is connected to the oscilloscope a trace resembling that in Figure 5 is displayed. You can find the time period by noting the setting on the time base and measuring the peak-to-peak or trough-to-trough distance on the display grid.

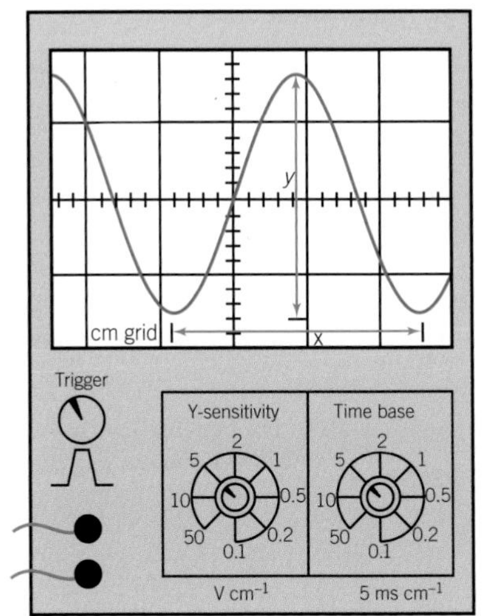


▲ **Figure 5** *Using an oscilloscope*

### Synoptic link

You have met the relationship between time period and frequency in Topic 1.4, Polarisation of electromagnetic waves.

## Worked example: Finding the frequency of a signal using an oscilloscope

Find the frequency of the signal displayed in Figure 5.

**Step 1:** Determine the time base set on the oscilloscope.

The illustration shows that the time base is set to $5\,\mathrm{ms\,cm^{-1}}$. 1 cm on the $x$-axis represents a time interval of 5 ms.

**Step 2:** Measure the distance between two crests, or between two troughs.

Looking on the grid you can see that this distance is 3.4 squares on the cm grid.

**Step 3:** Use the time base setting to find the time period.

Each cm represents a time interval of 5 ms – therefore, the time period is $5\,\mathrm{ms\,cm^{-1}} \times 3.4\,\mathrm{cm} = 17\,\mathrm{ms} = 17 \times 10^{-3}\,\mathrm{s}$.

**Step 4:** Use the equation frequency $= \dfrac{1}{\text{time period}}$ to find the frequency.

$$\text{Frequency} = \frac{1}{17 \times 10^{-3}\,\mathrm{s}} = 59\,\mathrm{Hz}$$

## Phase and phasors

**Phase** describes the stage in a wave cycle. *At the top of the wave* is a statement about phase. When two points are at the same stage in the cycle they are **in phase**. You can see from Figure 3 that points

## Study tip

**Radian measure**

Although angles are commonly measured in degrees (°), the standard unit is the **radian**. Phase angle is often given in radians. Make sure you are comfortable with both units and that you can convert from one to the other.

## Synoptic link

More information on the conversion between angles and radians can be found in the Maths Appendix.

P and Q are in phase, so are points R and S. They are in the same stage of the wave cycle at the instant of time represented by the snapshot.

### Phase difference

Two waves doing the same thing at the same moment are in phase. They have no phase difference. Two waves doing exactly opposite things at the same moment are in **antiphase**. These are special cases; if waves are neither exactly in phase nor exactly in antiphase they are said to be **out of phase**.

Phase and phase difference can be measured by a **phase angle**. We can use a rotating arrow, a **phasor**, to show where the wave is in its cycle. The arrow turns through $2\pi$ radians (360°) as the wave goes through one cycle. Figure 6 shows this. The vertical displacement of the tip of the clock arrow above or below the midpoint represents the displacement of the wave at that instant in time. If two waves are in phase the difference in phase angle is zero. Waves in antiphase have a phase difference of $\pi$ radians.

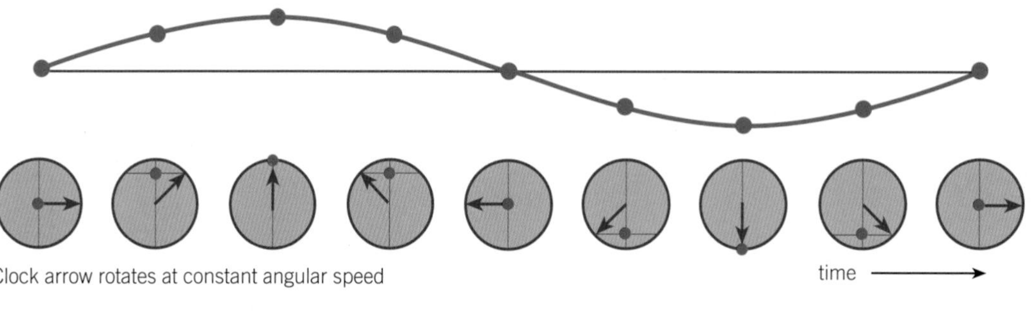


**Phase angle**

| | | | | | | | | | |
|---|---|---|---|---|---|---|---|---|---|
| degrees | 0 | 45 | 90 | 135 | 180 | 225 | 270 | 315 | 360 |
| radians | 0 | $\frac{\pi}{4}$ | $\frac{\pi}{2}$ | $\frac{3\pi}{4}$ | $\pi$ | $\frac{5\pi}{4}$ | $\frac{3\pi}{2}$ | $\frac{7\pi}{4}$ | $2\pi$ |

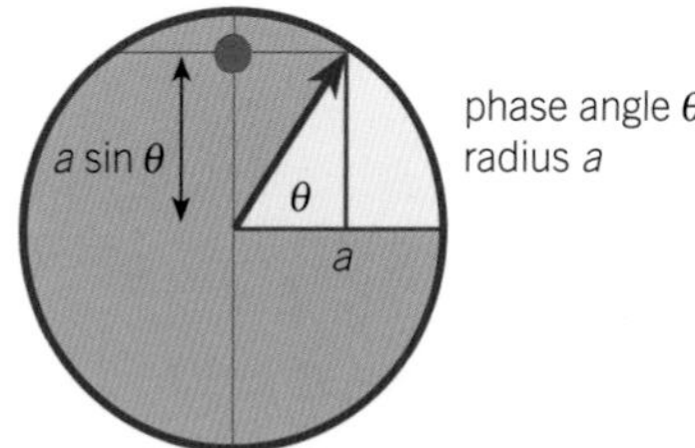


Clock arrow rotates $2\pi$ in periodic time $T$

angle $\theta = 2\pi\left(\frac{\text{time}}{T}\right)$

$T = \frac{1}{f}$ ($f$ = frequency)

angle $\theta = 2\pi ft$

displacement $= a \sin\theta = a \sin 2\pi ft$

▲ **Figure 6** *Phase and angle*

We can calculate the displacement when two waves superpose by adding the individual wave displacements together. You can see this in Figure 7. The picture of a phasor arrow is a useful one here. The length of the phasor arrow represents the amplitude of the wave. By adding the phasor arrows together, tip-to-tail, we can find the amplitude of the resultant phasor which is the length of the gap between the tail of the first phasor arrow and the tip of the last. Put another way, the arrow sum of the individual phasor components. Look at the two waves with a phase difference of $\frac{\pi}{2}$ radians in

## Synoptic link

The addition of phasor arrows is similar to the addition of vectors that you will meet in Topic 8.2, Vectors.

Figure 7. The length of the resultant phasor arrow (in red) tells us the amplitude of the two waves superposed and the angle tells us where the wave is in its cycle.

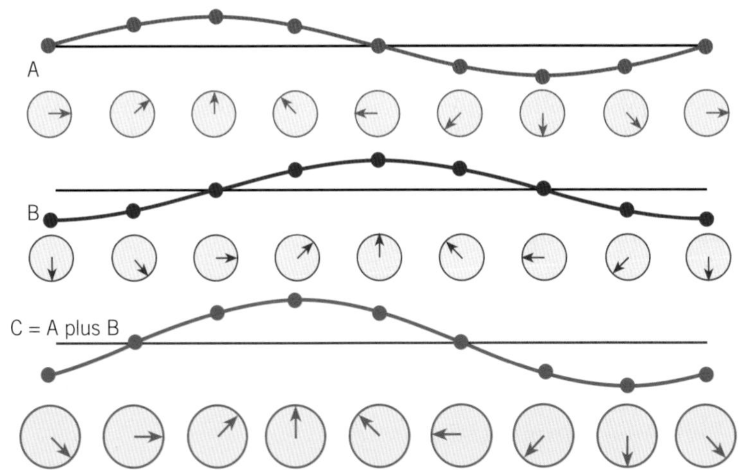

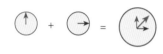

Rotating arrows add up

arrows add tip-to-tail 

For any phase difference, amplitude of resultant = arrow sum of components

▲ **Figure 7** *Oscillations with $\frac{\pi}{2}$ (90°) phase difference*

## Standing waves

A ripple moving across the surface of water is an example of a progressive wave – you can see the crest of the wave moving (or progressing) in time. When two progressive waves of the same frequency are travelling in opposite directions (say, along a string), the waves can appear to stop moving. When this happens, a standing or stationary wave is formed.

Standing waves are an example of superposition. These waves can also be produced when sound waves and electromagnetic waves superpose.

**Hint**

Although a string moves (up and down) when a standing wave is produced, the positions of maximum oscillation and minimum oscillation do not change.

### Forming a standing wave on a string

When a string is plucked, waves:

- move along the string in opposite directions
- reflect at the ends of the string
- superpose as they pass through one another.

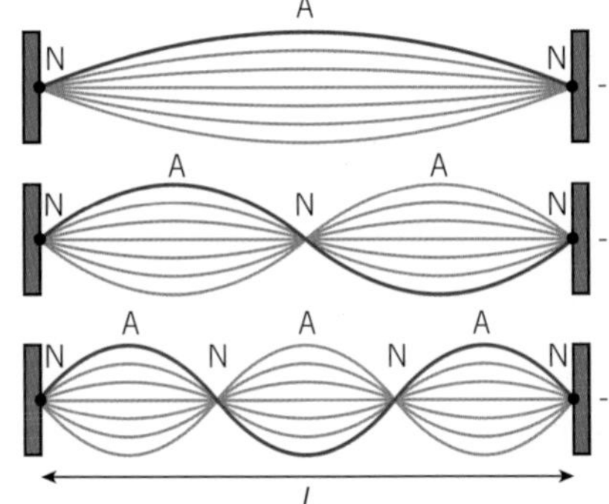


▲ **Figure 8** *Nodes (N) and antinodes (A) on a standing wave on a string for $\lambda = 2L$, $\lambda = L$, and $\lambda = \frac{2}{3}L$*

There will be maximum amplitude of oscillation at points along the string where the waves meet in phase. These are called **antinodes**. Where the waves travelling along the string meet in antiphase there is zero amplitude. These points are called **nodes**.

Adjacent nodes are half a wavelength apart. The longest standing wave has a wavelength of twice the length of the string. This has the lowest frequency of vibration, called the fundamental frequency. The next possible standing wave has a wavelength equal to the length of the string. You can see this in Figure 8.

## Calculating wavelength

To calculate wavelength we use the equation

wave velocity $v$ ($\mathrm{m\,s^{-1}}$) = frequency $f$ (Hz) × wavelength $\lambda$ (m)

### Worked example: Using $v = f\lambda$

A ukulele string has a vibrating length of 345 mm. The lowest (fundamental) frequency of its vibration is 440 Hz.

**a** Calculate the wavelength of the standing wave.

As the wavelength of the fundamental frequency = twice the length of the string, $\lambda = 2 \times 345\,\mathrm{mm} = 690\,\mathrm{mm}$

**b** Calculate the velocity of the wave in the string.

**Step 1:** Select the appropriate formula.

$$v = f\lambda$$

**Step 2:** Convert values into the correct unit before substituting into the formula.

$$690\,\mathrm{mm} = 0.69\,\mathrm{m}$$

**Step 3:** Substitute values.

$$v = 440\,\mathrm{Hz} \times 0.69\,\mathrm{m} = 304\,\mathrm{m\,s^{-1}}\ (3\text{ s.f.})$$

### Practical 4.1d(iii): Standing waves on a rubber cord

This practical gives you an opportunity to see standing waves. The apparatus is set up as in Figure 9. The rubber cord has been stretched to about twice its original length. Set a signal generator at 10 Hz and gradually increase the frequency. The first standing wave pattern you see has an antinode in the middle of the length of the string and nodes at either end. At double this frequency you will find a second standing wave pattern, which has two humps (see Figure 9). Increase the frequency further and you may see a third standing wave pattern.

The velocity of the progressive waves that form the standing wave can be found by measuring the wavelength of the standing wave, taking the value of the frequency from the signal generator and using the equation $v = f\lambda$.

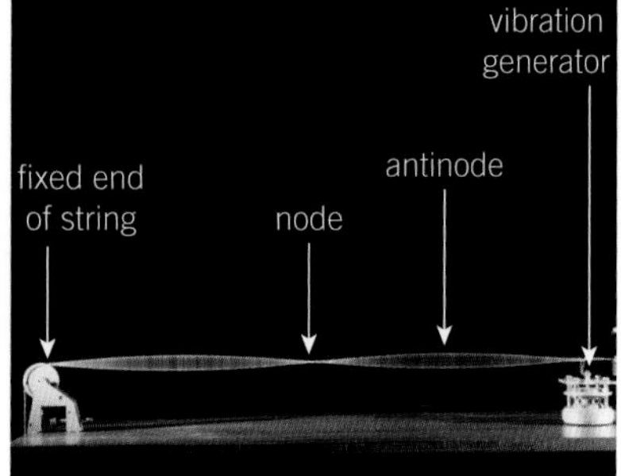


▲ Figure 9

## Standing waves in air

Standing waves can also be formed in air columns – that's how wind instruments such as bagpipes, trumpets, and recorders form their notes.

- A sound wave travels along the tube.
- The wave reflects at the end of the tube.
- Waves travelling up and down the tube superpose with each other.

Sound can be reflected from a closed end of a tube as well as from an open end, so there are two distinct sets of standing waves in tubes. Some instruments have two open ends, a flute, for example. Others effectively have one open and one closed end, for example, a clarinet. Where the waves travelling in opposite directions meet in phase an antinode is formed. A node is formed where the waves meet in antiphase. Figure 10 shows some of the possible standing waves from closed and open pipes. You will see that there is always an antinode at an open end of a tube.

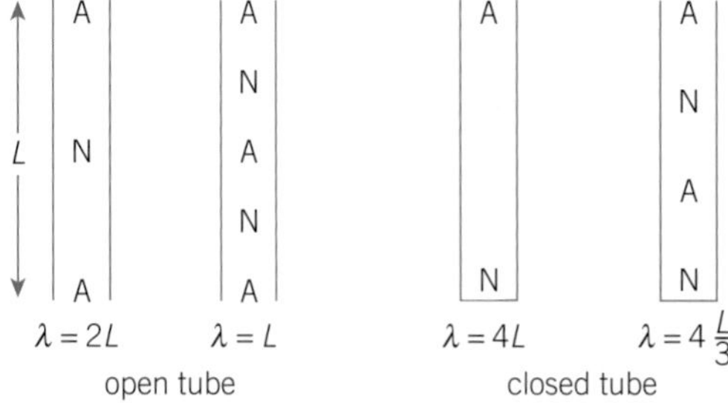


▲ **Figure 10** *Positions of nodes (N) and antinodes (A) in open and closed tubes*

You can also observe standing waves in sound and in microwaves by placing the source of the waves in front of a wall or reflector. Particular separations between the source and reflector will give rise to standing waves as the reflected waves superpose with those from the source.

### Practical 4.1d(v): Determining the speed of sound in air using a tube

You can find the speed of sound in air by using a *resonance tube*, a hollow tube with one end immersed in water. A source of sound waves, such as a tuning fork of known frequency or a loudspeaker controlled by a signal generator, is required for this experiment.

By holding a tuning fork at the top of the tube (Figure 11), you can find $L_1$, the shortest length of tube that increases the amplitude of the sound from the tuning fork. For this length there will be a node at the water surface and an antinode a little above the top of the tube. In other words, the length of the tube plus an *end correction*, $k$, is equal to one-quarter of the wavelength of the sound wave.

When there is about three times as much tube out of the water the amplitude will increase once more. At this position the length of tube $L_2$ (plus end correction $k$) is equal to three-quarters of the wavelength of the note.

By measuring $L_1$ and $L_2$ you can work out the wavelength of the sound wave.

$$\frac{\lambda}{4} = L_1 + k \quad \text{and} \quad \frac{3\lambda}{4} = L_2 + k$$

Therefore $L_2 - L_1 = \frac{\lambda}{2}$

Once you have found the wavelength the velocity of the wave can be calculated using the equation $v = f\lambda$.

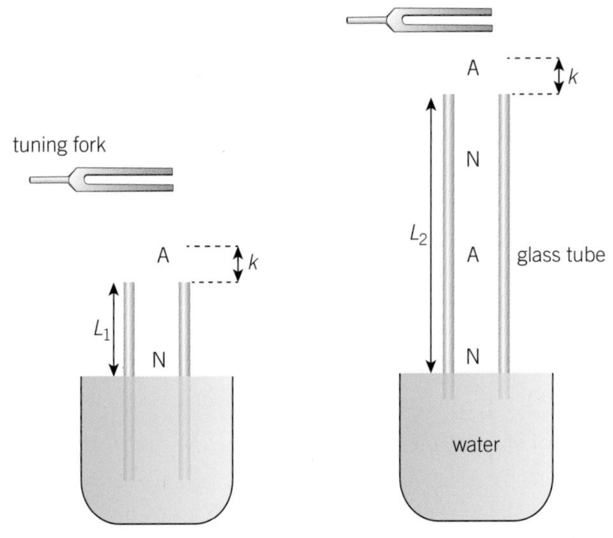


▲ **Figure 11** *Resonance tube experiment*

## Summary questions

**1** Sketch displacement–displacement graphs of:

**a** two equal-amplitude waves in phase; *(2 marks)*

**b** two waves of different amplitudes in antiphase; *(2 marks)*

**c** two equal-amplitude waves with a phase difference of $\frac{\pi}{2}$ radians. *(2 marks)*

**d** Sketch the superposition pattern of the waves you have sketched in **b**. *(1 mark)*

**2** A loudspeaker points directly at a wall 3 m away. It emits a sound wave of frequency 680 Hz and a standing wave is formed.

**a** Describe how a standing wave is formed. Include an explanation of the terms 'node' and 'antinode'. *(3 marks)*

**b** Calculate the distance between each node if the speed of sound is $340\,\text{m}\,\text{s}^{-1}$. *(2 marks)*

**3** An organ pipe can produce very low frequency sound waves. Calculate the length of pipe, closed at one end, that is needed to produce a sound wave of 30 Hz when the speed of sound is $340\,\text{m}\,\text{s}^{-1}$. *(2 marks)*

**4** A transmitter emits microwaves of frequency 15 GHz. The speed of microwaves in air is $3.0 \times 10^8\,\text{m}\,\text{s}^{-1}$.

**a** At what rate does the phasor representing the wave rotate? *(1 mark)*

**b** How far does the wave travel in one rotation of its phasor? *(2 marks)*

**c** Describe how a student could measure the wavelength of the microwaves using a microwave transmitter, microwave detector, reflector, and a metre rule. *(5 marks)*

# 6.2 Light, waves, and refraction

Specification reference: 4.1a(iii), 4.1b(i), 4.1 c(ii), 4.1d(ii)

## Refraction

Put a couple of drinking straws in a glass of water, look at them from the side, and they appear disjointed and magnified. A similar effect is seen when you look at the straws from above. This is an example of refraction. Waves change speed when they change medium – the material they are travelling through. This change of speed makes the light rays bend and change direction. When you look down on the straws in the water the light you see is refracted and bent at the boundary between the water and the air, giving the impression of a disjointed straw.

### Learning outcomes

Describe, explain, and apply:

→ the refraction of light at a plane boundary using the wave model and the changes in speed at the boundary

→ the term refractive index

→ the equations for calculating the refractive index of a material

→ how the refractive index for a transparent block can be determined.

### Refractive index

The behaviour of light has puzzled many thinkers across the centuries. The first reliable estimate for the speed of light came in the 17th century but many at that time continued to think that light travelled at infinite speed.

A vacuum is a region of space that contains no matter. Scientists now know that light travels at 299 792 458 m s$^{-1}$ in a vacuum. This figure is rounded to $3.00 \times 10^8$ m s$^{-1}$ in this book, but the very precise measurement is extremely important in many areas of physics. In fact, the precise value is used to define the value of the metre. Light travels at very nearly the same speed in air (which is mostly vacuum, when you think about it) but it is very slightly slower in air because it interacts with the electrons in the atoms in the air. The difference is small because air is not very dense so interactions between light and the electrons in the atoms are few and far between. Light travels considerably slower in glass, at around $2 \times 10^8$ m s$^{-1}$, depending on the type of glass. Glass is much denser than air so the number of interactions per metre between light and the electrons in the atoms of glass is much greater.

▲ **Figure 1** *Drinking straws in a glass*

The ratio of the speed of light in a one medium to the speed of light in another medium is called the **refractive index**.

$$\text{refractive index} = \frac{\text{speed of light in medium 1}}{\text{speed of light in medium 2}}$$

This can be written as refractive index, $n = \dfrac{c_{\text{1st medium}}}{c_{\text{2nd medium}}}$

If the first medium is a vacuum the equation becomes

$$\text{Refractive index of material} = \frac{\text{speed of light in a vacuum}}{\text{speed of light in material}}$$

This is known as the absolute refractive index.

### Synoptic link

You will consider the interactions between light and electrons in Chapter 7, Quantum behaviour.

### Hint

**Refractive index**

When you see the refractive index of a material given, this figure is for light travelling from a vacuum or air (1st medium) into the material (2nd medium).

### Worked example: Refractive index of quartz

The speed of light in air is $3.00 \times 10^8$ m s$^{-1}$. The speed of light in quartz is $1.95 \times 10^8$ m s$^{-1}$.

Calculate the refractive index of quartz.

**Synoptic link**

You have met the wave-front and ray models of light in Topic 1.1, Bending light with lenses.

**Step 1:** Select the appropriate equation.

$$n = \frac{c_{\text{1st medium}}}{c_{\text{2nd medium}}}$$

**Step 2:** Substitute values.

$$n = \frac{3.00 \times 10^8\,\text{m s}^{-1}}{1.95 \times 10^8\,\text{m s}^{-1}} = 1.54$$

## Refraction of light and Snell's law

Refraction can be modelled using both wave-fronts and light rays.

Figure 2 shows a representation of a ray of light travelling from air into glass. The ray bends towards the **normal**, an imaginary line at 90° to the surface of the glass (the air-glass boundary). The diagram also shows the angles of incidence and refraction, measured from the normal.

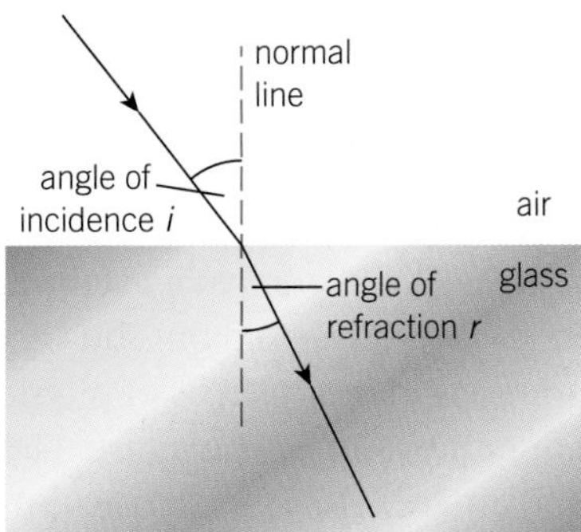


▲ **Figure 2** *Ray representation of refraction at a plane boundary*

The ratio $\frac{\sin i}{\sin r}$ is the refractive index of the material.

This statement is known as Snell's law after its discoverer, the Dutch mathematician Willebrord Snell, sometimes rather wonderfully known by the Latinised version of his name, Willebrordus Snellius.

From the two statements of the refractive index we can write

$$\frac{\sin i}{\sin r} = \frac{c_{\text{1st medium}}}{c_{\text{2nd medium}}}$$

### Worked example: Using Snell's law

The refractive index of pure ice is 1.31. The speed of light in ice is $2.29 \times 10^8\,\text{m s}^{-1}$. The speed of light in air is $3.00 \times 10^8\,\text{m s}^{-1}$. Calculate the angle of refraction in ice for an incident angle of 40°.

**Step 1:** Select the appropriate equation.

$$\frac{\sin i}{\sin r} = \frac{c_{\text{1st medium}}}{c_{\text{2nd medium}}}$$

**Step 2:** Rearrange the equation.

$$\sin r = \frac{\sin i \times c_{\text{2nd medium}}}{c_{\text{1st medium}}}$$

**Hint**

**Dispersion**

The refractive index varies with wavelength of the light – that is why prisms and raindrops (forming rainbows) disperse light into the colours of the spectrum.

**Step 3:** Substitute values and evaluate.

$$\sin r = \frac{\sin 40 \times 2.29 \times 10^8\,\text{m s}^{-1}}{3.00 \times 10^8\,\text{m s}^{-1}} = 0.49$$

**Step 4:** Find the angle from $\sin^{-1}$.

$$r = \sin^{-1} 0.49 = 29°$$

## Practical 4.1d(ii): Determining the refractive index for a transparent block

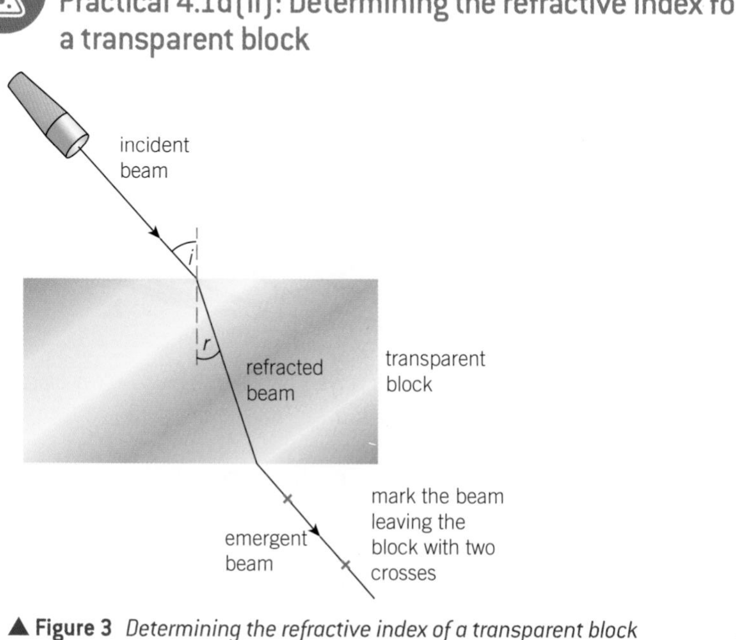


▲ **Figure 3** *Determining the refractive index of a transparent block*

You can find the refractive index of a glass block by finding the path of rays as they travel through the block. You need to measure the angle of refraction for a range of angles of incidence, and then calculate sin *i* and sin *r* for each pair of values. If you plot a graph of sin *i* (*y*-axis) against sin *r* (*x*-axis) you can find the refractive index from the gradient of the line.

The uncertainties in the measurements of *i* and *r* will introduce an uncertainty in sin *i* and sin *r*. When you plot the graph there may be an uncertainty in the gradient of the line, and therefore in the value of refractive index that you calculate.

## Huygens and wavelets

The 17th century Dutch mathematician Christiaan Huygens provided an explanation of the behaviour of light, including refraction, using the idea that light travelled as a wave.

Huygens imagined light spreading out as tiny *wavelets*. He pictured every point on a wave-front as a source of circular wavelets – just like the circular ripples you get when you let a drop fall into a bowl of water. Huygens suggested that where these wavelets meet in phase they combine to form a new wave-front. Everywhere else the wavelets

▲ **Figure 4** *Christiaan Huygens (1629–1695)*

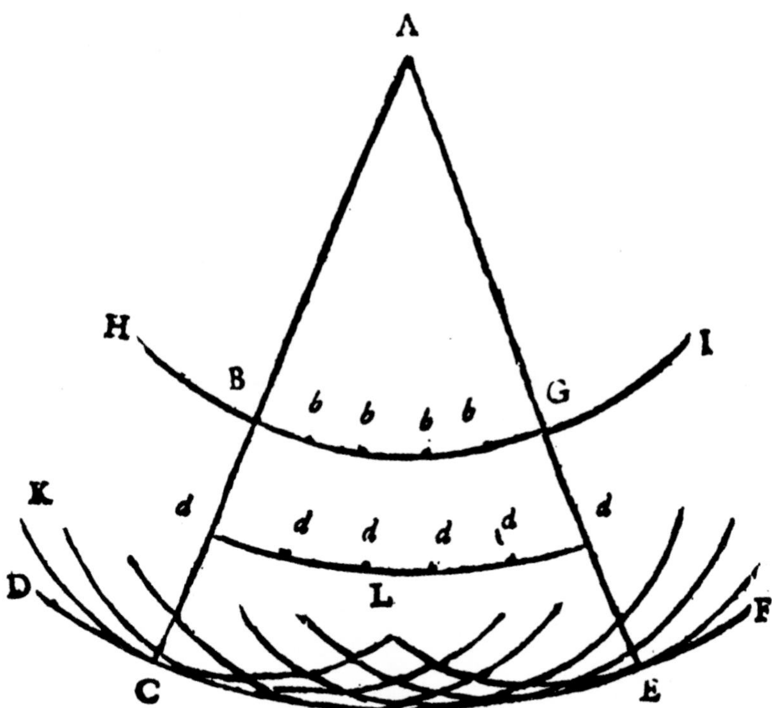


▲ **Figure 5** *Huygens' diagram showing that wave-front DCEF is made of wavelets on wave HBGI starting from all the places like b, b, b*

from one part of the wave will be in antiphase with those from another part of the wave and will cancel out. This simple picture is far-reaching. You have to think of wavelets from the wave as spreading everywhere, but adding up to nothing except where the wave actually goes, another example of superposition.

We can use the model of light as a succession of wave-fronts to explain why light bends when it moves from one material to another at an angle other than along the normal.

We begin by considering what happens when a wave-front enters glass from air parallel to the normal line, as shown in Figure 6. Huygens suggested that waves slow down when they enter the glass so the waves behind *catch up*. The wavelength is shorter in the glass but there is no change of direction.

When light enters a different material at an angle to the normal the wave-fronts become kinked at the boundary because the section of the wave-front travelling through air travels faster than the section of

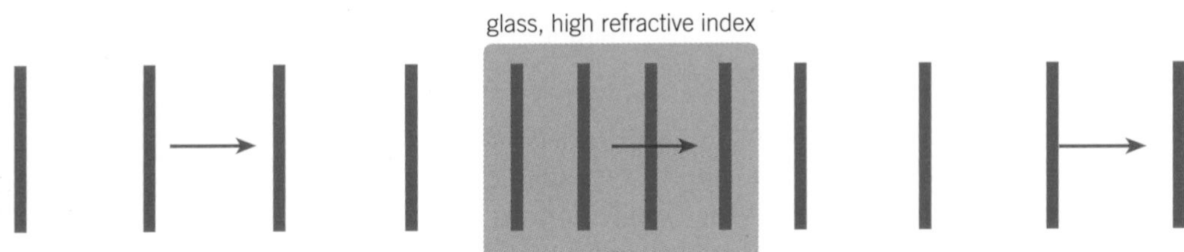


▲ **Figure 6** *Wave-front representation of light travelling from air into glass*

the wave-front in the material. This effect is shown in Figure 7. In time $\Delta t$ the wave-front moves distance RQ in medium 1 and distance PS in medium 2.

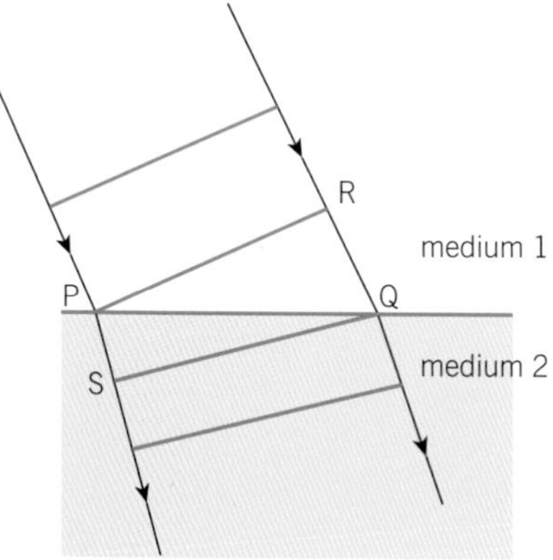


▲ **Figure 7** *Wave-fronts changing direction at a boundary*

We can use this model and a little geometry to explain why the ratio of speeds is the same as the ratio of the sines of the angles of incidence and refraction.

$$\text{As speed} = \frac{\text{distance}}{\text{time taken}}$$

$$\frac{\text{speed in air}}{\text{speed in glass}} = \frac{\frac{RQ}{\Delta t}}{\frac{PS}{\Delta t}} = \frac{RQ}{PS}$$

$$RQ = \sin i \times (PQ)$$

$$PS = \sin r \times (PQ)$$

$$\frac{RQ}{PS} = \frac{\sin i \times PQ}{\sin r \times PQ} = \frac{\sin i}{\sin r}$$

Snell's experimental law matches Huygens' analysis based on light wave-fronts slowing down when they travel from air into another transparent material.

## Summary questions

1. The velocity of light in air is $3.0 \times 10^8\ \text{m s}^{-1}$. Diamond has a refractive index of 2.419. Calculate the velocity of light in diamond. Explain your choice of the number of significant figures you give in your answer. *(3 marks)*

2. The refractive index of amber is 1.55. Calculate the angle of refraction in amber for an angle of incidence of 45°. *(3 marks)*

3. Figure 8 shows waves approaching a boundary to a material in which they speed up.
   **a** Copy and complete the diagram to show the faster-moving waves in the material. *(2 marks)*
   **b** State what happens to the wavelength and frequency of the waves as they enter the material. *(2 marks)*

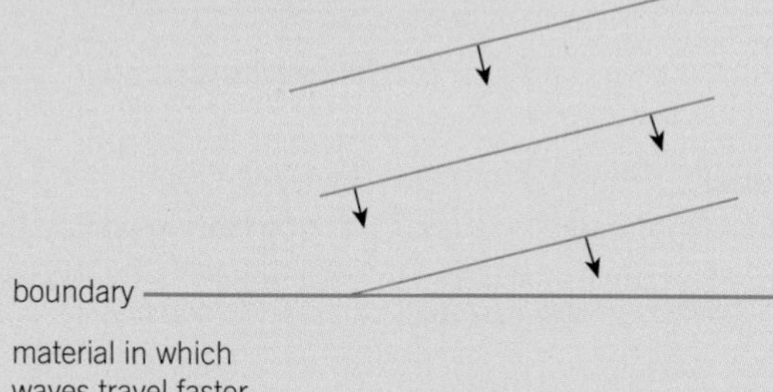


▲ **Figure 8**

4. A Pyrex boiling tube containing glycerol is immersed in a beaker of glycerol. Pyrex and glycerol have the same refractive index. Explain why the length of the tube in the glycerol becomes invisible. *(3 marks)*

5 White light passing through a glass prism is dispersed into the colours of the spectrum.

a State what the refractive index of a material tells you about the behaviour of light as it enters or leaves the material. *(1 mark)*

b Explain why the dispersion of light through a prism shows that different wavelengths of light travel at different speeds in glass. *(2 marks)*

c Explain why the dispersion of light shows that the refractive index of a material is wavelength dependent. *(2 marks)*

▲ **Figure 9** *Light dispersed into colours by a prism*

6 Table 1 shows the results from an experiment to determine the refractive index of a glass block. The student taking the results estimated the uncertainty in the readings of the angle of refraction as ± 2°.

▼ **Table 1**

| Angle of incidence $i$/° | Angle of refraction $r$/° |
|---|---|
| 0 | 0 |
| 10 | 5 |
| 20 | 14 |
| 30 | 20 |
| 40 | 26 |
| 50 | 32 |
| 60 | 41 |

a Draw a table of $\sin i$ and $\sin r$ for the results, include the uncertainty in $\sin r$. *(4 marks)*

b Draw a graph of $\sin i$ against $\sin r$. Use this graph to find the refractive index of the material. Show your working clearly and include an estimate of the uncertainty in your value. *(6 marks)*

# 6.3 Path difference and phase difference

## Specification reference: 4.1b(i), 4.1d(iii)

**Learning outcomes**

Describe, explain, and apply:

- → the terms: interference, coherence, path difference
- → the superposition experiment using microwaves to calculate wavelengths.

Designers and architects have to take superposition into account when designing concert halls. Some rooms have acoustic *dead spots* where the sound from the stage is of lower volume than expected. This happens when sound waves reflected from the walls and ceiling superpose and the amplitude of the wave is reduced. This effect can also happen in classrooms. The voice of a teacher can be just the right frequency to produce a reduction in amplitude where the sound waves superpose in areas of the room. This has been linked to poor performance of students!

This superposition effect is shown very clearly when two loudspeakers emit a note of the same frequency and amplitude. Walking in front of the loudspeakers you will notice positions where the sound is loud and others where it is quieter. If you position yourself at a region of quiet where you hear little sound and ask for one of the speakers to be turned off, you will find that you hear a louder sound. Switching the speaker back on lowers the volume you hear once more.

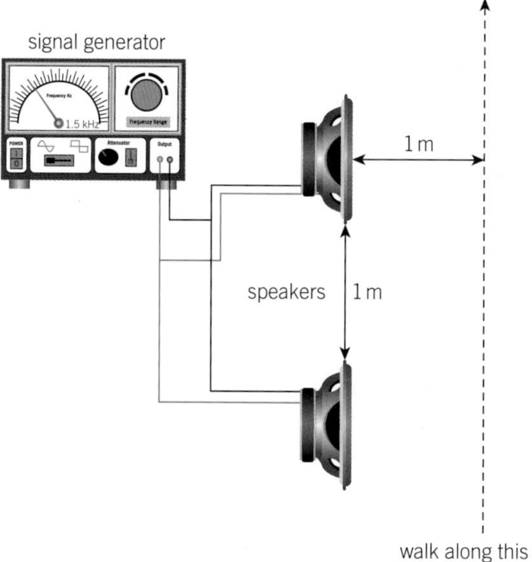


▲ **Figure 1** *Hearing superposition*

## Interference of waves

The term **interference** is often used to describe the effect of the superposition of waves. That is, the superposition of waves produces an interference pattern, such as the alternation of loud and quiet positions in the space in front of the speakers in Figure 1. In positions where waves from the two speakers meet in phase a louder sound is heard. Where they meet in antiphase the sound is quieter. If the waves from two speakers are of the same amplitude where they meet in antiphase they will cancel completely, producing silence.

**Synoptic link**

You have met the term *phase* in Topic 6.1, Superposition of waves.

## Path difference

You can see from the diagram in Figure 2 that the path length, the distance from speaker 1 to the microphone, is $6\lambda$, whereas the path length between speaker 2 and the microphone, is $8\lambda$. The **path difference**, the difference in path lengths, is $8\lambda - 6\lambda = 2\lambda$. These waves will meet in phase, increasing the amplitude of the superposed wave-form (this is a superposition maximum). If the path difference is, say, $1\frac{1}{2}\lambda$, the waves would meet in antiphase, and would completely cancel out if the amplitude of each wave was the same (this is a superposition minimum). Waves meeting with zero path difference will be in phase. You hear a louder sound when you stand at equal distances from the speakers in the experiment we have been describing.

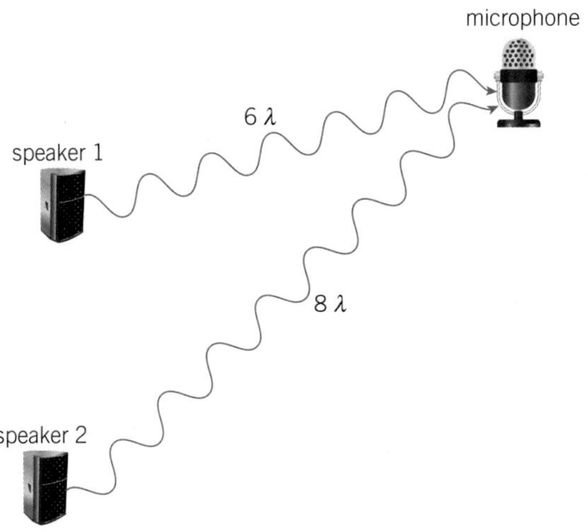


▲ **Figure 2** *An arrangement for waves meeting in phase*

We can state the conditions for waves to meet in phase and in antiphase as

- waves meet in phase when the path difference = $n\lambda$, where $n$ is an integer
- waves meet in antiphase when the path difference = $(n + \frac{1}{2})\lambda$.

### Worked example: Superposition maxima and minima

The two loudspeakers in Figure 2 emit a note of wavelength of 77 cm (corresponding to a frequency of 440 Hz). Suggest three values of path difference which will produce superposition maxima and three values of path difference where superposition minima will occur.

**Step 1:** Identify the correct equations to use.

Superposition maxima occur at a path difference of $n\lambda$, where $n$ is an integer.

Superposition minima occur at a path difference of $(n + \frac{1}{2})\lambda$.

**Step 2:** Substitute $n = 1$, 2, and 3 into the two equations.

Superposition maxima will occur at $n\lambda$.

$1 \times 77\,\text{cm} = 77\,\text{cm}$, $2 \times 77\,\text{cm} = 154\,\text{cm}$, and $3 \times 77\,\text{cm} = 231\,\text{cm}$

Superposition minima will occur at $(n + \frac{1}{2})\lambda$.

$1.5 \times 77\,\text{cm} = 115.5\,\text{cm}$, $2.5 \times 77\,\text{cm} = 192.5\,\text{cm}$,
$3.5 \times 77\,\text{cm} = 269.5\,\text{cm}$

## Coherence

A *stable* superposition pattern is one in which the position of the maxima and minima don't change over time. For example, in the loudspeaker experiment, if you find a minimum position and stand there for a period of time the volume of sound will not change, unless someone moves the speakers or changes the frequency of the note.

Stable superposition patterns can only occur when there is a constant phase difference between waves from the two sources. Waves with a constant phase difference are **coherent**.

The first two sets of waves in Figure 3 are coherent. They are not in phase but they do have a constant phase difference. The second two sets of waves are incoherent.

coherent waves with constant phase difference

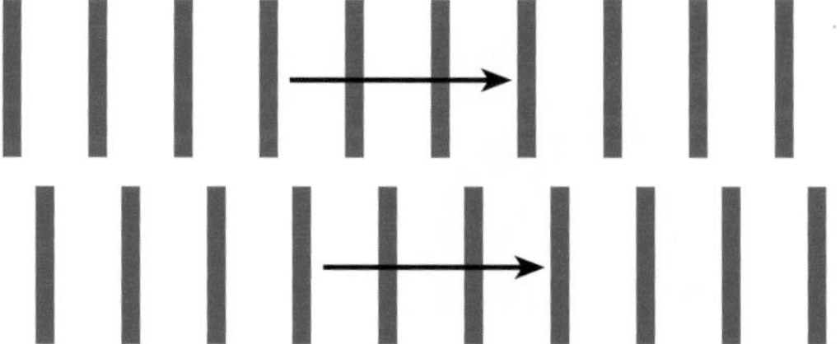

incoherent wave bursts with changing phase difference

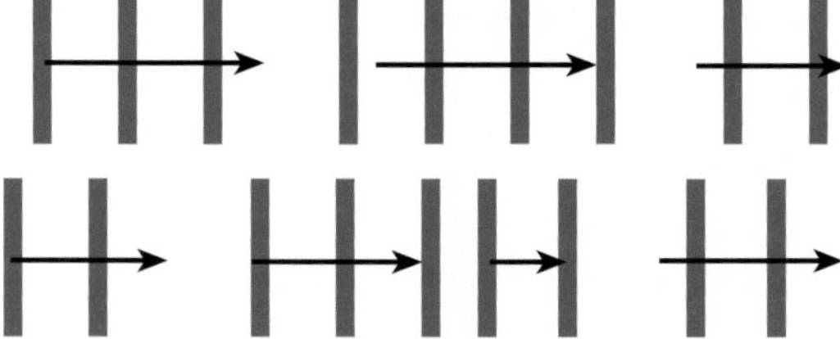

▲ **Figure 3** *Coherence*

## Superposition and interference

You will meet the terms *constructive interference* and *destructive interference*. These are both names for the effects of superposition. Constructive interference occurs when waves meet in phase leading to a larger superposition amplitude (a maximum). Destructive interference occurs when waves meet in antiphase leading to a lower superposition amplitude (a minimum).

## Active noise-reducing headphones

Noise-reducing (often sold as *noise-cancelling*) headphones allow the wearer to listen to his or her choice of music without needing to turn the volume up to overcome low frequency background noise.

Some of the background noise is blocked simply by the material of the headphones reducing the energy of sound waves passing into the ear. Active noise-reducing headphones use superposition to further reduce the effects of background noise.

A microphone in the ear piece detects the background noise that is not stopped by the material of the headphone ear piece.

Electronic circuitry *flips* the wave-form of the noise so that it is in antiphase ($\pi$ radians phase difference) with the background noise.

The flipped noise is fed into the headphone speakers. The noise and the flipped noise cancel out.

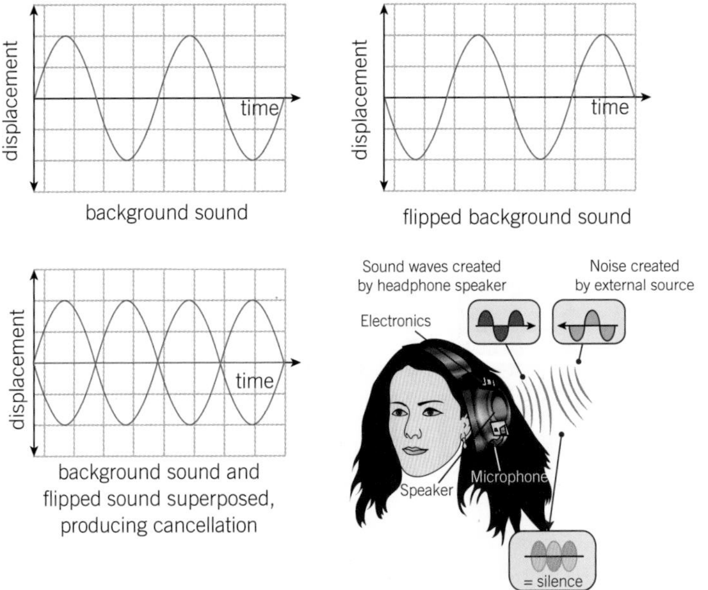


▲ **Figure 4** *Principle of active noise-reducing headphones*

1. Why is it so important that the electronic circuitry is able to flip the background sound extremely quickly?
2. Why do you think these headphones are better at reducing low frequency sounds such as engine noise rather than speech?

## Practical 4.1d(iii): Interference with microwaves

This practical is an example of using superposition to measure the wavelength of waves emitted from a source, in this case a microwave source.

The microwave transmitter T, receiver R, and mirror M are positioned as shown in Figure 5. Waves from the transmitter reach the receiver along the path TR and the path TMR.

As you move the mirror towards and away from the line TR you will detect a series of maxima and minima. Once you've found a position when the receiver shows a maximum value you can measure the distances TR and TMR. You can then find the path difference TMR – TR in terms of $\lambda$. As this path difference is for a maximum we can write TMR – TR = $n\lambda$.

When the mirror is moved further away from the line TR the signal at the receiver will decrease to a minimum and then rise again to a maximum as the mirror is moved further. The path difference for this second maximum is found ($n = 2$).

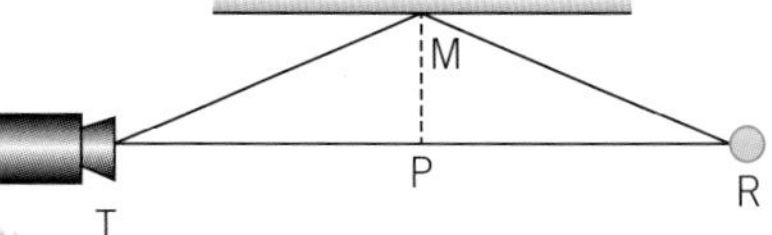


▲ Figure 5

### Study tip

Make sure that you know one method of measuring the wavelength of microwaves, visible light, and sound.

## Worked example: Interference with microwaves

An experiment is conducted in which a microwave transmitter T, receiver R, and mirror M are positioned as shown in Figure 5. Waves from the transmitter reach the receiver along the path TR and the path TMR. Measurements can be taken to a precision of 0.005 m.

distance TR = 0.900 m  distance MP = 0.270 m

distance TP = 0.450 m

**a** Calculate the length of the path TMR.

We use Pythagoras' theorem here to calculate the distance between the transmitter and the mirror. We then double this to find TMR.

$$\text{TMR} = 2 \times \text{TM} = 2 \times \sqrt{\text{TP}^2 + \text{MP}^2}$$
$$= 2 \times \sqrt{(0.450\,\text{m})^2 + (0.27\,\text{m})^2} = 1.050\,\text{m}$$

**b** Calculate the path difference for waves arriving at the receiver from the two paths. The receiver detects a maximum in this position.

Path difference = TMR – TR = 1.050 – 0.900 = 0.15 m

**c** The mirror is slowly moved, reducing the distance MP to 0.240 m. The amplitude of the microwaves detected at the receiver falls to a minimum and rises back to a maximum when MP is 0.240 m. Calculate the new length of the path TMR.

We use Pythagoras' theorem once again to find the length of the path TMR.

$$\text{New path TMR} = 2 \times \sqrt{(0.45\,\text{m})^2 + (0.240\,\text{m})^2} = 1.02\,\text{m}$$

**d** Calculate the path difference when MP = 0.240 m

Path difference = TMR − TR = 1.020 − 0.900 = 0.12 m

**e** The amplitude of the microwaves detected at the receiver falls to a minimum and rises back to a maximum when MP changes from 0.270 m to 0.240 m. Use this information to calculate the wavelength of the microwaves.

When distance MP is reduced from 0.270 m to 0.240 m the received signal goes through one cycle of maximum–minimum–maximum. This shows that the path difference TMR − TR has been reduced by one wavelength.

Therefore, the wavelength of microwaves = 0.15 m − 0.12 m = 0.03 m.

## Summary questions

1 Explain the meaning of the terms *path difference* and *phase difference*. *(2 marks)*

2 A laser is a coherent source of light. Explain what *coherent* means in this context. *(1 mark)*

3 Look at Figure 6. When the microphone is in position A an interference maximum is detected.

  a State the path difference necessary for an interference maximum. *(1 mark)*

  b Calculate the path difference between waves from speaker 2 to the microphone and speaker 1 to the microphone. *(1 mark)*

  c The microphone is slowly moved to position B. As the microphone is moved the signal falls to a minimum and then rises to a maximum at B. The path difference between the speakers and the microphone at position B is 0.50 m.
  Calculate the wavelength of the sound from the speakers.
  Describe how you reached your answer. *(2 marks)*

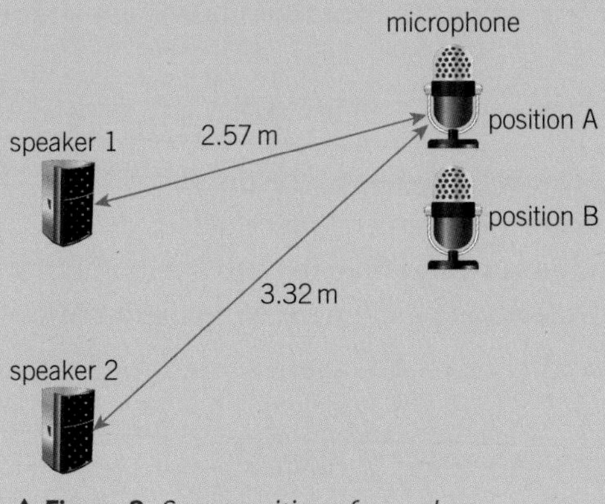


▲ **Figure 6** *Superposition of sound*

# 6.4 Interference and diffraction of light

Specification reference: 4.1a(ii), 4.1a(iv), 4.1a(v), 4.1b(i), 4.1c(iii), 4.1d(iii), 4.1d(iv)

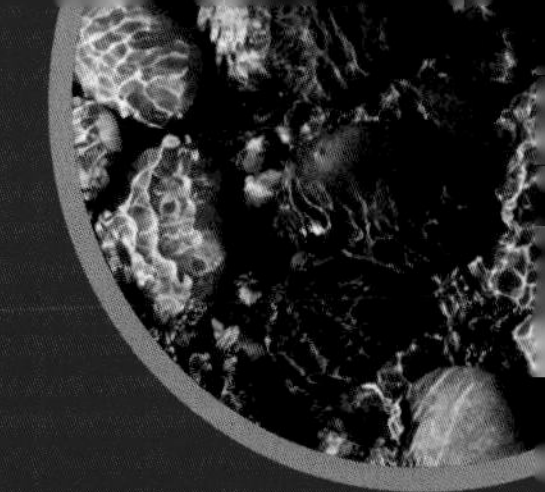

In the 17th century the poet John Milton described the flames of Hell in these words:

> *Yet from those flames*
>
> *No light; but rather darkness visible*

You may think that it is impossible for light to appear dark, but just as sound waves can cancel if they meet in antiphase, light meeting light in antiphase can produce a dark band where the two waves meet. This is just another example of superposition.

## Learning outcomes

Describe, explain, and apply:

- the diffraction of waves through a narrow aperture, a double slit, and a diffraction grating
- the term diffraction
- the equation $n\lambda = d \sin \theta$ to calculate path differences for double slits and diffraction gratings
- superposition experiments using light
- how the wavelength of light can be determined using a double slit and a diffraction grating.

## Diffraction

When waves pass through a gap of roughly the same width as their wavelength the waves spread out. This is an example of **diffraction**. This can be seen in the aerial photograph of waves entering a harbour, shown in Figure 1.

The amount the waves spread out depends on the width of the gap compared to the wavelength of the waves passing through. For a given wavelength, the smaller the gap the greater the spreading caused by diffraction. Diffraction does not alter the wavelength, speed, or frequency of the waves.

▲ **Figure 1** *Waves diffracting as they enter a bay*

You have experienced sound diffraction without realising it! When someone is talking to you in the middle of an open space the person talking does not need to be facing you for you to hear what is being said. This is because the sound is diffracted from the person's mouth and spreads out. Think how wide your mouth is when speaking – a few centimetres. The wavelength of human speech is of the order of one metre so it is no surprise that the sound diffracts around the speaker as the wavelength is so much bigger than the gap the waves are passing through.

It is more difficult to observe diffraction in light. Light does not seem to bend round corners or spread out through slits. The reason that it is not observed in our everyday lives is that the wavelength is of the order of $10^{-7}$ m, so a very narrow gap is needed for an appreciable amount of spreading to occur.

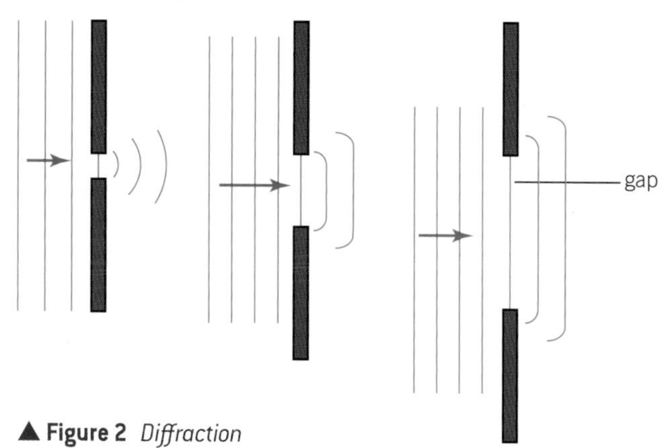


▲ **Figure 2** *Diffraction*

### Young's double slit experiment

The theory that light travels as waves gained a lot of support from an experiment published by Thomas Young (the same man as the Young modulus is named after) in 1807. He passed light

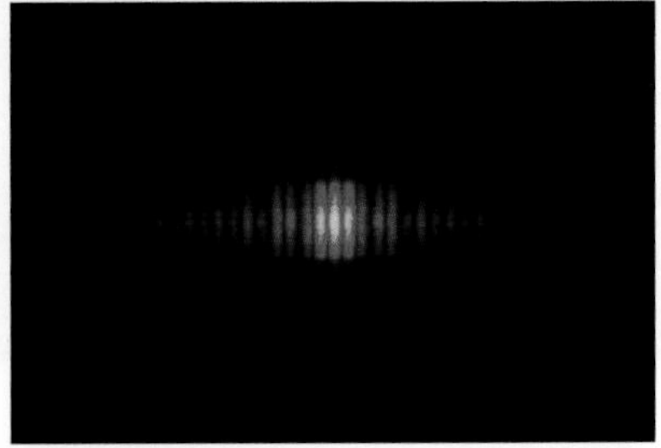

▲ **Figure 4** *Fringe pattern obtained from two slits*

through two pinholes very close together and observed a pattern of dark and bright fringes.

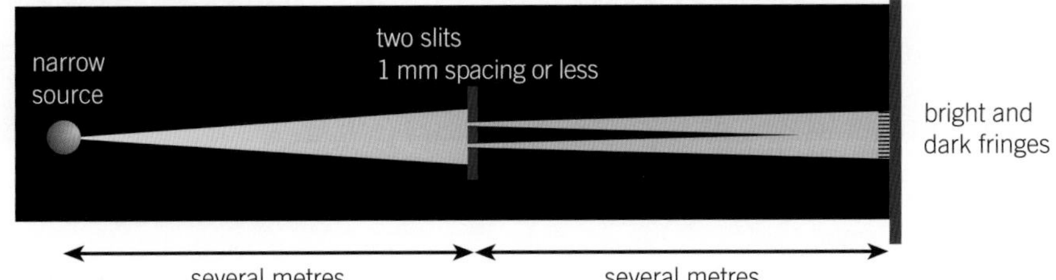


▲ **Figure 3** *Young's double slit experiment*

Modern versions of this experiment use slits rather than pinholes. The observed pattern is due to the superposition of light from the two slits – this is Young's double slit experiment. Where light from both slits meets at the screen in phase, a bright fringe is observed. Light meeting in antiphase produces a dark patch. Figure 5 shows wave-fronts of light from two slits which meet in phase and out of phase at a distant screen.

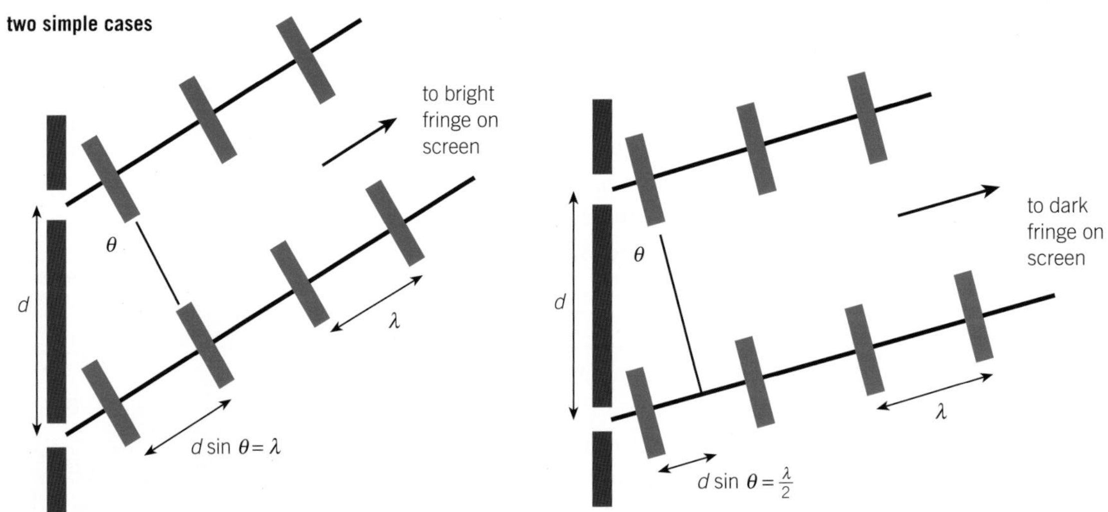


▲ **Figure 5** *Producing bright and dark fringes*

### Synoptic link

You have seen from Topic 6.3, Path difference and phase difference, that a bright fringe means that there is a path difference of $n\lambda$ between the light from the two slits.

You may wonder why diagrams of the double slit experiment show parallel rays of light meeting on a distant screen to produce fringe patterns. Of course this is only an approximation as parallel lines never meet. It is entirely reasonable to make this approximation as the distance between the slits and the screen is so large compared to the separation of the slits.

Figure 6 shows a ray representation of the double slit experiment.

The situation in Figure 6 produces a bright fringe. The path difference must therefore be a whole number of wavelengths. In this case we have indicated that the path difference is one wavelength.

$$\text{As } \sin\theta = \frac{\text{opposite}}{\text{hypotenuse}}$$

$$\sin\theta = \frac{\lambda}{d}$$

Therefore, for waves with a path difference of one wavelength, $\lambda = d\sin\theta$, where $d$ is the slit separation and $\theta$ is the angle made at the slits between one bright fringe and the next.

Figure 7 shows the situation for a path difference of two wavelengths.

For a path difference of $n\lambda$ we can write $n\lambda = d\sin\theta_n$.

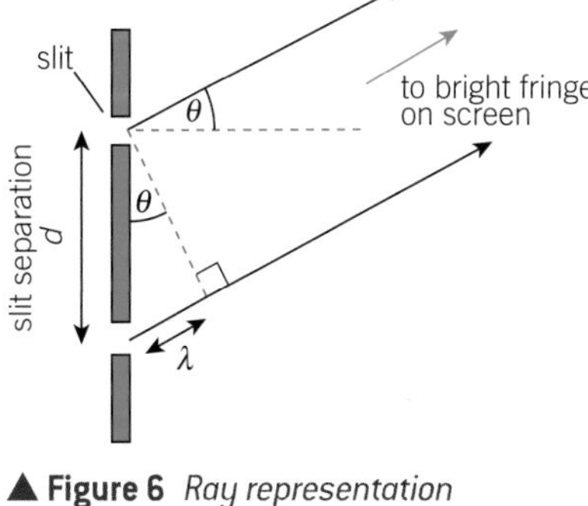


▲ **Figure 6** *Ray representation*

### Order of maxima

The order of a maximum shows the number of wavelengths path difference between light from two adjacent slits and the screen (or detector). Therefore, the zeroth order ($n = 0$) is the maximum when the path difference between two adjacent slits and the screen is zero. The first order ($n = 1$) is the maximum for a path difference between the slits and the screen of one wavelength. The second order ($n = 2$) is the maximum produced by a path difference of two wavelengths and so on.

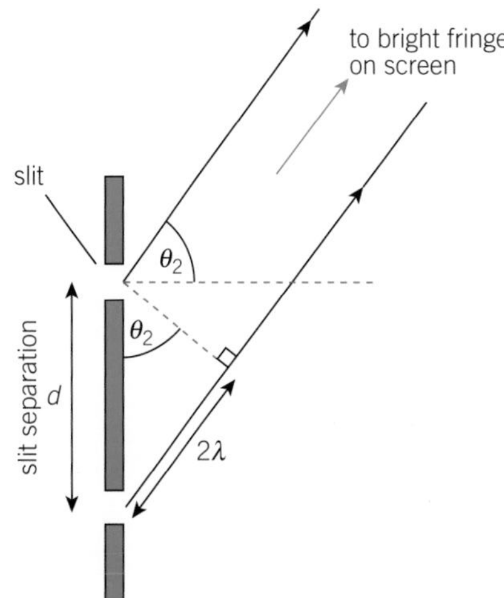


▲ **Figure 7** *Ray representation, two wavelength path difference*

## Worked example: Calculating angles from orders of maxima

Green light of wavelength 500 nm is incident on two narrow slits. The slit separation is $0.5 \times 10^{-3}$ m.

**a** Calculate the angle of the first order ($n = 1$) maximum.

**Step 1:** Select the appropriate equation.

$$n\lambda = d\sin\theta_n \text{ which rearranges to give } \sin\theta_n = \frac{n\lambda}{d}$$

**Step 2:** Substitute values and evaluate, remembering that 1 nm = $1 \times 10^{-9}$ m.

$$\sin\theta_1 = \frac{1 \times 500 \times 10^{-9}\,\text{m}}{0.5 \times 10^{-3}\,\text{m}} = 1 \times 10^{-3}$$

**Step 3:** Find the angle from the value of the sin.

$$\theta_1 = \sin^{-1} 1 \times 10^{-3} = 0.06°$$

**b** Calculate the angle of the fourth order ($n = 4$) maximum.

**Step 1:** Substitute values into $\sin\theta_n = \frac{n\lambda}{d}$, where $n = 4$.

$$\sin\theta_4 = \frac{4 \times 500 \times 10^{-9}\,\text{m}}{0.5 \times 10^{-3}\,\text{m}} = 4 \times 10^{-3}$$

$$\theta_4 = \sin^{-1} 4 \times 10^{-3} = 0.23°$$

These angles are very small. This is why the screen must be a good distance away from the slits to allow the bright and dark fringes to be observed and measured.

We can use the equation $n\lambda = d\sin\theta_n$ to determine the wavelength of the light passing through the pair of slits. Figure 8 shows the geometry of rays meeting at a distant screen.

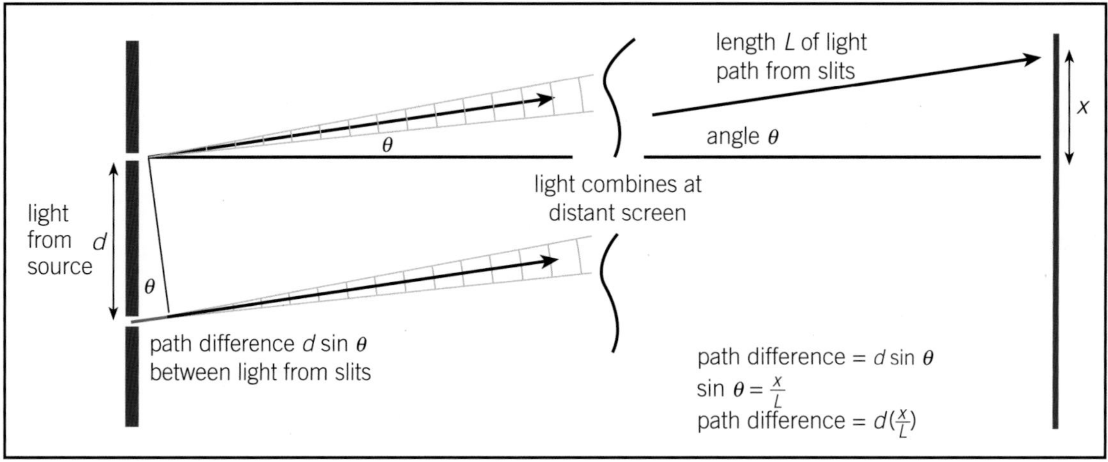


Approximations – angle $\theta$ very small; paths effectively parallel; distance $L$ equal to slit–screen distance. Error less than 1 in 1000

▲ **Figure 8** *Geometry*

## Synoptic link

Small angle approximations are used in the assumption for Young's double slit experiment. This is covered in the Maths Appendix.

The diagram shows that $\tan\theta = \frac{x}{L}$.

For small $\theta$, $\tan\theta \approx \sin\theta$.

By combining $\tan\theta = \frac{x}{L}$ and $\sin\theta = \frac{\lambda}{d}$ you find $\frac{x}{L} = \frac{\lambda}{d}$.

Therefore $\lambda = \frac{xd}{L}$.

### Worked example: Young's double slit experiment

A double slit experiment is set up using a laser. The screen is 3.00 m away from the slits, which have a separation of 0.4 mm. The spacing between bright fringes on the screen is 5.0 mm. Calculate the wavelength of light used.

**Step 1:** Select and rearrange the appropriate equation.

$$\lambda = \frac{xd}{L}$$

**Step 2:** Substitute values and evaluate.

$$\lambda = \frac{5.0 \times 10^{-3}\,\text{m} \times 4 \times 10^{-4}\,\text{m}}{3.00} = 6.7 \times 10^{-7}\,\text{m (2 s.f.)}$$

## The diffraction grating

A diffraction grating is a multiple slit version of the two slit system you have been thinking about. The equation $n\lambda = d\sin\theta_n$ also holds for gratings.

Using a grating of many slits increases the brightness of the image on the screen because more light gets through. Gratings also give a sharper fringe pattern. Diffraction gratings can also spread white light out into its component colours because each wavelength of light will produce a maximum at a different angle.

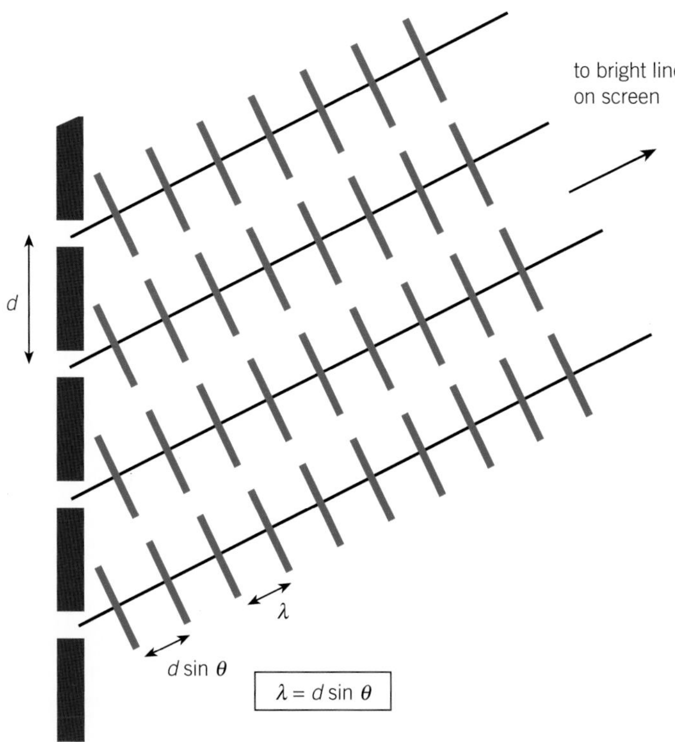


▲ **Figure 9** *Waves from many sources all in phase*

Diffraction gratings are often described by their number of lines per mm. To calculate the separation of each line (the equivalent of the slit separation), convert the number of lines per mm into lines per metre and take the reciprocal.

$$\text{line separation (m)} = \frac{1}{\text{number of lines per metre}}$$

## Worked example: Diffraction gratings

A diffraction grating has 300 lines per mm. Calculate the angle to the first order maximum for:

**a** red light of wavelength 670 nm;

**Step 1:** Select and rearrange the appropriate equation.

$$\sin\theta_1 = \frac{\lambda}{d}$$

**Step 2:** Rewrite the equation above in terms of number of lines per metre, not $d$.

Since line separation $d = \dfrac{1}{\text{number of lines per metre}}$

$\sin\theta = \lambda \times \text{lines per m}$

**Step 3:** Substitute values and evaluate.

$$\sin\theta = 6.7 \times 10^{-7}\,\text{m} \times 300\,000\,\text{m}^{-1}$$

$$\sin\theta = 0.201 \text{ therefore } \theta = 11.6°$$

**b** blue light of wavelength 460 nm.

Work through Steps 1 and 2 as above. Step 3 becomes

$$\sin\theta = 4.6 \times 10^{-7}\,\text{m} \times 300\,000\,\text{m}^{-1} = 0.138 \text{ therefore } \theta = 7.9°$$

## Practical 4.1d(iii) and (iv): Determining the wavelength of light using superposition of light (Young's double slits or a diffraction grating)

There are many ways of determining the wavelength of light using interference effects. The key measurements are:

- the slit separation
- the distance between the slits and the screen
- the fringe separation.

The basic experimental set-up using a laser is shown in Figure 10. The screen should be a few metres away from the double slit.

The greatest source of uncertainty in the double slit experiment is the measurement of the slit separation. If you are using a diffraction grating you will be given the number of lines per mm which will allow you to calculate a confident and precise value of the slit separation.

Performing the experiment with simple apparatus is a worthwhile challenge as you can still get a reasonable estimate for the wavelength of the light you are using and you will be using a method very similar to Thomas Young at the beginning of the 19th century.

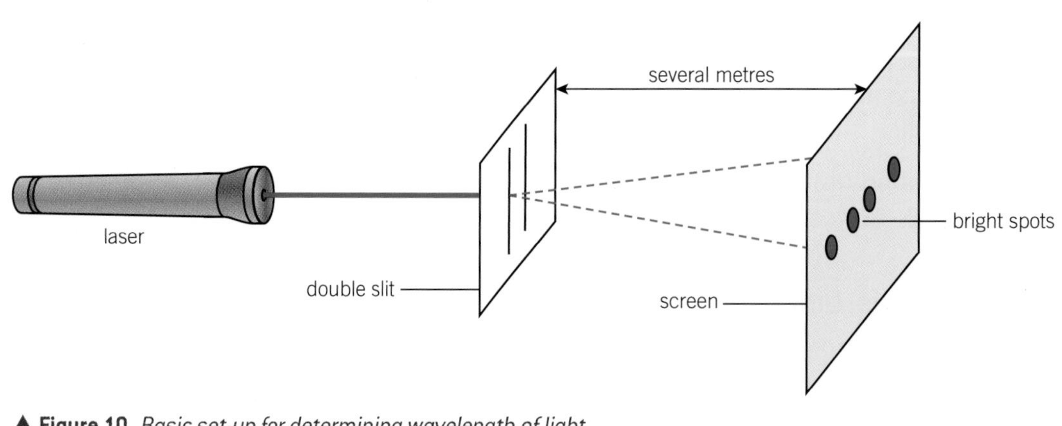


▲ **Figure 10** *Basic set-up for determining wavelength of light*

### Diffraction through a single slit

We can use phasor thinking to explain diffraction through a narrow aperture, as shown in Figure 11.

**Single slit**

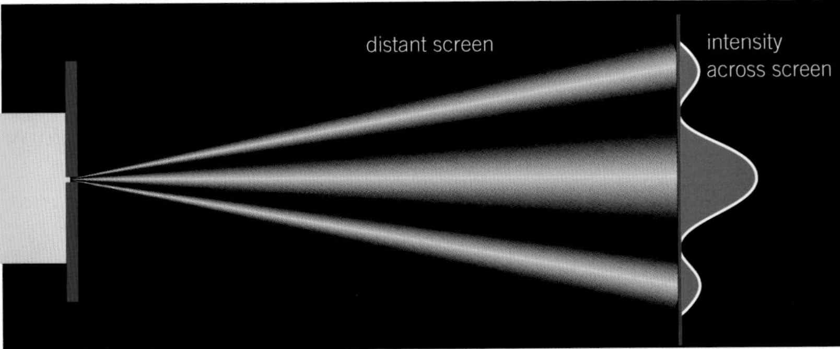


▲ **Figure 11** *Single slit diffraction*

As a wave passes through a slit we imagine that each point in the wave-front is a source of new wavelets – just like Huygens' idea. But we are going to imagine these as rotating phasor arrows.

Where the phasor arrows are all in the same direction we get a large resultant, creating a maximum.

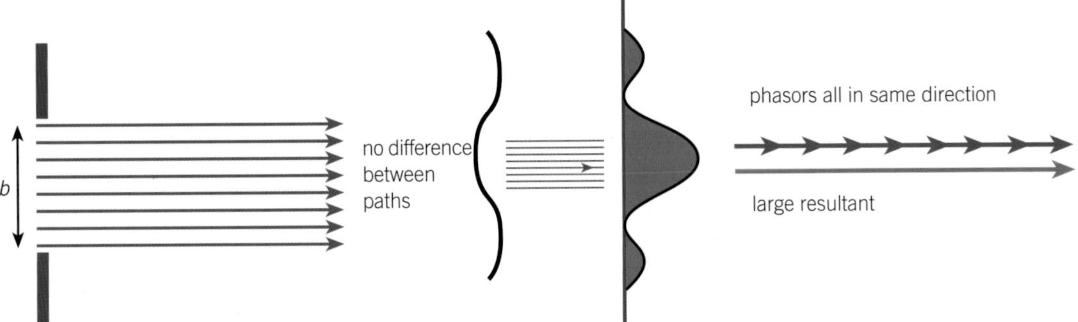


▲ **Figure 12** *Simplified case – distant screen with paths nearly parallel, no path difference*

At an angle $\theta$, the path difference across the whole slit is $\lambda$. Each phasor will reach the screen a little out of phase with the phasor from the adjacent point. All the phasors will add together to give a zero resultant. This is a diffraction minimum – it occurs when $\sin\theta = \frac{\lambda}{b}$.

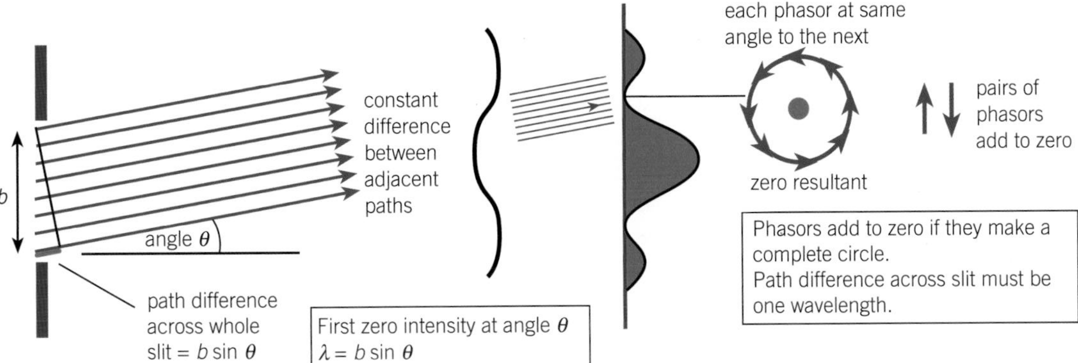


▲ **Figure 13** *Simplified case – distant screen with paths nearly parallel, constant path difference*

The angle ($\theta$ in Figure 13) to the first diffraction minimum gives a measure of how much the waves spread out through the gap.

▲ **Figure 14** *Close up of a tail feather of a peacock*

## The tail of the peacock

You began this chapter with a picture of a peacock. So, how are the colour changes produced as it moves past? Light strikes the feathers of the bird and is reflected from tiny ridges on the feathers. These ridges act like a diffraction grating. Instead of light passing through a uniform group of slits it is reflected from a uniform group of ridges. There is a path difference between waves from each ridge so at any angle one colour will predominate because light of that wavelength will be meeting at the eye in phase. There are many examples of iridescence in nature and although the details of the effect vary they all depend on the principle of superposition.

### Hint

Remember that $\sin\theta \leq 1$. This means that the highest order maximum will be when $\sin\theta = 1$. In other words, *number of orders of maxima possible* $= \frac{\textit{slit separation}}{\textit{wavelength}}$.

## Summary questions

1 a Describe what is meant by diffraction. State whether there is any change in the speed, wavelength, or frequency of waves when they diffract. *(3 marks)*

b A diffraction grating has 80 lines per mm. Light that is perpendicularly incident on the grating produces a first order maximum at an angle of 2.2°. Calculate the wavelength of the light. *(3 marks)*

2 Light of wavelength $5.4 \times 10^{-7}$ m passes through a diffraction grating of 300 lines per mm.

a Calculate the angle to the second order maximum. *(3 marks)*

b Calculate how many orders of maxima are possible for this set-up. *(2 marks)*

3 Light of wavelength 600 nm passes through a single slit of width $1 \times 10^{-3}$ m.

a Find the angle $\theta$ to the first order minimum where $\sin\theta = \frac{\lambda}{b}$, where $b$ is the slit width. *(2 marks)*

b Explain why the result from **a** shows that diffraction would not be observed by the unaided eye in such an arrangement. *(2 marks)*

4 A class performs a Young's double slit experiment. The gap between the slits is $0.5 \pm 0.1$ mm. The distance between the slits and the screen is $3.40 \pm 0.01$ m. The fringe spacing is measured to be $3.5 \pm 0.5$ mm.

a Use these measurements and uncertainties to calculate the biggest and smallest values for the wavelength of light. *(4 marks)*

b Explain which measurement contributes most to the overall uncertainty. *(2 marks)*

# Physics in perspective

## Shedding light on light

### The debate about the nature of light

Many thinkers have puzzled over the nature of light. In the 17th and 18th centuries the so-called *corpuscularists* favoured a particle-like description, whereas the *undulationists* supported a wave-like description. Christiaan Huygens was an undulationist — his famous illustration of the candle (Figure 1)shows waves emanating from every part of the flame. However, the arguments between the undulationists and the corpuscularists were not always clear-cut.

Isaac Newton (1642–1727) wrote in his book *Optiks,* "Are not the rays of light very small bodies emitted from shining substances?". This suggests that Newton was firmly in the corpuscularist camp, but he went on to suggest that rays have vibrations associated with them of 'several bignesses' or, in modern terms, different wavelengths.

Newton knew of many phenomena that are now explained by waves, and invented a way of measuring them accurately – known as Newton's rings. Newton's rings can be observed when a lens that is convex on one side and flat on the other is placed on a flat glass slide and illuminated from above. A pattern of concentric rings is observed. If the incident light is monochromatic the pattern shows bright and dark rings, with a dark spot in the centre where the surfaces touch. Similar effects are seen when two microscope slides are placed together.

▲ **Figure 1** *Huygens' candle*

Newton did not decisively support one model or the other, but in the years following his death it was generally thought that the great mathematician and experimenter had strongly supported the particle model. This may have made it more difficult for alternative ideas to gain acceptance.

### Thomas Young explains interference

Before 1833 you could not be a scientist, for the term had not been invented. Before this date there were only a few individuals who would now be considered 'professional' scientists. So when Thomas Young stood in front of the Royal Society in London in a series of lectures during the early 1800s, the fact that a medical doctor was lecturing about the nature of light was rather less surprising than it seems now. Nevertheless, Thomas Young was an extraordinary individual of many talents – a true polymath.

▲ **Figure 2** *Amongst other interests, Thomas Young (1773–1829) was a linguist, an archaeologist, a poet, and a doctor*

Young used a device of his own invention, a ripple tank, to show how interference effects can be produced in water waves. It is thanks to Young that ripple tanks are used in physics lessons to this day, and that we talk about 'interference' of waves — a phrase he used to describe the superposition effects observed in water.

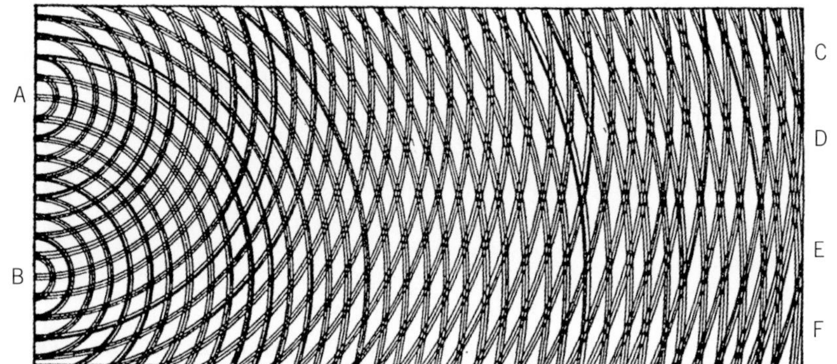


▲ **Figure 3** *Young's illustration of the double slit pattern*

Young compared interference effects in water with those in light, but interference is not easily observed in light. It was not enough for Young to suggest that light could interfere like water waves — he had to give evidence for it. In 1803, he provided the evidence in his now-celebrated double-slit experiment.

At the time, however, some of Young's audiences were less-than-impressed, and it took the work of Auguste Fresnel (1788–1829), a French engineer, to really change scientific opinion.

### Interference explains Newton's rings

Figure 4 shows the basis of Young's explanation of Newton's rings. Some of the incident light is reflected from the curved surface of the lens (path 1). Light that passes through the lens is reflected from the surface of the slide (path 2). The superposition of the two rays produces bright fringes when the waves are in phase and dark lines when the waves are in antiphase.

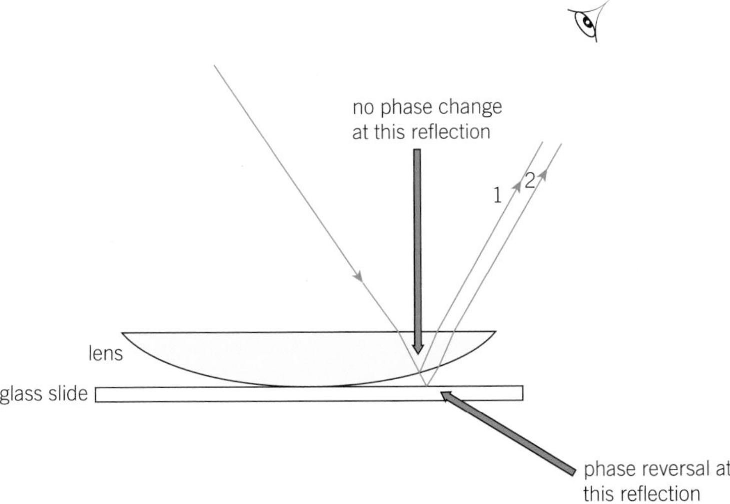


▲ **Figure 4** *Light reflecting from two surfaces*

When reflection occurs at a boundary with a material of higher refractive index, there is a phase change of π radians — when light travelling through air reflects at a glass surface, there will be a phase change.

Reflection at a boundary with a material of lower refractive index produces no phase change – when light travelling through glass reflects from the air boundary, there is no phase change.

A dark spot is observed where the two surfaces are very close. There is no path difference, but the phase of light reflecting from the slide is reversed, and so it is in antiphase with the light reflecting from the curved surface of the lens.

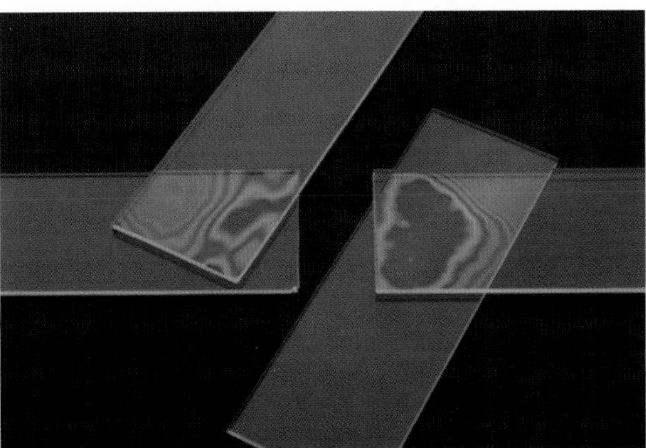

▲ **Figure 5** *Newton's rings effect from two microscope slides*

## Using interference in Solar Cells

Understanding interference helps design non-reflective coatings. These are frequently used on lenses and solar cells.

The silicon used in solar cells reflects about 30% of the light incident upon it. This means that 30% of the energy incident on the cell cannot be used to generate electricity. A layer of transparent silicon monoxide acts as non-reflective coating.

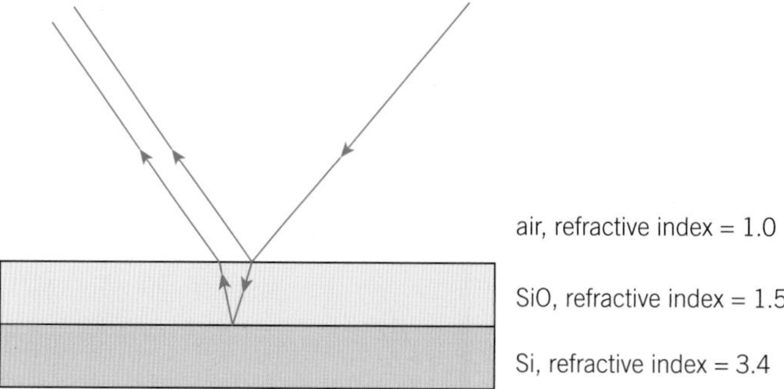


▲ **Figure 6** *Anti-reflection coating*

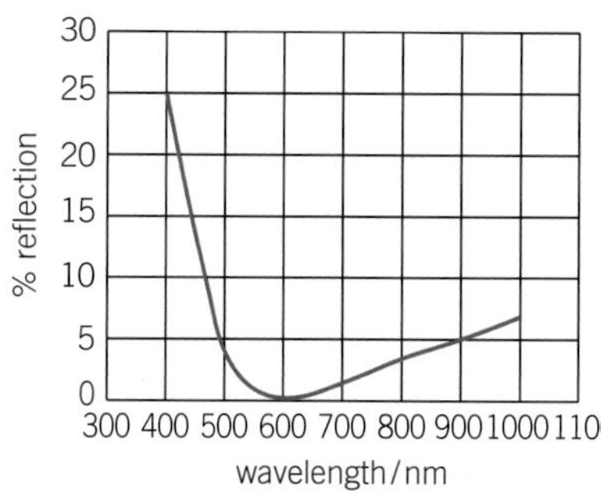


▲ **Figure 7**

There is a phase change of π radians for light reflecting at the air-silicon monoxide boundary *and* at the silicon monoxide–silicon boundary. If the wavelength of light in the silicon monoxide layer is equal to four times the thickness of the layer, the two reflected rays will cancel. The reflective losses can be reduced to about 10% by this method, therefore increasing the efficiency of the solar cell.

## Summary questions

**1** This question is about the development of the wave model of light.

**a** Why is diffraction of visible light not easily observed?

**b** Young used a Newton's ring set-up to determine the wavelength of red light as 650 nm. What is the minimum air gap required between the lens and the slide to produce a bright fringe?
Remember to consider the phase reversal on reflection from the slide.

**c** Young and Huygens both considered light to be a longitudinal wave. Explain why the phenomenon of the polarisation of light suggests that light is a **transverse** wave.

**2** This question is about anti-reflective coating on a solar cell.

**a** In Figure 6, explain why both the light reflecting from the silicon monoxide and the light reflecting from the silicon experience a phase change of π radians.

**b** Using ideas about path difference and phase difference, explain why reflected light from the two surfaces cancels out when the thickness of the silicon monoxide layer is equal to one quarter of the wavelength of the light in the material.

**c** Silicon monoxide has a refractive index of 1.5. Yellow light of wavelength 580 nm in air is incident on the silicon monoxide layer.

**i** Calculate the wavelength of the yellow light in the silicon monoxide.

**ii** Calculate the thickness of silicon monoxide required to produce cancellation of the reflected light from the two surfaces.

**d** The graph in Figure 7 shows how the percentage reflection varies with wavelength with a coating of thickness of $\lambda/4$ for yellow light. Explain why the coating minimises reflection for yellow light but reflection of violet light ($\lambda \sim 400$ nm) is barely changed at all.

## Practice questions

1 Light enters glass from air. Which of the following statements is/are correct?
Statement 1: The wavelength is shorter in the glass than in air.
Statement 2: The velocity is slower in the glass than in air.
Statement 3: The frequency is lower in the glass than in air.

**A** 1, 2, and 3 are correct

**B** Only 1 and 2 are correct

**C** Only 2 and 3 are correct

**D** Only 1 is correct

2 Figure 1 shows light entering glass from air. The angles of incidence and refraction are given. Use information from the diagram to calculate the velocity of light in the glass.

Velocity of light in air = $3.0 \times 10^8\,\mathrm{m\,s^{-1}}$.

(*2 marks*)

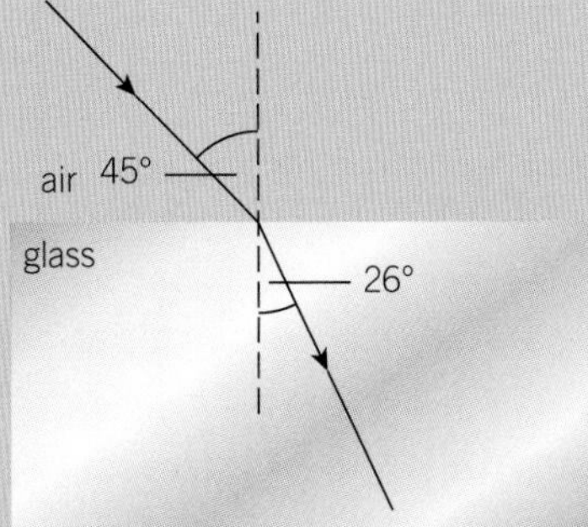


▲ **Figure 1**

3 A coherent beam of red light is perpendicularly incident on a diffraction grating.

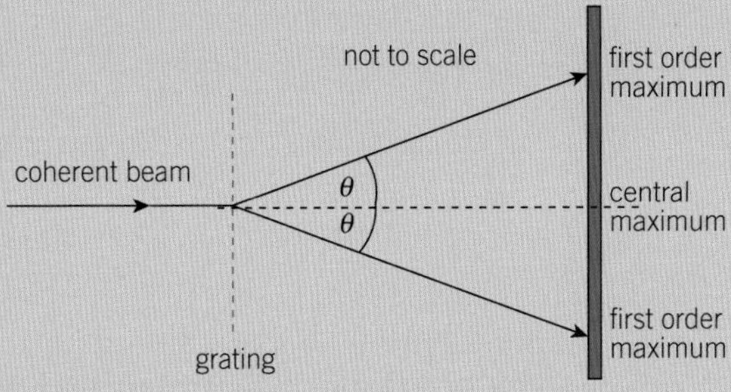


▲ **Figure 2**

a Explain what is meant by a *coherent* beam of light. (*1 mark*)

b The grating has 80 lines per mm. The angle to the first order maximum is 3°. Calculate the wavelength of the light. (*2 marks*)

4 Make a sketch copy of the graph of the oscillation shown in Figure 3. Add a second waveform of the same amplitude with a phase difference of $\frac{\pi}{2}$ radians. (*2 marks*)

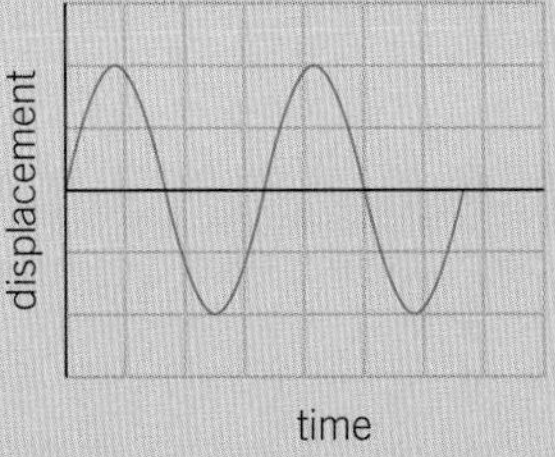


▲ **Figure 3**

5 Figure 4 represents water waves passing through a gap.

a Copy Figure 4 and show the next three wave-fronts in the water beyond the gap. (*2 marks*)

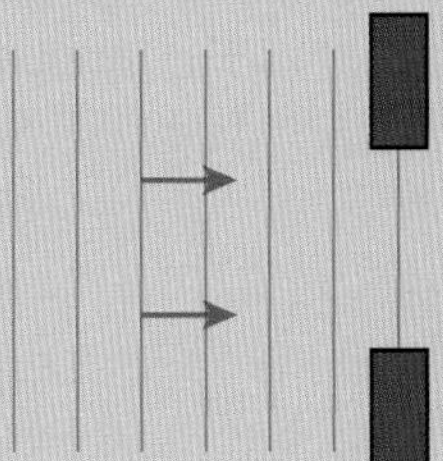

▲ **Figure 4**

b The velocity of the waves is proportional to $\sqrt{d}$ where $d$ is the depth of the water.

The depth of water changes from $d$ to $2d$. Calculate how this change of depth affects the wavelength of the water waves, assuming that the frequency remains constant. (*2 marks*)

c State the effect the change in wavelength has on the pattern of the waves passing through the gap. (*1 mark*)

6 A DVD can be used as a reflection grating. Light of wavelength 670 nm is incident on the DVD. It is reflected from a series of tracks as shown in Figure 5. A first order maximum is observed at 31°.

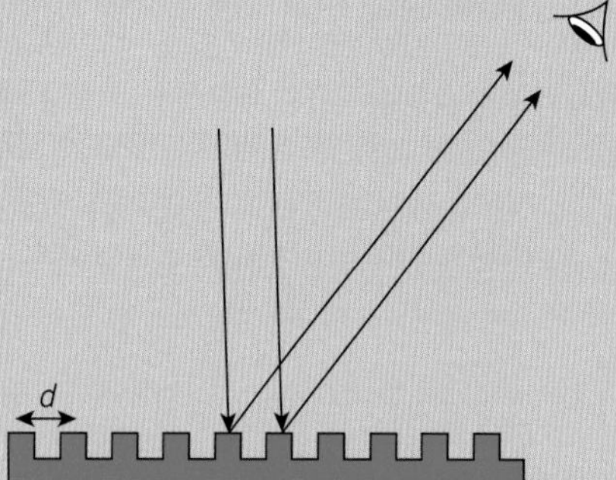


▲ Figure 5

a Use the equation $\sin\theta = n\lambda/d$ to calculate the distance $d$ between tracks. (*2 marks*)

b Explain why only one maximum is observed for light of this wavelength reflecting from the DVD. (*2 marks*)

7 This question is about standing waves in a tube closed at one end, such as a clarinet. When the player blows into the instrument a reed vibrates at the closed end. A standing wave is formed in the tube. The tube is 0.6 m long.

a (i) Copy Figure 6 and indicate the position of the node(s) and antinode(s) for the longest wavelength standing wave. (*1 mark*)

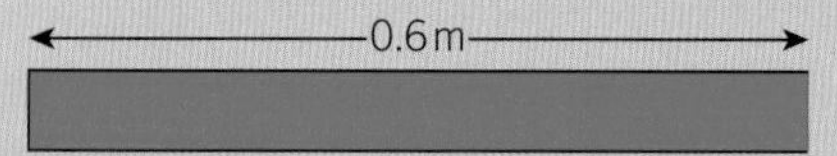


▲ Figure 6

(ii) State the wavelength of the longest-wavelength standing wave. (*1 mark*)

(iii) Explain how standing waves form in a tube. (*3 marks*)

b (i) Draw a diagram indicating the nodes and antinodes in the tube when it produces a note of three times the frequency of the lowest possible note. (*1 mark*)

(ii) The speed of sound at room temperature is about 340 m s$^{-1}$. Calculate the frequency of the sound produced in the tube from the standing wave in **(b)(i)**. (*2 marks*)

c The speed of sound increases with temperature. State and explain the effect this has on the frequency of the note produced by the clarinet. (*3 marks*)

8 This question is about an experiment to determine the wavelength of blue light using a pair of slits. A student makes the following measurements:

Slit separation = $5 \times 10^{-4} \pm 1 \times 10^{-4}$ m
Distance from slits to screen = $2.5 \pm 0.01$ m

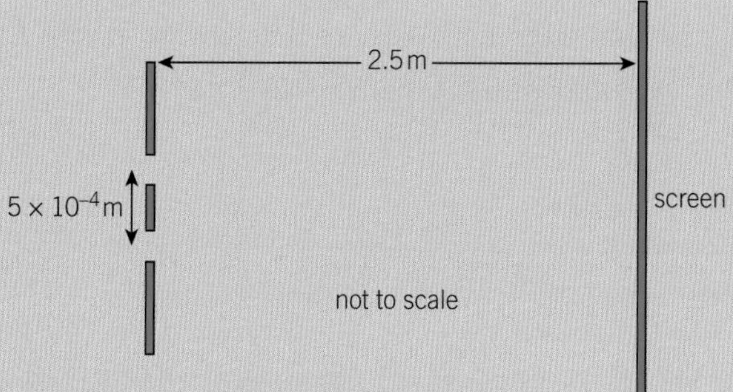


▲ Figure 7

The student measures 10 bright fringes in a distance of 24 ± 1 mm on the screen.

a (i) The student states that she only needs to consider uncertainty in the slit separation when calculating the wavelength of light. Explain her reasoning. (*2 marks*)

(ii) Calculate the wavelength of light using the student's results. Include a calculated estimate of uncertainty in your final answer. (*4 marks*)

The student suggests reducing the uncertainty in the experiment by doubling the slit spacing and doubling the distance between the slits and the screen.

**b** (i) Use your answer to **(a)(ii)** to calculate an estimate of the expected fringe spacing for the new slit separation and slit-screen distance. *(2 marks)*

(ii) Discuss whether these changes have reduced the uncertainty in the value for wavelength and suggest any disadvantages of the new set up. *(3 marks)*

**9** This question is about the superposition of sound. Two speakers, A and B, produce a steady, coherent note. As the microphone moves along the line XY it detects a series of amplitude maxima and minima.

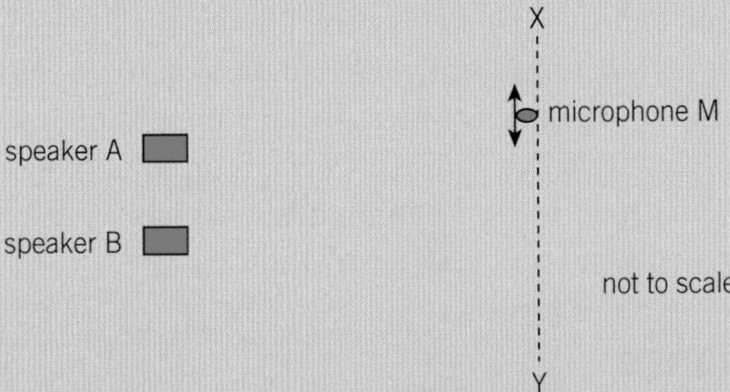


▲ **Figure 8**

**a** Use ideas about path difference and phase difference to explain why the microphone detects a series of maxima and minima. *(4 marks)*

**b** The microphone detects a maximum when path difference BM – AM = 2.4 m. It detects the next maximum when the path difference is 3.2 m. State the wavelength of the sound from the speakers. *(1 mark)*

**c** The speed of sound $v$ in air varies with temperature as $v = k\sqrt{(T+273)}$ where $T$ is the air temperature in °C. Calculate the percentage difference in the velocity of sound when the air temperature rises from 10° C to 20° C and explain the effect this change will have on the pattern of maxima and minima from the speakers. The frequency of the sound is not changed. *(4 marks)*

**10** This question is about standing waves on guitar strings.

Figure 9 shows a guitar whose strings are 0.65 m long.

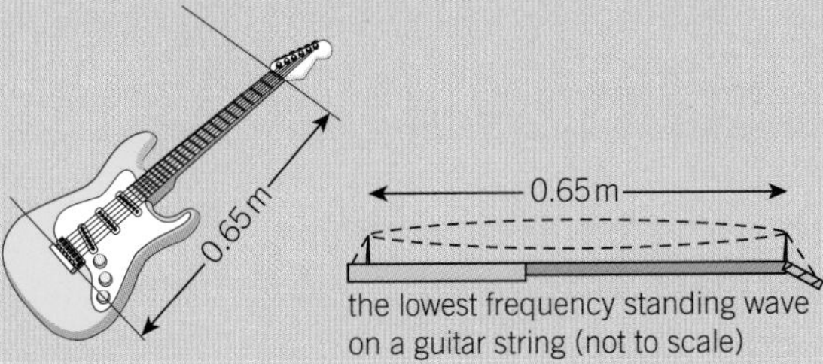


▲ **Figure 9**

**a** Explain why the wavelength of the standing wave shown in Figure 9 is 1.3 m. *(1 mark)*

**b** The lowest frequency standing wave on the thickest guitar string is at 82 Hz.

Show that the speed of the wave travelling along the string is about $100\,\text{m}\,\text{s}^{-1}$. *(2 marks)*

**c** (i) The speed $v$ of waves along a string is given by the equation

$$v = \sqrt{\frac{T}{\mu}}$$

Where $T$ is the tension in the string, and $\mu$ is the mass of a metre length of the string.

Use this equation to calculate the tension $T$ in the thickest guitar string where $\mu = 8.4 \times 10^{-3}\,\text{kg}\,\text{m}^{-1}$. *(2 marks)*

(ii) All strings on the guitar have the same tension and length. Use the equation above to explain why the fundamental frequency of the thinnest string is higher than the fundamental frequency of the thickest string. *(1 mark)*

**d** Explain clearly how waves travelling along a string can produce standing waves on the string *(3 marks)*

*OCR Physics B Paper G492 Jan 2012*

# 7 QUANTUM BEHAVIOUR

## 7.1 Quantum behaviour

Specification reference: 4.1a(vi), 4.1b(i), 4.1c(iv), 4.1 d(vi)

**Learning outcomes**

Describe, explain, and apply:

- evidence that photons exchange energy in quanta
- the equation $E = hf$ applied in light-emitting diodes (LEDs), the photoelectric effect, and line spectra
- the terms: electronvolt, intensity, work function, threshold frequency
- the determination of the Planck constant using different coloured LEDs.

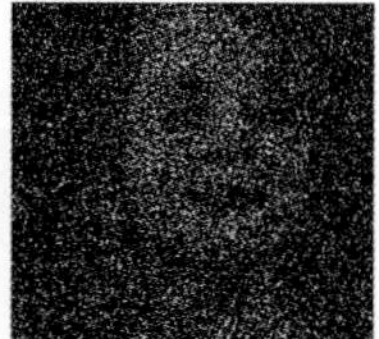

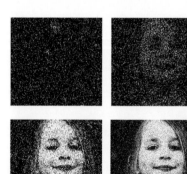

▲ **Figure 1** *Six images representing exposures with increasing numbers of photons. With very little light (few photons), the picture is a random pattern of exposed dots, but it becomes more detailed and recognisable as more photons arrive.*

You have seen that by modelling light as a wave you can explain refraction, diffraction, and interference, but there are some phenomena that cannot be explained using this model.

What do you think happens when you take a photograph in progressively dimmer light? You may think that the photograph will simply appear fainter, but it does not. In fact, the photograph breaks up into randomly arranged exposed patches on the film or digital sensor, suggesting that light is arriving randomly in small packets. Turn up the brightness and the lumps of light arrive at a greater rate, producing the smooth-looking picture you expect. These packets, or quanta of light, are called **photons**. This effect is shown in Figure 1, which represents the same picture taken with successively more photons.

### Photons and chance

The images are a good illustration of how photons hit a detector. Where the picture is bright, there is a good chance of photons arriving at that location. Where it is dim, the chance is low. These are probabilities, not certainties, so it is possible (but not likely) for a photon to appear where the probability is low, or not appear where the probability is high.

A smooth picture is built up through many, many events happening one by one at random with a certain probability. It is a bit like going out in the rain – you cannot tell when or where the next raindrop will fall, but if you stay in the rain you will get wet. Strong sources of electromagnetic radiation are like heavy rain, weak sources are like light rain – the random nature of the arrival of photons (or raindrops) is much more obvious when the rate of arrival is low. If you meet someone whose clothes are spattered with water, you could conclude that they were in a rain shower – meet someone who is soaked from head to foot and you cannot tell if they have been in a very heavy shower of rain (particles) or been hit by a wave!

#### Quanta

The idea that electromagnetic radiation can be emitted in energy quanta, discrete packets of energy, was suggested by the German physicist Max Planck in 1900. Albert Einstein extended the concept in 1905 when he suggested that electromagnetic radiation is not only emitted in energy **quanta**, but is also absorbed as quanta (photons). The relationship is very simple. Electromagnetic radiation of frequency $f$ is emitted and absorbed in quanta of energy $E$ where

$$\text{energy } E \text{ (J)} = \text{Planck constant } h \text{ (J s)} \times \text{frequency } f \text{ (Hz)}$$

$h$, the Planck constant, is equal to $6.6 \times 10^{-34}$ J s.

As speed of light $c = f\lambda$, we can write $f = \frac{c}{\lambda}$ and substitute this into the photon–energy equation to give $E = \frac{hc}{\lambda}$.

### Worked example: Photons from a light-emitting diode

An LED emits blue light of wavelength 470 nm at a power of 250 mW. Calculate the number of photons emitted by the LED per second.

**Step 1:** Identify the correct equation to use.

Since the wavelength is given, and we can find values of $c$ and $\lambda$, we can use the equation $E = \frac{hc}{\lambda}$.

**Step 2:** Substitute values into equation.

$$E = \frac{hc}{\lambda} = \frac{6.6 \times 10^{-34}\,\text{J s} \times 3 \times 10^{8}\,\text{m s}^{-1}}{470 \times 10^{-9}\,\text{m}} = 4.2\ldots \times 10^{-19}\,\text{J}$$

This is the energy of a photon of wavelength 470 nm.

**Step 3:** Identify the energy transferred per second.

Power = energy transfer per second

As power = 250 mW

Energy transfer per second by the LED = $250 \times 10^{-3}$ J

**Step 4:** Calculate the number of photons emitted per second.

$$\text{Number of photons emitted per second} = \frac{\text{total energy transferred per second}}{\text{energy of single photon}}$$

$$= \frac{250 \times 10^{-3}\,\text{J}}{4.2\ldots \times 10^{-19}\,\text{J}} = 5.9 \times 10^{17} \text{ (2 s.f.)}$$

A single LED releases $5.9 \times 10^{17}$ photons each second. This large number helps explain why light appears to be continuous (like a wave motion) rather than a random release of lumps of energy.

## Measuring energy in electronvolts

The joule is not a convenient energy unit to use when discussing the energy of photons or sub-atomic particles, as the energy values considered are so small, as you can see in the example above.

The **electronvolt** (eV) is often used as an alternative unit. One electronvolt is the energy transferred when an electron moves through a potential difference of one volt.

As $W = VQ$, and the magnitude of the charge on an electron is $1.6 \times 10^{-19}$ C

Energy transferred, $W = 1.6 \times 10^{-19}\,\text{C} \times 1\,\text{J C}^{-1} = 1.6 \times 10^{-19}\,\text{J}$

Therefore, $1\,\text{eV} = 1.6 \times 10^{-19}\,\text{J}$.

### Synoptic link

You have met electrical power and the relationship between charge, energy, and potential difference in Topic 3.1, Current, p.d., and electrical power.

### Worked example: Converting from J to eV

A photon has energy $2.4 \times 10^{-19}$ J. What is its energy in eV?

**Step 1:** $1\,\text{eV} = 1.6 \times 10^{-19}\,\text{J}$

**Step 2:** Energy in eV $= \frac{2.4 \times 10^{-19}\,\text{J}}{1.6 \times 10^{-19}\,\text{J eV}^{-1}} = 1.5\,\text{eV}$

## The photoelectric effect and Einstein's equation

One of the unresolved problems in physics at the beginning of the 20th century concerned the photoelectric effect – the emission of electrons when light of a sufficiently high frequency strikes a metal surface.

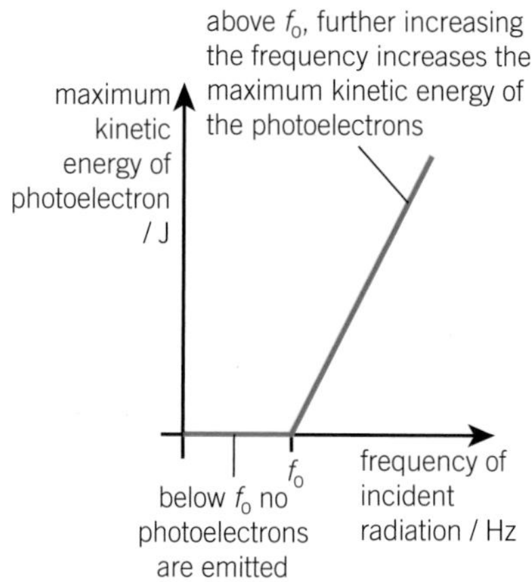


▲ **Figure 2** *Graph showing the relationship between the maximum kinetic energy of the photoelectrons and the frequency of incident light*

The **intensity** of light is the amount of energy transferred per metre squared per second. In everyday language we think of the intensity as the brightness of a light source. The wave model of light suggested that there would be a delay in the emission of photoelectrons from the metal surface when low intensity light strikes the surface. It was also suggested that a more intense source of light striking the metal surface would result in the emission of photoelectrons with greater kinetic energy.

However, it was found that the kinetic energy of the ejected electrons (photoelectrons) was *not* affected by the intensity of the light striking the metal surface. A more intense light produces greater numbers of photoelectrons but the maximum energy of the electrons depends *only* on the frequency of the light. A low intensity source produces fewer photoelectrons, but there is no measurable delay in the emission. If the frequency is lower than a certain **threshold frequency** $f_0$, no electrons are released no matter how bright the light source.

### Einstein's theory

In 1905 Albert Einstein published ideas that changed the way we think about energy and matter. He produced five papers in that year, an incredible achievement considering that he had a full-time job and was, effectively, an amateur scientist at the time. One of the papers gave an explanation for the photoelectric effect, work that led to Einstein receiving a Nobel Prize in 1921. In his ground-breaking work he pictured light as interacting with matter as particles. This allowed him to explain the results of the photoelectric effect in terms of individual photons interacting with individual electrons on the metal surface. This is completely different from the wave model which pictures the energy of the incident light spread evenly over the metal surface.

It is interesting to note that 1905 also saw the publication of possibly the most famous equation in physics: $E = mc^2$, but it was Einstein's work on the photoelectric effect that led to his award of the Nobel Prize.

▲ **Figure 3** *Einstein in 1905*

**Questions**

1 How does Einstein's picture of light as photons interacting individually with electrons explain the observation that no photoelectrons are emitted below a minimum frequency of incident light?

2 Einstein's explanation of the photoelectric effect was accepted by most of the scientific community following very careful experiments by the American physicist Robert Millikan in 1916. Why do you think Einstein's explanation was not accepted by all physicists as soon as it became widely known?

It takes energy to remove an electron from the metal surface. The amount of energy required to do so is known as the **work function** of the metal, $\phi$. Einstein's equation states that the maximum kinetic energy of the ejected electrons is equal to the energy of an individual photon minus the work function of the metal.

$$E_{k(\max)} = hf - \phi$$

When one photon strikes the surface of the metal it transfers energy $hf$ to an *individual* electron. Some of this energy goes to releasing the electron from the surface ($\phi$), any remaining energy is transferred as the kinetic energy of the electron.

**Hint**

The work function is the energy required to remove an electron from the *surface* of the metal. Some photons interact with electrons (a little) below the surface, which require more energy to be released. As we cannot be sure where the electrons emitted came from, we must describe their energies as *maximum* kinetic energy.

## Worked example: The photoelectric effect

The energy required to release an electron from the surface of zinc (the work function of zinc) is $6.9 \times 10^{-19}$ J.

**a** Calculate the minimum frequency of light that will release electrons from the surface.

**Step 1:** Select the appropriate equation.

Minimum photon energy $E = 6.9 \times 10^{-19}\,\text{J} = hf$

**Step 2:** Rearrange the equation and evaluate.

$$f = \frac{E}{h} = \frac{6.9 \times 10^{-19}\,\text{J}}{6.6 \times 10^{-34}\,\text{J s}} = 1.0 \times 10^{15}\,\text{Hz (2 s.f.)}$$

The wavelength of light of this frequency is $3 \times 10^{-7}$ m, in the ultraviolet region of the spectrum.

**b** Ultraviolet light of frequency $1.7 \times 10^{15}$ Hz is incident on the zinc. Calculate the maximum kinetic energy of the photoelectrons emitted from the surface.

**Step 1:** Select the appropriate equation.

$$E_{k(\max)} = hf - \phi$$

**Step 2:** Substitute values and evaluate.

$$\begin{aligned} E_{k(\max)} &= (6.6 \times 10^{-34}\,\text{J s} \times 1.7 \times 10^{15}\,\text{s}^{-1}) - 6.9 \times 10^{-19}\,\text{J} \\ &= 1.122 \times 10^{-18}\,\text{J} - 6.9 \times 10^{-19}\,\text{J} \\ &= 4.3 \times 10^{-19}\,\text{J (2 s.f.)} \end{aligned}$$

An electron with this kinetic energy travels at about $1 \times 10^{6}\,\text{m s}^{-1}$.

### Using the photoelectric effect to determine the Planck constant

The gradient of the graph in Figure 4 is the Planck constant. The work function can be calculated by finding the threshold frequency (the intercept on the frequency axis) and multiplying by the Planck constant (gradient of the graph).

## Worked example: The photoelectric effect

The graph in Figure 4 shows the maximum kinetic energy of photoelectrons emitted from a metal surface when illuminated with light of different frequencies.

**a** Use the graph to determine the Planck constant.

**Step 1:** The Planck constant is the gradient of the line. Select suitable pairs of $x$- and $y$-values (see graph).

**Step 2:** Evaluate gradient.

$$\text{Gradient} = \frac{5.4 \times 10^{-19}\,\text{J} - 0.6 \times 10^{-19}\,\text{J}}{13.2 \times 10^{14}\,\text{Hz} - 6.0 \times 10^{14}\,\text{Hz}} = \frac{4.8 \times 10^{-19}\,\text{J}}{7.2 \times 10^{14}\,\text{Hz}} = 6.7 \times 10^{-34}\,\text{J s (2 s.f.)}$$

Note that this value is not quite the expected value of $6.6 \times 10^{-34}$ J s. This is because of the inaccuracy in reading values from the graph.

**b** Use the accepted value of the Planck constant to calculate the work function of the metal.

**Step 1:** Select the appropriate equation.

Work function = threshold frequency × the Planck constant

**Step 2:** Find the value for the threshold frequency from the $x$-intercept of the graph.

Threshold frequency = $5.0 \times 10^{14}$ Hz

**Step 3:** Substitute and evaluate.

Work function = $6.6 \times 10^{-34}\,\text{J s} \times 5.0 \times 10^{14}\,\text{Hz} = 3.3 \times 10^{-19}\,\text{J}$

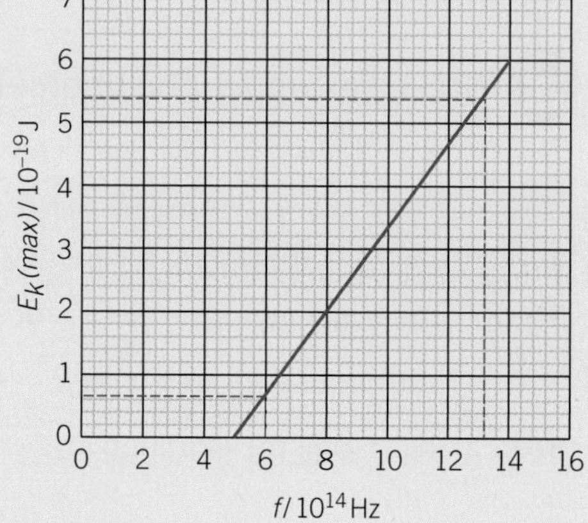


▲ **Figure 4**

## Line spectra

Electrons in atoms have different energies and can gain and lose energy by changing **energy levels**. We can picture these energy levels rather like rungs of a ladder. Atoms release photons when an electron falls from a higher energy level (a higher rung) to a lower level (lower rung). The released photon has energy equal to the difference between the energy levels. Figure 5 shows the six energy differences possible for electrons moving between four energy levels.

As $E = hf$, and $c = f\lambda$, higher energy photons will have higher frequency and therefore shorter wavelength.

When a compound or element is heated, electrons in the atoms move into higher energy levels. Photons are released when the electrons fall back to lower levels. An example of this is materials glowing 'red-hot' when they are heated to a sufficiently high temperature.

Atoms of each element have their own set of energy levels. This means they will emit photons of different energies – light of different wavelengths.

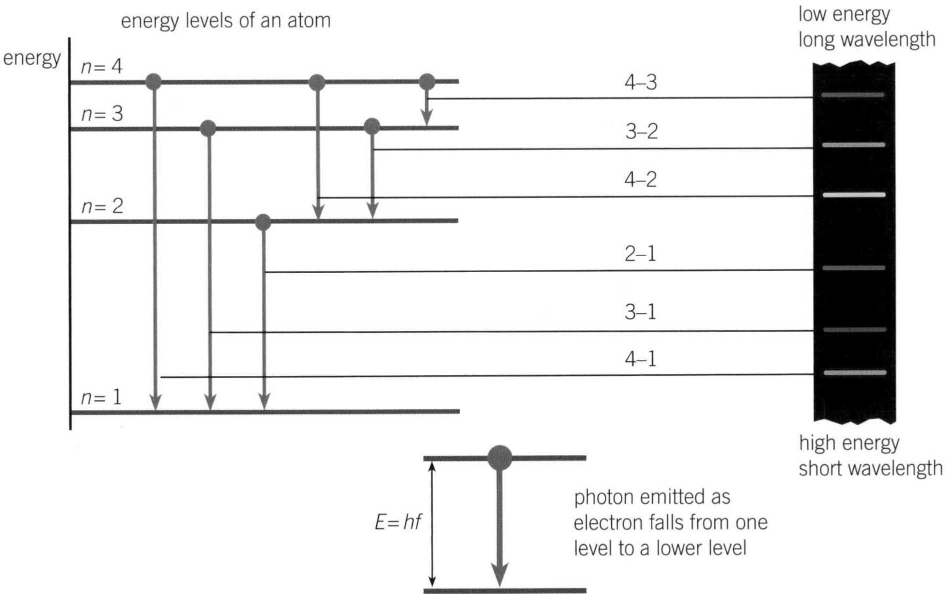


▲ **Figure 5** *Spectral lines and energy levels*

A diffraction grating can be used to spread incident light into its component wavelengths. Line spectra resemble multi-coloured bar codes with each line produced by electrons dropping from a higher energy level to a lower one within the visible spectrum. Atoms from each element have their own pattern of lines – their own bar code.

Figure 6 shows line spectra of several elements. The tungsten filament at the top gives a continuous spectrum which is one reason why many people still prefer 'old-fashioned' filament lamps. The yellow-orange light from sodium gives the familiar colour of many streetlights.

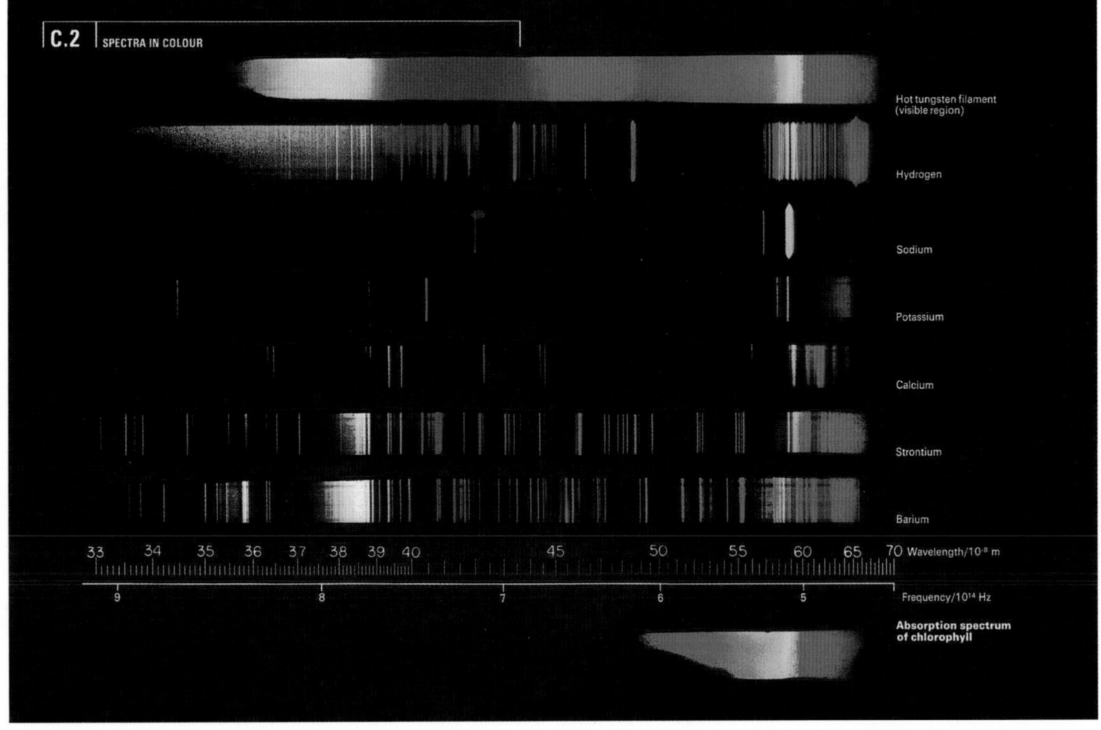


▲ **Figure 6** *Line spectra of several elements*

## Light-emitting diodes (LEDs)

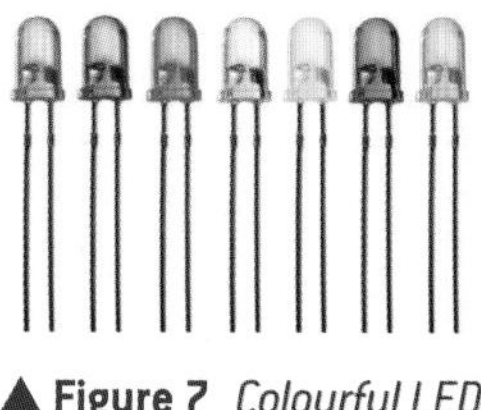

▲ **Figure 7** *Colourful LEDs*

LEDs are semiconducting devices that only allow current to travel in one direction. As an electron crosses from one side of the diode to the other it falls to a lower energy level and emits a photon.

The energy change between levels can be calculated by multiplying the p.d. across the diode when it just begins to glow (the 'striking' potential) by the charge on an electron. This value also depends on the semiconducting material. Diodes made from different semiconducting materials will release photons of different energy, that is, light of different wavelengths.

### Worked example: Calculating the wavelength of light emitted from an LED

A diode using zinc selenide as the semiconductor material has a striking potential of 2.5 V. Calculate the wavelength of the light emitted at this potential difference.

Charge on an electron = $1.6 \times 10^{-19}$ C

**Step 1:** Select the appropriate equation to obtain the energy transfer.

$$\text{Energy change } W = qV$$

**Step 2:** Substitute and evaluate.

$$\text{Energy change} = 1.6 \times 10^{-19}\,\text{C} \times 2.5\,\text{V} = 4.0 \times 10^{-19}\,\text{J}$$

**Step 3:** Select the appropriate equation to find the wavelength of light for a photon energy of $4.0 \times 10^{-19}$ J.

$$E = hf = \frac{hc}{\lambda} \;\therefore\; \lambda = \frac{hc}{E}$$

**Step 4:** Substitute and evaluate.

$$\lambda = \frac{6.6 \times 10^{-34}\,\text{J s} \times 3.0 \times 10^{8}\,\text{m s}^{-1}}{4.0 \times 10^{-19}\,\text{J}} = 5.0 \times 10^{-7}\,\text{m (2 s.f.)}$$

This is in the blue region of the spectrum.

### Practical 4.1d(vi): Determining the Planck constant using LEDs

The striking p.d.s for LEDs emitting different wavelengths of light (photons with different energies) are measured. These measurements need some care – you need to look carefully, ideally in a darkened room, so that you can determine the p.d. at which the LED just begins to glow.

The striking p.d. for each diode is recorded and the energy of the photons emitted is found using the equation

$$\text{Energy of a photon emitted by the LED} = e \times V$$

where $e$ is the charge on an electron and $V$ is the striking p.d. The frequency of the light emitted by each diode is calculated from the wavelengths given by the manufacturers using the equation

$$\text{Frequency} = \frac{\text{wave speed}}{\text{wavelength}} \text{ where wave speed} = 3 \times 10^{8}\,\text{m s}^{-1}$$

A graph of photon energy ($y$-axis) against frequency ($x$-axis) is drawn. The Planck constant can be found from the gradient of the graph.

Although you may be given LEDs of specified wavelengths, they actually emit light over a small *range* of wavelengths. This can add to the uncertainty in the experiment and, together with the difficulty of measuring the striking p.d., should be remembered when considering the uncertainty bars on the graph.

## Summary questions

**1** Single photons of frequencies $f = 3.0$ GHz, $f = 6.0 \times 10^{14}$ Hz, and $f = 1.2 \times 10^{18}$ Hz are absorbed. Calculate the energy transferred in each case. Give your answers in J and eV. *(3 marks)*

**2** Choose one of these phenomena:

Line spectra Photoelectric effect LEDs

Describe and explain your chosen phenomenon using the idea of the exchange of energy in quanta. *(6 marks)*

**3** Radiation of wavelength 320 nm strikes a metal surface, releasing photoelectrons with a maximum kinetic energy of $1.4 \times 10^{-19}$ J. Use the equation $E_{k(max)} = hf - \phi$ to calculate $\phi$ (the work function of the metal) and the threshold frequency. *(4 marks)*

**4** A student investigates the striking potentials of LEDs emitting different colours. Her results are given in Table 1.

▼ **Table 1**

| Wavelength of light / nm | Frequency of photon / $10^{14}$ Hz | Striking potential / V | Photon energy / $10^{-19}$ J |
|---|---|---|---|
| 470 +/− 30 | 6.4 +/− 0.4 | 2.6 +/− 0.2 | 4.2 +/− 0.4 |
| 503 +/− 30 | | 2.5 +/− 0.2 | |
| 585 +/− 30 | | 2.1 +/− 0.2 | |
| 620 +/− 30 | | 2.0 +/− 0.2 | |

**a** Explain why the photon energy = striking potential × charge on an electron. *(1 mark)*
**b** Copy and complete Table 1. *(2 marks)*
**c** Plot a graph of photon energy (J) against photon frequency (Hz). Include uncertainty bars on your graph. *(6 marks)*
**d** Calculate a value for the Planck constant. Include an estimate of the uncertainty of the result and explain how you arrived at this value. *(4 marks)*

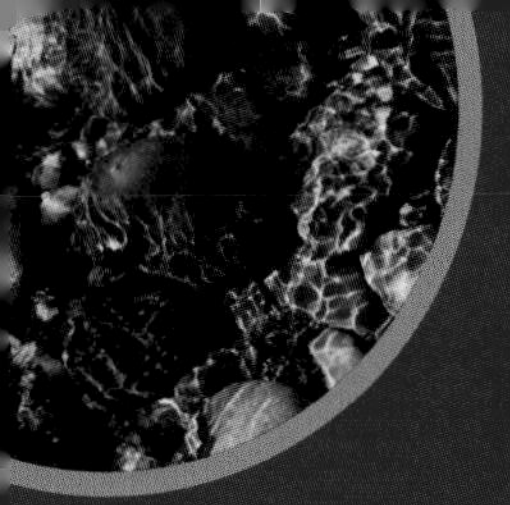

# 7.2 Quantum behaviour and probability

Specification references: 4.1a(vii), 4.1b(i)

**Learning outcomes**

Describe, explain, and apply:

- → quantum behaviour in terms of probability of arrival obtained by adding phasor arrows
- → the terms: probability, intensity.

▲ **Figure 1** *Using digital devices*

Think of modern communication technology and you will be thinking of devices that use the understanding that physicists have gained about quantum behaviour. Televisions, phones, and tablet computers all rely on understanding the behaviour of light and electrons at the quantum level. In this topic we will consider a way of understanding this behaviour.

## The rule for quantum behaviour

Quantum objects such as photons have their own way of behaving – it isn't wave behaviour and it isn't particle behaviour. The rule for quantum behaviour is very simple – explore all possible paths. This behaviour produces wave-like effects such as interference and particle-like effects such as the way images build up. So, how can we predict or explain phenomena such as the fringes observed in Young's double-slit experiment, an experiment that helped develop the wave model of light?

There are rules for the behaviour of photons.

- A photon is emitted by a source and is detected at a certain place and time.
- Imagine the photon taking every possible path from the source to arrive at that place and time. The longer the path, the earlier the time at which the photon will have to be emitted.
- A path includes the emission, travel, and detection of a photon. For each path a combined phasor arrow can be calculated. The phasor arrow can be thought of as rotating with frequency $f = \frac{E}{h}$ (from $E = hf$) for a time equal to the time it takes the photon to travel the length of the path. We call this the *trip time*.
- The angle at which the phasor ends up can be determined for each path.
- Add the phasor arrows for all possible paths, tip-to-tail, to get the resultant phasor.

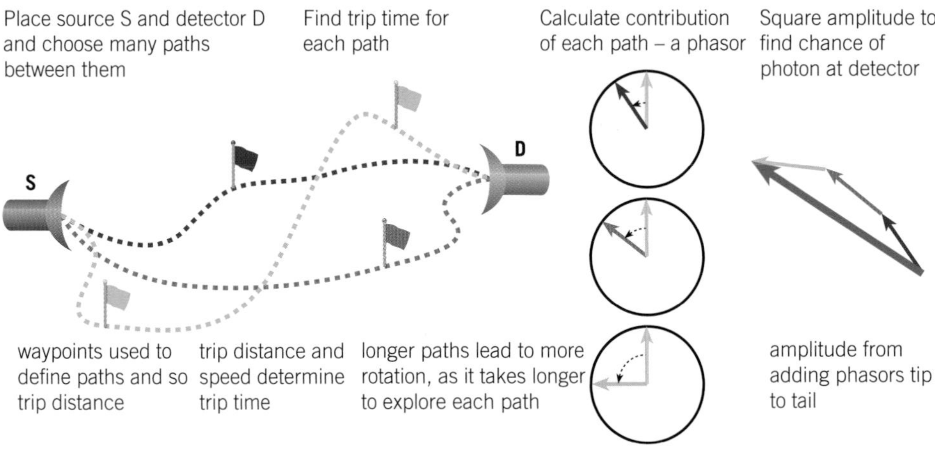
Place source S and detector D and choose many paths between them
Find trip time for each path
Calculate contribution of each path – a phasor
Square amplitude to find chance of photon at detector
S
D
waypoints used to define paths and so trip distance
trip distance and speed determine trip time
longer paths lead to more rotation, as it takes longer to explore each path
amplitude from adding phasors tip to tail
Repeating for more paths produces more accurate calculations. With limited time, choose wisely.

▶ **Figure 2** *Carrying out quantum calculations*

### Worked example: Frequency of a phasor arrow

Calculate the rate of rotation for the phasor arrow associated with a photon of energy $5.4 \times 10^{-19}$ J.

**Step 1:** Select the appropriate equation.

$$f = \frac{E}{h}$$

**Step 2:** Substitute values and evaluate.

$$f = \frac{5.4 \times 10^{-19}\,\text{J}}{6.6 \times 10^{-34}\,\text{J s}} = 8.2 \times 10^{14}\,\text{Hz}$$

## Phasor amplitude, probability, and intensity of light

The length (or amplitude) of the resultant phasor arrow at a particular location gives a measure of the **probability** of a photon arriving at that point – the longer the arrow (larger the amplitude), the higher the probability of arrival. This is an important statement as it shows that where photons end up is not a certainty, but a probability. This explains the random nature of the arrival of photons discussed previously.

Consider these statements about light striking a screen.

**a** The rate at which energy arrives at a point on a screen is the **intensity** of the light at that point.

**b** Intensity is proportional to the probability of the arrival of a photon.

**c** The intensity of light is proportional to the square of the length of the resultant phasor arrow at that point.

Statements **b** and **c** lead to the relationship

*probability of arrival of a photon* $\propto$ *square of the length of resultant phasor arrow*

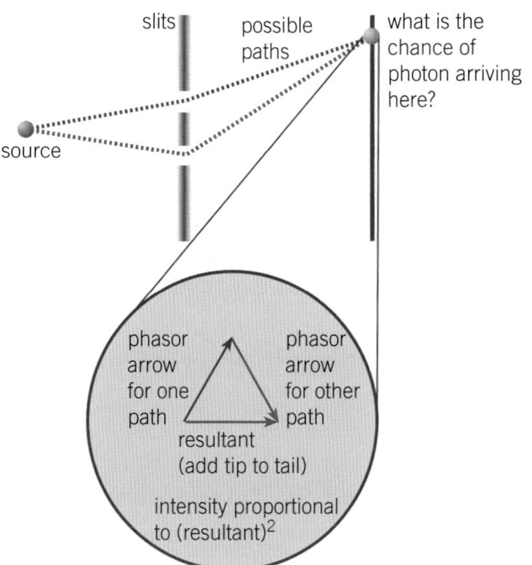


▲ **Figure 3** *Double-slit photon experiment*

### Hint

We are using the term 'probability' to mean how likely it is that a photon will arrive at a particular place during a given time interval. In this section you will not be required to calculate individual probabilities, but to give a method of comparing the likelihood of a photon arriving at one place compared to another.

## Worked example: Phasor arrows and probability

The resultant phasor arrows for the points A and B on the screen of a Young's double-slit experiment are shown in Figure 4. Resultant phasor A is 2 units in length and resultant phasor B is 6 units in length. Calculate the probability of arrival of a photon at point B compared to point A.

**Step 1:** Select the appropriate relationship.

$$\text{probability of arrival} \propto \text{square of the length of resultant phasor arrow}$$

**Step 2:** Substitute values in the relationship.

$$\text{probability of photon arriving at B} \propto 6^2$$
$$\text{probability of photon arriving at A} \propto 2^2$$

**Step 3:** Use a constant of proportionality to turn the relationships into equations.

$$\text{probability of photon arriving at B} = k6^2$$
$$\text{probability of photon arriving at A} = k2^2$$

**Step 4:** Divide the equation for B by the equation for A. The constant of proportionality cancels.

$$\frac{\text{probability of photon arriving at B}}{\text{probability of photon arriving at A}} = \frac{6^2}{2^2} = \frac{36}{4} = 9$$

A photon is nine times more likely to arrive at point B than point A.

If a continuous stream of photons passed through the slits, the fringe at point B would be nine times the intensity of that at A.

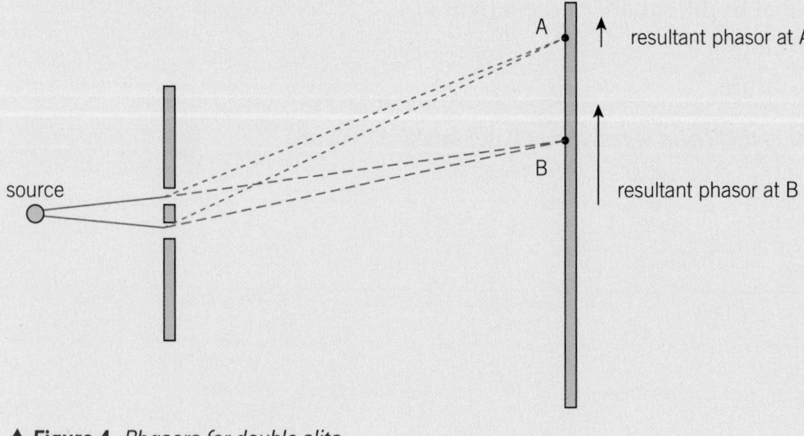


▲ **Figure 4** *Phasors for double slits*

## Examples of explaining wave behaviour using phasors

### Reflection

Photons obey the law of reflection – they come off the surface of a mirror at the same angle at which they reach it. This can be shown using phasors in Figure 5 (next page). By allowing photons to explore all possible paths from the source S, including ones which are obviously 'wrong', each trip will take a different time and affect the number of rotations of each phasor arrow. Longer trip time = more rotations. Adding the phasor arrows together tip-to-tail at the detector D produces the resultant phasor arrow shown.

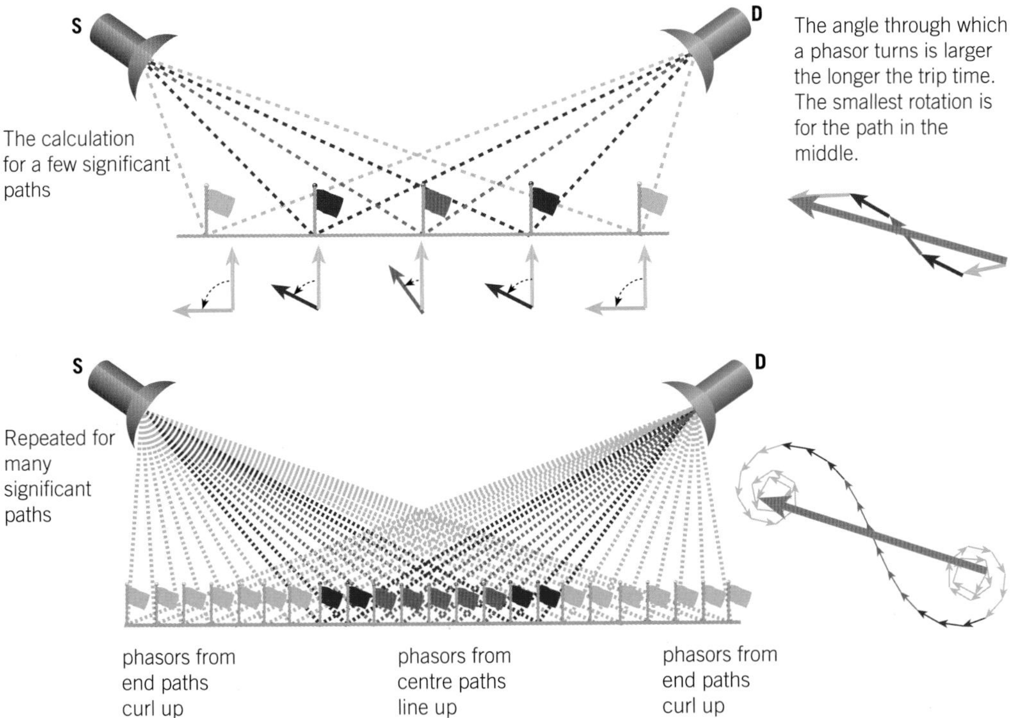
S
D
The calculation for a few significant paths
The angle through which a phasor turns is larger the longer the trip time. The smallest rotation is for the path in the middle.
S
D
Repeated for many significant paths
phasors from end paths curl up
phasors from centre paths line up
phasors from end paths curl up

▲ **Figure 5** *Reflection – explorations over a surface*

Look at the coiled pattern produced when the phasor arrows are added together. The phasors from the ends (following the 'obviously wrong' paths) reach the detector at very different **phases** and tend to curl up. These paths do not contribute much to the resultant amplitude. The paths taken near the equal-angled path (the path that follows the law of reflection) have little phase difference and line up. Most of the probability of photons arriving at the detector comes from these paths. The rules of quantum behaviour predict the law of reflection.

**Synoptic link**

You have met refraction in terms of waves slowing down in glass in Topic 6.2, Light, waves, and refraction.

## Refraction

Photons slow down in glass so a phasor arrow for a photon travelling a given distance in glass will make more rotations than the arrow for a photon travelling the same distance through air.

An example showing how light from a source is focused by a converging lens is shown in Figure 6.

Photons travelling straight through the lens to the detector take the shortest possible route but are slowed down in the glass. Photons taking the longest route (to the tip of the lens in the figure) are not slowed down by the glass. The point at which the lens focuses the light is simply the point at which all the phasor arrows have made the same number of rotations.

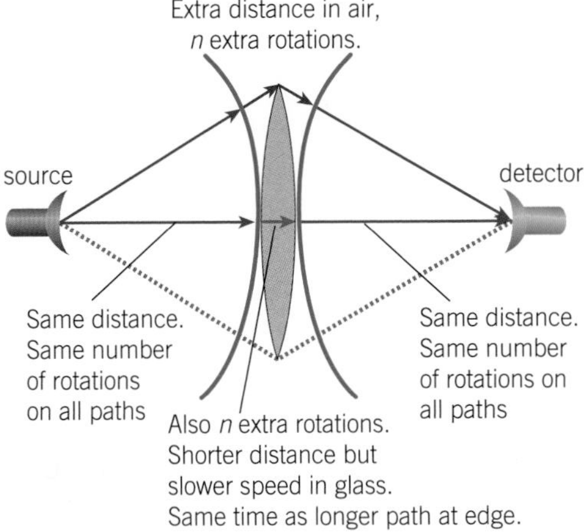
Extra distance in air, *n* extra rotations.
source
detector
Same distance. Same number of rotations on all paths
Same distance. Same number of rotations on all paths
Also *n* extra rotations. Shorter distance but slower speed in glass. Same time as longer path at edge.

▲ **Figure 6** *Refraction with a lens*

**Hint**

Remember that frequency is another way of describing the number of rotations per second.

**Hint**

Remember that $v = \frac{s}{t}$ and the velocity of photons in air is $c$.

## Summary questions

1 Calculate the number of rotations per second for the phasor associated with a photon of energy $3.3 \times 10^{-19}$ J. *(2 marks)*

2 A photon of frequency $6.0 \times 10^{14}$ Hz explores a path 6.0 m long.

**a** Calculate its trip time. *(2 marks)*

**b** Calculate the number of rotations of its phasor arrow during the trip. *(2 marks)*

3 Show how three phasor arrows of equal length can add together to produce a zero resultant. *(1 mark)*

4 A parabolic mirror brings light from a distant source to a sharp focus as shown in Figure 7.

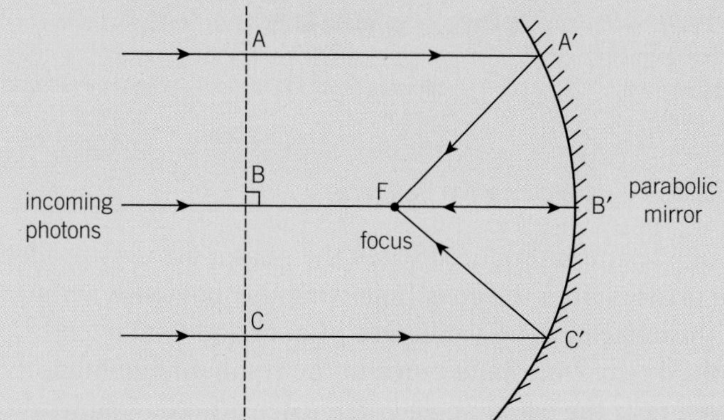


▲ **Figure 7**

**a** State the relationship between the trip times for the photons exploring paths AA′F, BB′F, and CC′F. *(1 mark)*

**b** State the phase relationship at F between photons from the three paths. Explain your reasoning. (Assume that the photons were in phase along line ABC). *(1 mark)*

**c** Use your answers to **a** and **b** to explain why the intensity of reflected light is much lower at positions other than F. *(3 marks)*

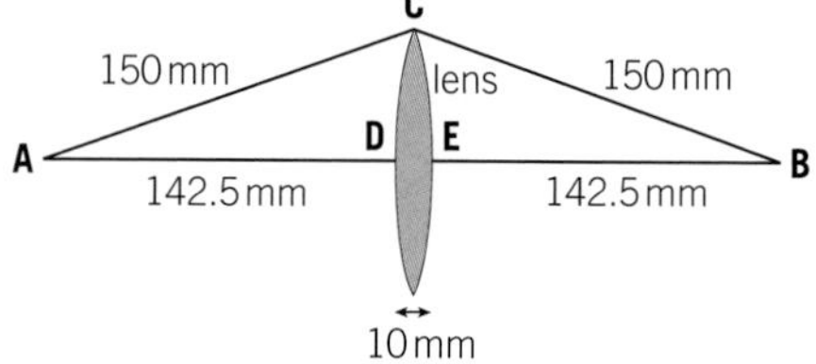


▲ **Figure 8**

5 A converging lens is focusing light, as shown in Figure 8.

**a** Express the speed of light in mm $s^{-1}$. *(1 mark)*

**b** How long does it take light to follow the route ACB? *(1 mark)*

**c** How long must the light spend between D and E in the glass? *(1 mark)*

**d** What is the speed of light in the glass? *(1 mark)*

**e** What is the refractive index of the glass? *(2 marks)*

**f** The phasors following route ACB make the same number of turns as those following route ADEB. Explain why this is the case. *(2 marks)*

# 7.3 Electron diffraction

## Specification references: 4.1a(viii), 4.1c(v)

In 1897 the Cambridge physicist J.J. Thomson showed that so-called 'cathode rays' consist of charged particles. He had discovered the first sub-atomic particle, the electron, a discovery that led to his Nobel Prize in 1906. Thirty-one years later the prize was given for demonstrating that electrons showed wave-like behaviour – that electrons can diffract. The recipients of the 1937 award were the American Clinton Davisson and G.P. Thomson, the son of J.J. Thomson.

### Learning outcomes

Describe, explain, and apply:

→ evidence from electron diffraction that electrons show quantum behaviour

→ the de Broglie relationship
$$\lambda = \frac{h}{p}$$

## Electron diffraction

Davisson and Thomson independently demonstrated that a thin layer of atoms can act as a natural diffraction grating, the two-dimensional array of atoms producing a diffraction pattern of concentric circles. This shows that the electrons are showing wave-like behaviour. Particle behaviour would result in a central bright region with no concentric circles. Experiments show that electrons diffract through slits in the same manner as light.

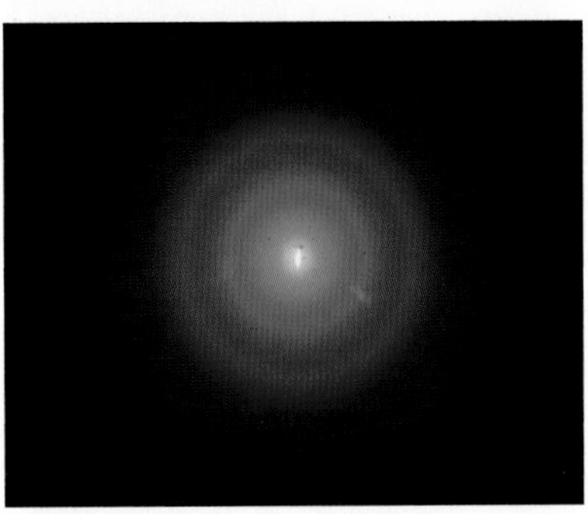

▲ **Figure 1** *Electron diffraction using a thin sheet of graphite*

### de Broglie's equation

In the 1920s the French physicist de Broglie developed an equation which suggested that an electron has a wavelength associated with it. This wavelength $\lambda$ is related to the Planck constant $h$, and the momentum of the electron $p$ by the equation

$$\lambda = \frac{h}{p}$$

and momentum is given as $mv$, where $m$ is the mass of the particle (in this case the electron) and $v$ is its velocity. This equation can therefore be rewritten as

$$\lambda = \frac{h}{mv}$$

### Hint

The Planck constant $h$ is fundamental to picturing the wave-like nature of particles, just as it is fundamental to picturing the particle-like nature of light ($E = hf$).

### Worked example: Calculating the de Broglie wavelength of an electron

An electron has a velocity of $1.5 \times 10^6\,\mathrm{m\,s^{-1}}$. Calculate its de Broglie wavelength (mass of electron = $9.1 \times 10^{-31}\,\mathrm{kg}$).

**Step 1:** Select the appropriate equation – de Broglie wavelength.

$$\lambda = \frac{h}{mv}$$

**Step 2:** Substitute and evaluate.

$$\lambda = \frac{6.6 \times 10^{-34}\,\mathrm{J\,s}}{9.1 \times 10^{-31}\,\mathrm{kg} \times 1.5 \times 10^{6}\,\mathrm{m\,s^{-1}}} = 4.8 \times 10^{-10}\,\mathrm{m}\ (2\text{ s.f.})$$

### Rotating electron arrows

An electron is not exactly like a photon. Electrons have mass and so cannot travel at the speed of light like a photon. Instead, they can travel at different speeds slower than light. However, just like a photon, an electron explores all possible paths between two points and a phasor can be used to track the phase change along every path. Phasors for photons and electrons superpose in the same way.

### Rules for the quantum behaviour of electrons

- An electron of mass $m$ and speed $v$ is emitted by a source and is detected at a certain place and time.
- The longer the path, the earlier the time at which the electron will have to be emitted. This depends on the speed of the electron.
- A path includes the whole process of emission, travel, and detection of an electron. For each path a combined phasor arrow can be calculated. The phasor arrow can be thought of as making one turn for each distance $\frac{h}{p}$ along the path.
- Add up the phasor arrows for all possible paths, tip-to-tail, to get their resultant phasor.
- The probability of detection of an electron can be calculated from the square of the resultant phasor.

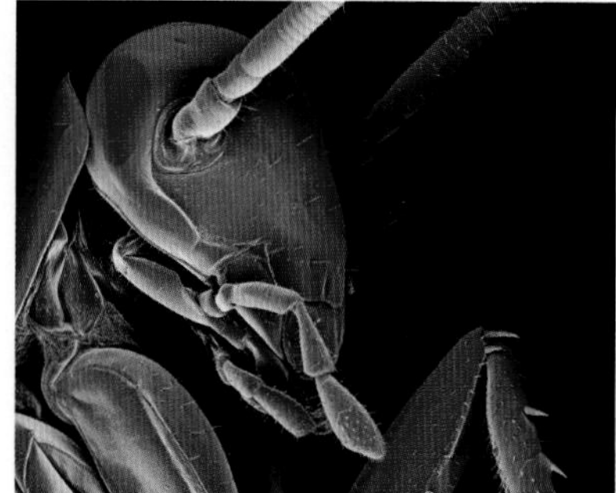

▲ **Figure 2** *Electron microscope image of the head of a cockroach.*

The image of the head of a cockroach in Figure 2 is taken with an electron microscope. Modern electron microscopes rely on the quantum behaviour of electrons.

## Summary questions

1 Calculate the de Broglie wavelength for electrons with momentum $5 \times 10^{-24}$ kg m s$^{-1}$ ($h = 6.6 \times 10^{-34}$ J s). *(2 marks)*

2 Copy and complete the table ($h = 6.6 \times 10^{-34}$ J s, mass of electron = $9.1 \times 10^{-31}$ kg).

| Momentum, $mv$ | Speed, $v$ | Wavelength, $\lambda$ |
|---|---|---|
| | | 10 nm |
| $1.0 \times 10^{-24}$ kg m s$^{-1}$ | | |
| | $2.0 \times 10^{6}$ m s$^{-1}$ | |

*(3 marks)*

3 $3 \times 10^{-19}$ J is roughly the energy of a photon of visible light (wavelength about 600 nm). The momentum of an electron with this energy is about $7 \times 10^{-24}$ kg m s$^{-1}$. Compare, using calculations, the wavelength of the electron to the wavelength of light associated with photons of the same energy. Explain, using your comparison, why electrons need to pass through far smaller gaps than photons for interference effects to be observed. *(3 marks)*

4 The images in Figure 2 suggest that electrons were showing both wave-like and particle-like behaviour in the experiment.

a The image is composed of many discrete spots. State why this suggests that electrons are interacting with the screen like particles. *(1 mark)*

b The image is a series of bright and dark fringes. State why this suggests that the electrons are showing wave-like behaviour. *(1 mark)*

c Explain the pattern on the screen using the phasor model of electron behaviour. *(4 marks)*

# Module 4.1 Summary

## Wave behaviour

### Superposition of waves

- how standing waves are formed
- the terms: phase, phasor, amplitude, superposition
- identify the wavelength of standing waves from diagrams
- practical task — using an oscilloscope to find the frequencies of waves
- practical task — using a rubber cord to demonstrate superposition
- practical task — determining the speed of sound by the formation of stationary waves in a resonance tube

### Light, waves, and refraction

- describe the refraction of light using the wave model and the changes in speed at the boundary between the two media
- the term refractive index, $n$
- Snell's Law: $n = \frac{\sin i}{\sin r} = \frac{\text{speed of light in medium 1}}{\text{speed of light in medium 2}}$
- absolute refractive index $= \frac{\text{speed of light in a vacuum}}{\text{speed of light in medium}}$
- Practical task — determining the refractive index of a transparent block

### Path difference and phase difference

- the terms: interference, coherence, path difference
- practical task — determining the wavelength of microwaves

### Interference and diffraction of light

- diffraction of waves through an aperture, a double slit and a diffraction grating
- using the equation $n\lambda = d \sin \theta$
- practical task — determining the wavelength of light using superposition of light

## Quantum behaviour

### Quantum behaviour

- evidence that photons exchange energy in quanta
- the equation $E = hf$ applied to line spectra, LEDs, and the photoelectric effect
- measuring energy in electronvolts
- the terms: work function, threshold frequency
- practical task — using different coloured LEDs to determine the Planck constant

### Quantum behaviour and probability

- quantum behaviour in terms of the probability of arrival obtained by adding phasor arrows — probability of arrival of a photon found by combining amplitude and phase for all paths
- probability of arrival of a photon $\propto$ square of the length of the resultant phasor arrow
- intensity at a point as the rate at which energy arrives at the point

### Electron diffraction

- electron diffraction as evidence that electrons show quantum behaviour
- the de Broglie relationship $\lambda = \frac{h}{mv}$

# Physics in perspective

## A deeper look at quantum behaviour

### What does it all mean?

A peculiar thing about quantum behaviour is that although everything agrees with experimental results, there is more than one picture of what lies behind the calculations, and no agreement between scientists about which is the 'right' picture. In this chapter we have chosen the story that seems to be the simplest — a story first devised by physicist Richard Feynman.

You must keep in mind that the picture we have painted is not a definitive description of *how things are*. The story of photons following all possible paths is an easy way of remembering how to do the calculations, but it's important to not take this interpretation too literally. For example, don't imagine photons trying paths one at a time, one after the other — if anything, they must be imagined trying all paths at once.

Feynman's "try all paths" idea was not altogether new — Huygens long ago sowed the seeds. Huygens thought of his wavelets as spreading out everywhere they could, and building up the new position of a wave by all arriving at that point in phase. The reason the wavelets are not observed everywhere is that in most places they add up to nothing at all.

### Shocking or not?

People originally found quantum behaviour peculiar. Niels Bohr, one of the founders of quantum theory wrote "Anyone who is not shocked by quantum theory has not fully understood it."

Quantum behaviour can be modelled by combining phasors (amplitude and phase in one spinning arrow) to calculate probabilities. When there are countless numbers of photons in a beam, the probability just determines the brightness, which varies smoothly from place to place. And the brightness varies just as if waves were superposing to give the result, because of the quantum adding-up of phasors. Wave-like behaviour is what you observe when the photons are numerous.

▲ **Figure 1** *Neils Bohr with Einstein. The two men could never agree about the probabilistic nature of the quantum world.*

This argument may feel uncomfortable, as it is explaining something almost relatively simple (wave behaviour) in terms of something more challenging (phasors). However, if we didn't explain anything in this way, then the simplest concepts would never be explained.

### Waves or particles?

The modern story told so far avoids the question of whether electrons and photons are truly waves or particles. Many quantum physics books attempt to scratch the surface of this tricky question, but in doing so only make it itch even more — the best solution at this point is to answer that they are neither waves nor particles. They are simply quantum objects.

So far we have used the electron as the only example of a quantum object with mass, but quantum behaviour doesn't stop with electrons. In 1999, a team working at the University of Vienna showed that 'buckyballs', spherical molecules composed of sixty carbon atoms, can diffract through a suitable grating in the same manner as electrons.

The mass of a buckyball is more than a million times that of an electron, but in carefully controlled conditions buckyballs can behave as quantum objects. In 2011, quantum behaviour was demonstrated in molecules composed of four hundred and thirty atoms with lengths of up to 6 nanometres. Experiments such as these are of more than academic interest — they may well help towards the development of quantum computers.

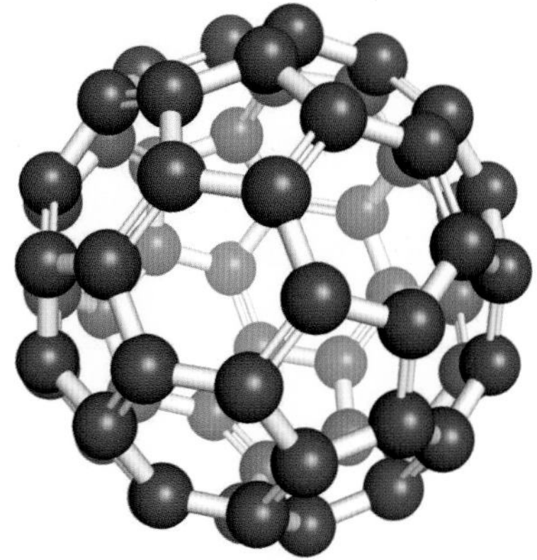

▲ **Figure 2** *A representation of a C-60 or carbon fullerene molecule, affectionately known as a buckyball*

## Summary questions

**1** It is suggested that Huygen's idea of wavelets superposing has similarities to the Feynman picture of phasors exploring all possible paths. Explain what these similarities are, comparing the addition of wavelets to the adding of phasor arrows. You can include diagrams in your answer.

**2** This question is about the quantum behaviour of buckyballs.

**a** Calculate the de Broglie wavelength of a buckyball of mass $1.2 \times 10^{-24}$ kg moving with a velocity of 220 m $s^{-1}$.

**b** Buckyballs of velocity 220 m $s^{-1}$ pass through a grating with slit separation of $1 \times 10^{-7}$ m. Show that the separation between interference maxima at a detector 1.3 m away from the grating is about $3 \times 10^{-5}$ m.

The diameter of a buckyball is about 1 nm.

**c** Suggest why the interference maxima and minima will blur together when the separation between maxima at the detector is less than this diameter.

**d** Calculate the velocity of the buckyballs at which the maxima and minima blur together.

## Practice questions

1 An electron travelling at $1 \times 10^6\,\text{m s}^{-1}$ has a de Broglie wavelength $\lambda_1$.

A proton has a mass of roughly 1800 times that of an electron.

What is the de Broglie wavelength of a proton travelling at $1 \times 10^6\,\text{m s}^{-1}$?

A $\frac{\lambda_1}{1800}$ B $1800\,\lambda_1$

C $\frac{1800}{\lambda_1}$ D $\lambda_1\sqrt{1800}$

2 A laser emits light of wavelength 400 nm. The energy is radiated at a rate of 18 mW ($18\,\text{mJ s}^{-1}$). Calculate the number of photons emitted per second by the laser. *(4 marks)*

3 An electron has kinetic energy = 1.6 keV.

Use the equation kinetic energy $= \frac{\text{momentum}^2}{2m}$ to calculate the momentum of the electron and use your value to calculate its de Broglie wavelength. *(4 marks)*

4 The energy required to release an electron from the surface of magnesium is 3.7 eV. Radiation of wavelength 170 nm is incident on the surface of the metal. Calculate the maximum kinetic energy of the photoelectrons released. *(3 marks)*

5 Calculate the velocity of an electron with a de Broglie wavelength of $6.6 \times 10^{-10}$ m. *(3 marks)*

6 This question is about photons and phasors. Figure 1 shows two paths of photons from slits $S_1$ and $S_2$ to a point on the screen $P_1$. The path difference $S_2P_1 - S_1P_1$ is $\frac{\lambda}{3}$ where $\lambda$ is the wavelength of the light.

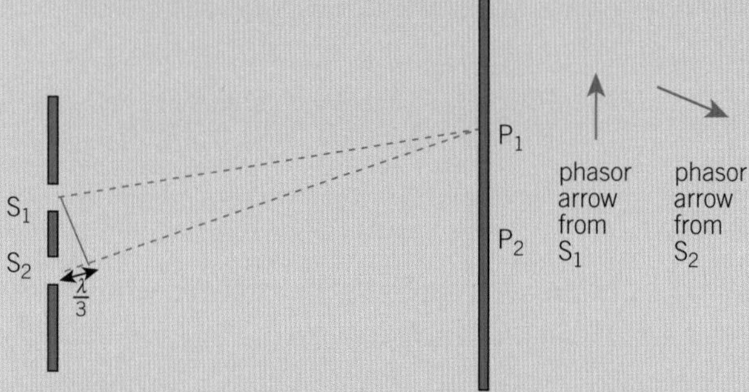


▲ Figure 1

a The angles of the phasor arrows arriving at $P_1$ from slits are shown on the right hand side of the diagram. State why the phase difference is $\frac{2\pi}{3}$ radians. *(1 mark)*

b Draw a scale diagram to find the length of the resultant phasor arrow. *(2 marks)*

c Phasors meet at P2 with no phase difference. Calculate the ratio $\frac{\text{length of the resultant arrow at } P_2}{\text{length of resultant arrow at } P_1}$. *(2 marks)*

d Calculate the ratio $\frac{\text{probability of photon arriving at } P_1}{\text{probability of photon arriving at } P_2}$. *(2 marks)*

e A third slit is added, as shown in Figure 2.

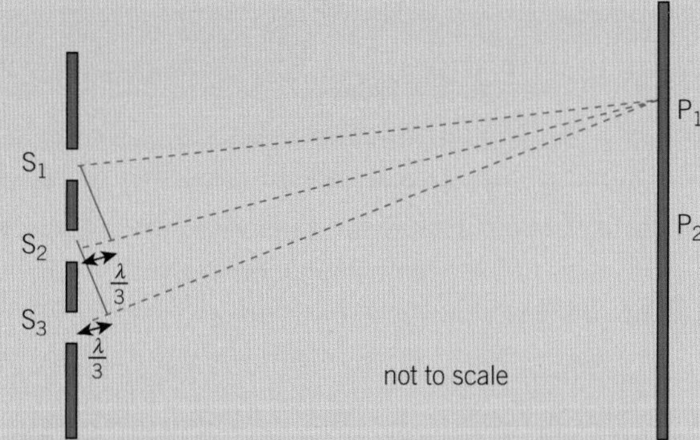


▲ Figure 2

Explain the effect this has on the probability of a photon arriving at P1. You may include a diagram of the phasor arrows at P1 in your answer. *(3 marks)*

7 This question is about the photoelectric effect. When light above the threshold frequency $f_0$ is incident on a metal surface, photoelectrons are emitted. The maximum kinetic energy of the photoelectrons is given by Einstein's equation:

$E_{k(max)} = hf - \phi$

where $\phi$ is the minimum energy required to remove an electron from the surface of the metal and $f$ is the frequency of the light incident on the surface.

Red light of frequency $4.5 \times 10^{14}$ Hz is incident on a metal surface. The maximum energy of the ejected photoelectrons is 0.2 eV.

**a** State how the number and energy of the emitted photoelectrons will change when the intensity of the light incident on the surface is doubled. Explain your answer in terms of photons interacting with electrons in the surface of the metal. *(4 marks)*

**b** Violet light of frequency $7.5 \times 10^{14}$ Hz ejects electrons with maximum energy of 1.4 eV. Use the data for red and violet light to calculate a value for the Planck constant $h$. *(4 marks)*

**c** Calculate the minimum frequency for the release of photoelectrons from this surface. *(2 marks)*

**8** In a simple wave model to explain the diffraction of waves at a gap, the gap of width $b$ is divided into three equal parts as shown in Figure 3a.

The centre of each part is treated as a source of waves.

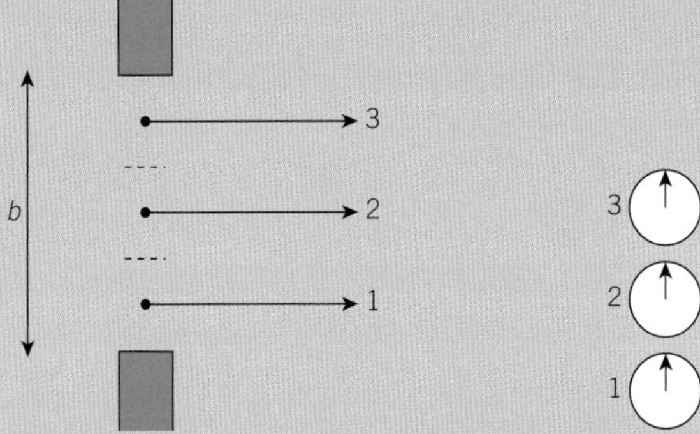


▲ Figure 3a ▲ Figure 3b

**a** The phasors for the waves from each of the three parts of the gap reaching a distant screen in the straight-on direction are shown in Figure 3b.

(i) The paths taken by the waves in Figure 3a are all equal in length. Explain how the phasors in Figure 3b confirm this. *(1 mark)*

(ii) Each phasor has an amplitude $A$. Write down the amplitude of the resultant phasor at the distant screen. *(1 mark)*

**b** At an angle $\theta$ to the straight-on direction, the path difference between neighbouring paths is $\Delta x$, as show in Figure 4a. For one particular value of $\theta$, the resultant intensity is **zero**.

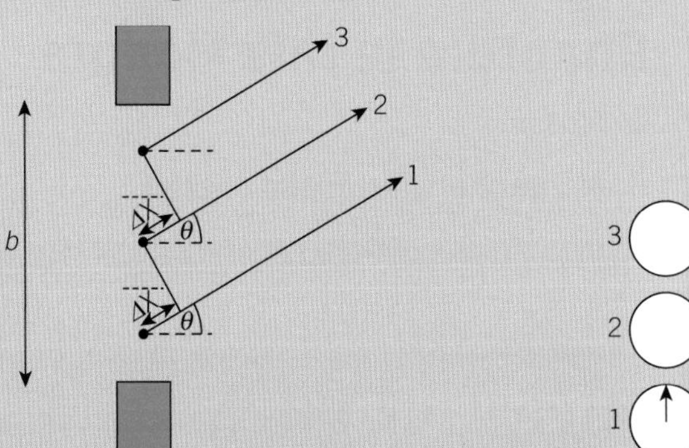


▲ Figure 4a ▲ Figure 4b

(i) Explain why the phasor for path 1 has rotated 120° more than the phasor for path 2, when $\Delta x = \frac{1}{3}\lambda$, where $\lambda$ is the wavelength of the waves. *(2 marks)*

(ii) Draw arrows on a copy of Figure 4b to represent the phasors for waves 2 and 3. Explain, using a diagram, why the three phasors have a zero resultant.

Label your phasors in the diagram 1, 2, and 3.

(iii) Use Figure 4a and the fact that $\Delta x = \frac{1}{3}\lambda$ to show that $\lambda = b\sin\theta$ where $b$ is the total width of the gap. Show your working clearly. *(2 marks)*

**c** Use the equation $\lambda = b\sin\theta$ to calculate the angle $\theta$ at which a minimum signal occurs when microwaves of wavelength 2.4 cm are incident on a gap of width 6.0 cm. *(2 marks)*

*OCR Physics B Paper G492 Jan 2011*

# MODULE 4.2

## Space, time, and motion

### Chapters in this Module



### Introduction

Newtonian dynamics, which is the study of the relationships between force and motion, forms the basis of all of Physics. In this module you will learn how physicists analyse and measure motion. You will see how the concepts of momentum, force, and energy provide a framework to explain phenomena, from the tiny interactions of molecules in a gas to the vast movements of stars and planets.

**Motion** develops the mathematical methods with which motion is described and analysed. Starting from graphs and definitions met at GCSE, the relationships between displacement, velocity, and acceleration are developed. Skills covered include the interpretation of gradients and areas under s-t and *v-t* graphs. Other analytical approaches to motion involve the use of vectors and the algebra of the kinematic equations of motion. Iterative modelling is introduced as a method of analysing change, preparing for its later use in both AS and A level work.

**Momentum, force and energy** is a logical consequence of chapter 8. Chapter 8 is about the *description* of motion — chapter 9 is about its *explanation*. The principle of conservation of momentum is used to analyse collisions and explosions. These concepts lead to Newton's laws of motion and force, providing the underlying concepts that explain changes in motion and the path of projectiles. This chapter will also cover work and the principle of conservation of energy, including calculations in kinetic energy and gravitational potential energy. Power is introduced as the rate at which work is done, or the rate at which energy is transferred.

## Knowledge and understanding checklist

**From your Key Stage 4 study you should be able to do the following. Work through each point, using your Key Stage 4 notes and the support available on Kerboodle.**

- ☐ Relate changes and differences in motion to appropriate distance-time and velocity-time graphs.
- ☐ Apply formulae relating distance, time, and speed for uniform motion, and for motion with uniform acceleration.
- ☐ Recall examples of ways in which objects interact (e.g., by gravity, by contact) and describe how such examples involve interactions between pairs of objects that produce a force on each object.
- ☐ Apply Newton's first law to explain the motion of objects and apply Newton's second law in calculations relating force, mass, and acceleration.
- ☐ Use vector diagrams to illustrate resolution of forces, a net (resultant) force, and equilibrium situations.
- ☐ Use the relationship between work done, force, and distance moved along the line of action of the force and describe the energy transfer involved.
- ☐ Explain the definition of power as the rate at which energy is transferred.
- ☐ Calculate energy efficiency for any energy transfer, and describe ways to increase efficiency.

## Maths skills checklist

**In this unit, you will need to use the following maths skills. You can find support for these skills on Kerboodle and through MyMaths.**

- ☐ **Change the subject of an equation, including nonlinear equations**, such as when using kinematic equations, or calculations involving Newton's second law and energy transfers.
- ☐ **Use an appropriate number of significant figures** in any calculation using data expressed to a certain number of significant figures.
- ☐ **Plot two variables from experimental or other data and use $y = mx + c$** when analysing the motion of falling objects.
- ☐ **Determine a rate of change from the gradient from a graph (including tangents)** in order to calculate velocities and accelerations.
- ☐ **Understand the possible physical significance of the area between a curve and the *x*-axis, and be able to calculate it or estimate it by graphical methods**, such as with curved displacement-time or force-time graphs.

# 8 MOTION

## 8.1 Graphs of motion

Specification reference: 4.2a(vii), 4.2b(i), 4.2b(ii), 4.2d(i)

### Learning outcomes

Describe, explain, and apply:

- how displacement, velocity, and acceleration can be measured
- the terms: displacement, speed, velocity, acceleration
- graphs of accelerated motion, including gradient and area under graph
- how the motion of trolleys can be investigated using light gates.

### Synoptic link

You will learn more about directional movement in Topic 8.2, Vectors.

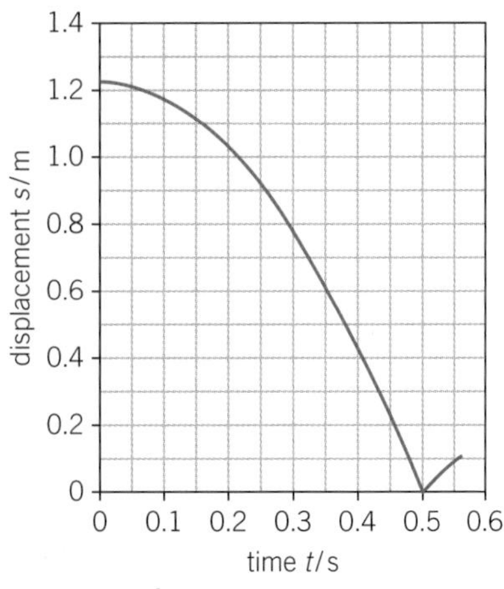


▲ Figure 1 *A dropped ball*

## Displacement and velocity

Imagine you go out for a run. How can you possibly have run a displacement of 0 m?

The terms *displacement* and *velocity* may seem to be just formal scientific words with exactly the same meaning as *distance travelled* and *speed*, but as you shall see later they give extra information about the direction of movement. In this case, the distance you travelled is the amount of ground you covered. The **displacement** is how far you are from your starting point, and in which direction. This would of course be zero if you end up back home.

In the same way, **speed** is just how fast you are going, whereas **velocity** also specifies the direction.

## Displacement–time graphs

The graph in Figure 1 shows the displacement of a ball dropped onto the floor. Displacement is measured from the floor, with the direction up being positive.

The graph shows the timeline of the fall. The ball was released at time 0 s, when its displacement, just over 1.2 m, is the height from which the ball was dropped. In the first 0.1 s the displacement changes quite slowly as the line is nearly horizontal. By the time the ball reaches the floor (displacement = 0 m), the steepness of the line shows that displacement is changing rapidly. The ball then bounces, but the rebound line is less steep, showing that it is moving more slowly.

To calculate the average velocity, you use the equation

$$\text{average velocity } v\ (\text{m s}^{-1}) = \frac{\text{change in displacement } s\ (\text{m})}{\text{total time taken } t\ (\text{s})}$$

A constant velocity is shown on a displacement–time graph by a straight line and the velocity is given by the gradient of the line. In Figure 1 the velocity is changing all the time, so you need to draw a tangent to the curve at the time when you need to find the velocity.

### Delta notation

Changes in the value of a parameter are common in physics, particularly in this chapter, so we will use the standard shorthand of the Greek letter delta ($\Delta$).

$\Delta s$ (read it as delta-es) means the change in $s$. If $s$ changes from 2.10 m to 1.64 m, then $\Delta s = -0.46$ m.

### Worked example: Calculating velocity from a displacement–time graph

Calculate the velocity of the ball at time $t = 0.20\,\text{s}$ in Figure 1.

**Step 1:** Draw a tangent to the curve at $t = 0.20\,\text{s}$.

**Step 2:** Draw the gradient triangle (the red lines in Figure 2).

Using such a large triangle reduces the percentage uncertainty in the readings of the change in $s$, $\Delta s$, and the change in $t$, $\Delta t$.

**Step 3:** Read $\Delta s$ and $\Delta t$ from the triangle.

$\Delta s$ = final $s$ − initial $s$ = 0.22 m − 1.40 m = −1.18 m

$\Delta t$ = final $t$ − initial $t$ = 0.60 s − 0.02 s = 0.58 s

**Step 4:** Calculate the velocity.

$$v = \text{gradient of tangent} = \frac{\Delta s}{\Delta t} = \frac{-1.18\,\text{m}}{0.58\,\text{s}} = -2.034\ldots\,\text{m}\,\text{s}^{-1}$$

**Step 5:** Round the answer to an appropriate number of significant figures, and do not forget the units.

$v = -2.03\,\text{m}\,\text{s}^{-1}$ (2 s.f.)

Two significant figures is acceptable in the answer – if you round to 1 s.f. you are losing information, and using 3 s.f. is not justified because you cannot read the graph to that level of precision.

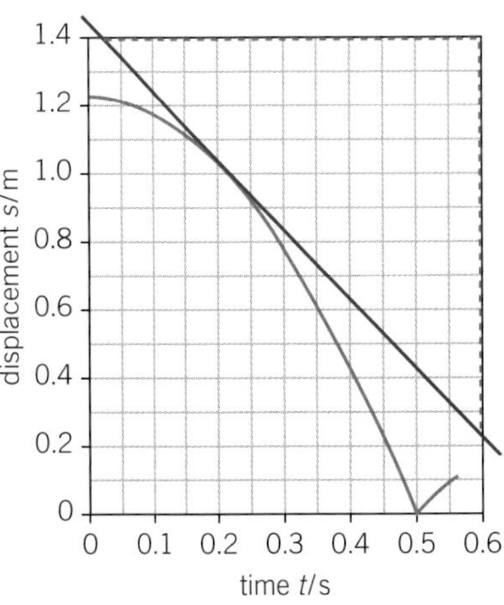


▲ **Figure 2** *Calculating velocity from a displacement–time graph*

**Hint**

$\Delta s$ and $v$ are both negative because the ball is falling downwards, and upwards has been defined as being positive.

**Hint**

**Tangents and gradients**

If you need to draw a tangent to a graph to find the gradient at a point, always draw as long a tangent as possible.

## Velocity–time graphs

Figure 3 shows a journey with varying velocity.

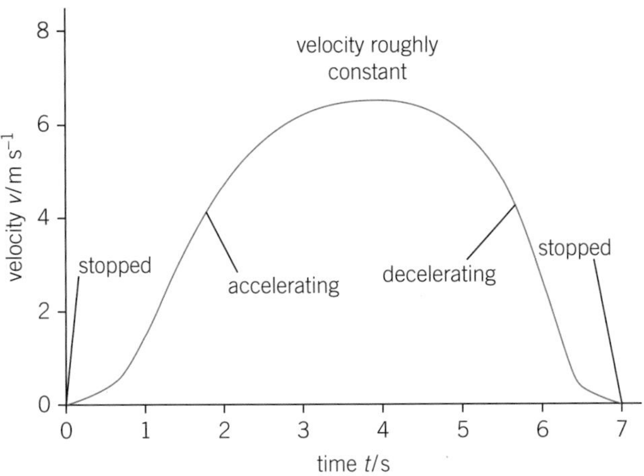


▲ **Figure 3** *Journey with varying velocity*

This graph shows the timeline for the journey of an object travelling in a straight line, and we can also use it to calculate the displacement. If the direction changed you would never be able to work out what the displacement was, as the object could have ended up back where it started.

You can see that the mean velocity is more than $2\,\mathrm{m\,s^{-1}}$ and less than $6\,\mathrm{m\,s^{-1}}$, probably between $3\,\mathrm{m\,s^{-1}}$ and $4\,\mathrm{m\,s^{-1}}$, so you can estimate the displacement by assuming a mean velocity of $3.5\,\mathrm{m\,s^{-1}}$ for the 7 s:

$$v = \frac{s}{t} \Rightarrow s = \text{mean velocity} \times \text{time} = 3.5\,\mathrm{m\,s^{-1}} \times 7\,\mathrm{s} = 24.5\,\mathrm{m}$$

But how can you use the graph to find the displacement more accurately? Figure 4 shows one tiny part of this journey when the velocity was $4.0\,\mathrm{m\,s^{-1}}$.

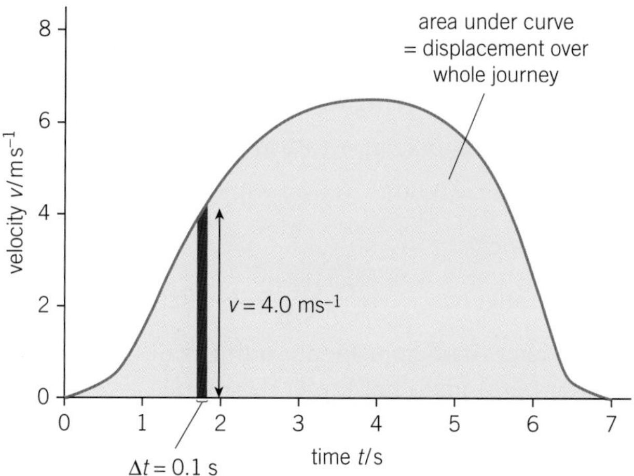


▲ **Figure 4** *Using a velocity–time graph to find displacement*

The area of the strip in Figure 4 = height of strip × width of strip

$= v \times \Delta t$

$=$ distance $s$ travelled in the short time $\Delta t$

$= 4.0\,\mathrm{m\,s^{-1}} \times 0.1\,\mathrm{s} = 0.4\,\mathrm{m}$

If a thin strip like this is drawn every 0.1 s over the whole graph from the beginning to the end of the motion, then adding all the areas together (the total area under the curve) gives the total distance travelled.

You can (correctly) observe that the thin strips do not exactly fit under the graph, as they make the curve into a saw-like edge. The solution – make the strips thinner, for example, 0.01 s, 1 ms, or 1 ns. This increases the accuracy of the estimate but it just takes longer to calculate for the increased number of strips.

In exactly the same way as the area under this velocity–time graph gives the displacement (providing the direction has stayed the same), so the area under a speed–time graph gives the distance travelled. Here a change in direction makes no difference.

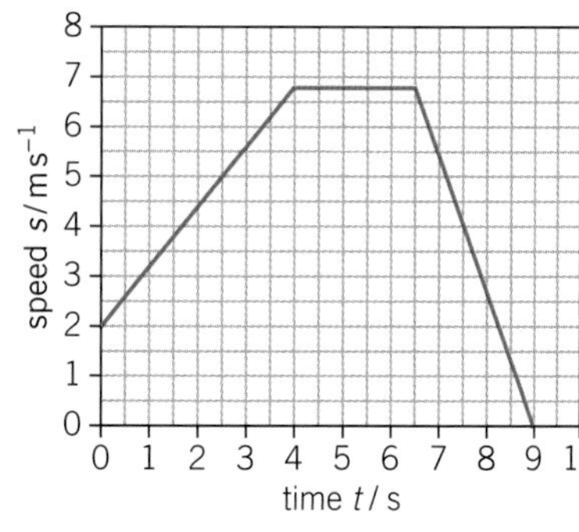


▲ **Figure 5** *Using a speed–time graph to find distance travelled*

### Worked example: Finding distance travelled from a speed–time graph

Find the distance travelled in the 9 s journey described by Figure 5.

**Step 1:** Divide the polygon into shapes whose area can be easily calculated. There are usually several ways to do this. Figure 6 shows one possibility.

**Step 2:** Sum the areas of all the shapes.

Total area = area 1 + area 2 + area 3 + area 4

$= (2\,\text{m}\,\text{s}^{-1} \times 4\,\text{s}) + \left(\frac{1}{2} \times [6.8 - 2.0]\,\text{m}\,\text{s}^{-1} \times 4\,\text{s}\right)$

$+ (6.8\,\text{m}\,\text{s}^{-1} \times 2.5\,\text{s}) + \left(\frac{1}{2} \times 6.8\,\text{m}\,\text{s}^{-1} \times 2.5\,\text{s}\right)$

$= 8\,\text{m} + 9.6\,\text{m} + 17\,\text{m} + 8.5\,\text{m} = 43.1\,\text{m}$

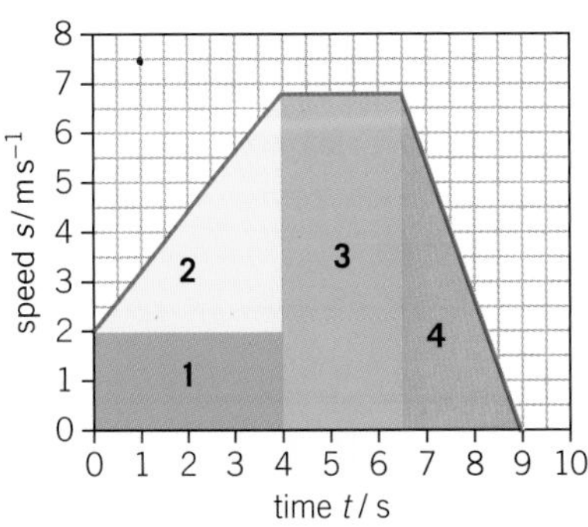


▲ Figure 6 *Speed–time graph divided into regular shapes*

## Calculating acceleration

In Figure 3, increasing velocity and decreasing velocity were described as **acceleration** and deceleration. These need definition to allow calculation.

$$\text{Just as velocity } (\text{m}\,\text{s}^{-1}) = \frac{\text{change in displacement (m)}}{\text{time taken (s)}}$$

$$\text{so acceleration } a\ (\text{m}\,\text{s}^{-2}) = \frac{\text{change in velocity } \Delta v\ (\text{m}\,\text{s}^{-1})}{\text{time taken } \Delta t\ (\text{s})}$$

$$a = \frac{\Delta v}{\Delta t}$$

In both cases, these are given by the gradient of a graph. $v = \frac{\Delta s}{\Delta t}$ is the gradient of the $s$–$t$ graph, and $a = \frac{\Delta v}{\Delta t}$ is the gradient of the $v$–$t$ graph.

**Hint**

Take care to include the units on the axes. This will help avoid *power of 10* errors by missing km, ms, and so on.

**Hint**

Remember that area under a *v-t* graph is in m, not $\text{m}^2$, because the units multiply like this: $\text{m}\,\text{s}^{-1} \times \text{s} = \text{m}$.

### Worked example: Using a velocity–time graph to calculate acceleration

Calculate the acceleration at time $t = 1.75\,\text{s}$ for Figure 3.

**Step 1:** Since the *accelerating* portion of the graph is not straight, a tangent must be drawn to find the acceleration.

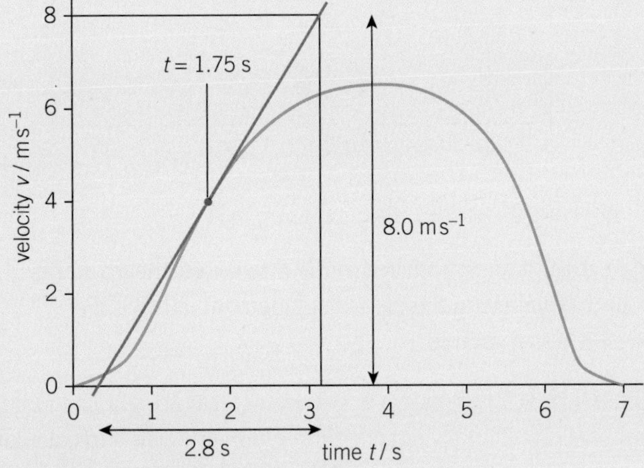


▲ Figure 7 *Acceleration*

**Step 2:** Draw a large gradient triangle and read the values of $\Delta v$ and $\Delta t$.

$\Delta v = 8.0 - 0 = 8.0\,\text{m}\,\text{s}^{-1}$ and $\Delta t = 3.1 - 0.3 = 2.8\,\text{s}$

## Hint

*Negative acceleration* is an alternative expression for deceleration.

**Step 3:** Calculate the acceleration by substituting appropriate values in the equation $a = \frac{\Delta v}{\Delta t}$, rounding the answer to an appropriate number of significant figures.

$$a = \frac{\Delta v}{\Delta t} = \frac{8.0\,\text{m s}^{-1}}{2.8\,\text{s}} = 2.857\ldots\,\text{m s}^{-2} = 2.9\,\text{m s}^{-2}\ (2\text{ s.f.})$$

### Practical 4.2d(i): Investigating the motion of a trolley using light gates

Figure 8 shows a dynamics trolley on a ramp tilted at an angle $\theta$.

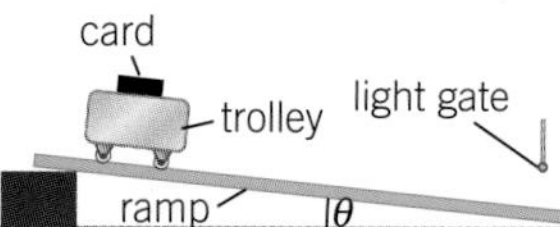


▲ **Figure 8** *Trolley on a ramp*

When released, the trolley will accelerate down the slope. Measurements of displacement $s$ and time $t$ from release are straight-forward, but measuring velocity $v$ is harder. The method shown in Figure 8 is to use a light gate containing an infrared beam that is interrupted by the card fastened to the moving trolley. The light gate is connected to a computer/data logger which will measure the time the card takes to pass through the beam ($\Delta t$). This then allows for the calculation of $v$ using $v = \frac{\Delta s}{\Delta t}$, where $\Delta s$ is the width of the card.

The motion of objects can also be investigated using ticker timers, data loggers, or video analysis.

## Summary questions

**1** A car moving along a straight road accelerates from rest at a uniform rate for 5 seconds, travels at a constant velocity for 5 seconds, and decelerates uniformly to rest over the next 10 seconds. Sketch the velocity–time and displacement–time graphs for this journey. Sketch the graphs one above the other, with the same scale on the time axis. *(5 marks)*

**2** Figure 9 is a speed–time graph for a car journey along a country road, where the direction changes frequently.

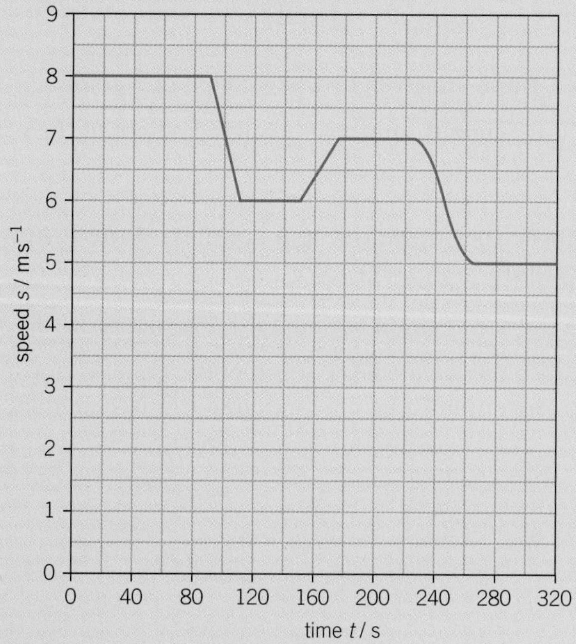


▲ **Figure 9** *Journey along a country road*

Calculate the distance travelled in the 5 minutes shown in the graph, and explain why the car's displacement in this time has a value less than this. *(5 marks)*

**3** The data below is for an object accelerating in a straight line at a non-uniform rate. Plot a velocity–time graph and use it to calculate the acceleration at time $t = 5$ s and the distance $s$ travelled in 10 seconds. *(5 marks)*

| $t$ / s | 0 | 2 | 4 | 6 | 8 | 10 |
|---|---|---|---|---|---|---|
| $v$ / m s$^{-1}$ | 0 | 3 | 8 | 15 | 24 | 35 |

# 8.2 Vectors

Specification reference: 4.2a(i), 4.2b(i), 4.2c(i), 4.2c(ii)

## What are vectors?

Vectors appear everywhere in the world around us, from ocean currents and air flow, to forces and fields.

Vectors are quantities that have both magnitude and direction. These can be represented by an arrow of the appropriate length pointing in the correct direction.

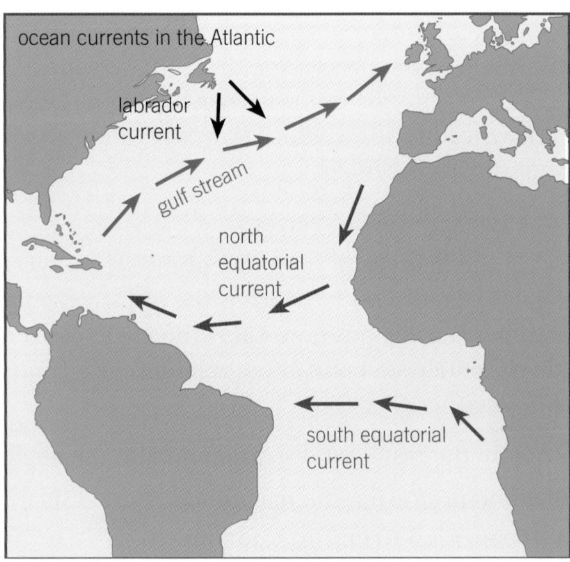


▲ **Figure 1** *Currents in the Atlantic Ocean*

Not all quantities have direction. You have met speed before and it does not tell you in which direction you are moving. Quantities like speed that do not have a direction are called **scalar** quantities.

## Distance and displacement

Distance is a scalar quantity as it has no direction, only magnitude (a numerical value together with its unit). If you wish to state not only how far away something is but in which direction, then you need a **vector** quantity which has the same magnitude as distance, but also has direction. In this case the vector quantity is displacement.

### Adding vectors

Each vector is represented by an arrow pointing in the appropriate direction. Choose an appropriate scale, and draw each arrow with a length proportional to its magnitude.

Draw the two vector arrows tip-to-tail. The resultant vector is given by the line from the tail of the first vector to the tip of the last.

### Learning outcomes

Describe, explain, and apply:

- → the use of vectors to represent displacement, velocity, and acceleration
- → the terms: vector, scalar
- → the resolution of a vector into two components at right angles to each other
- → the addition of vectors using graphs and algebra.

### Synoptic link

You have met scalar quantities such as speed and distance in Topic 8.1, Graphs of motion.

## Worked example: Adding displacements using graphical representations of vectors

A woman walks 25 m due east and then 15 m due north. What is her displacement from the starting position?

**Step 1:** Using graph paper, choose a suitable scale (see Figure 2).

In this example, one large square (1 cm) = 5 m. Put a compass rose in the corner to avoid mistakes (see Figure 2).

**Step 2:** Plot the displacements, one after the other.

It doesn't matter which you start with, the answer will be the same. The example shows the 25 m due east being plotted first. Then, from the tip of the first vector, draw the second vector with the correct magnitude in the correct direction. Remember to put an arrowhead to show which way it is going, and label the magnitude of the displacement.

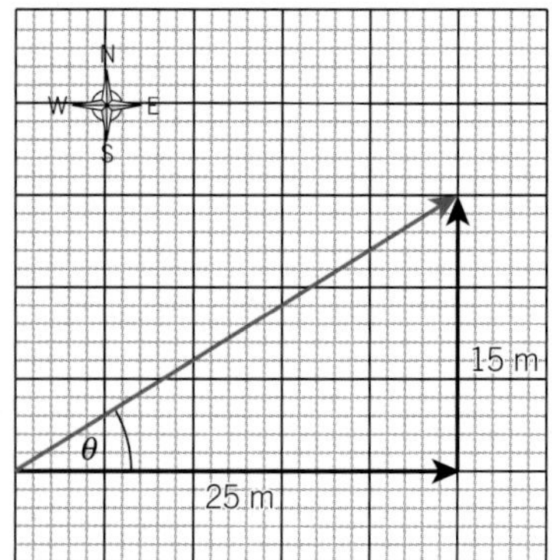


▲ Figure 2 *Adding displacements*

**Step 3:** Draw in the resultant vector.

Join the tail of the first vector to the tip of the second vector. This gives the resultant vector, which is the displacement of the woman from her starting position. Label the angle between the resultant displacement and one of the reference directions – here you can choose either N or E.

**Step 4:** Measure the resultant displacement and the angle.

Using a ruler and protractor, the magnitude of the displacement on the graph paper is 6.0 cm, corresponding to $\frac{6\text{ cm}}{1\text{ cm}} \times 5\text{ m} = 30\text{ m}$, and the angle is 30°.

The resultant displacement is therefore 30 m in a direction E 30° N (or a direction N 60° E, or a bearing of 060°).

Notice that the symbol $\theta$ (the Greek letter theta) has been used. This is the usual symbol chosen for an angle.

### Geometry versus trigonometry

You may have spotted that the displacement in this example is easier to calculate using Pythagoras' theorem and a bit of trigonometry. This has the advantage that you do not need to draw the vector addition triangle with great accuracy. A sketch will do, if it is clearly labelled.

## Worked example: Addition of vectors using trigonometry

By using trigonometry in the example above, Step 4 becomes, using Pythagoras' theorem:

$(\text{resultant magnitude})^2 = (25\text{ m})^2 + (15\text{ m})^2 = 850\text{ m}^2$

$\text{resultant magnitude} = \sqrt{850\text{ m}^2} = 29.1\ldots\text{ m} = 29\text{ m (2.s.f.)}$

You can use sin, cos, or tan to calculate $\theta$, as the triangle is right-angled and you know all three sides. Tan is usually easiest.

$$\tan\theta = \frac{\text{opposite}}{\text{adjacent}} = \frac{15\,\text{cm}}{25\,\text{cm}} = 0.60 \Rightarrow \theta \text{ is the inverse tangent of } 0.60$$

arctan (0.60) or $\tan^{-1}(0.60) = 30.9°... = 31°$ (2 s.f.)

The resultant displacement is therefore more accurately 29 m in a direction E 31° N (or a direction N 59° E, or a bearing of 059°).

## Study tip

Although trigonometry is more accurate than a scale drawing, you should learn to use both methods.

## Synoptic link

See the Maths Appendix for further uses of trigonometry in physics.

## Speed and velocity

In everyday speech, speed and velocity have the same meaning. In physics, they have an important difference. Speed is a scalar quantity, with no direction indicated. It is what is registered by a car speedometer. Velocity is a vector, where the direction is important, as shown in the weather map of Figure 3.

Figure 3 refers, quite correctly, to wind velocities, not speeds. This is because the arrows show the direction, and their size shows the magnitude. Wind speed maps that show only the magnitude are not common.

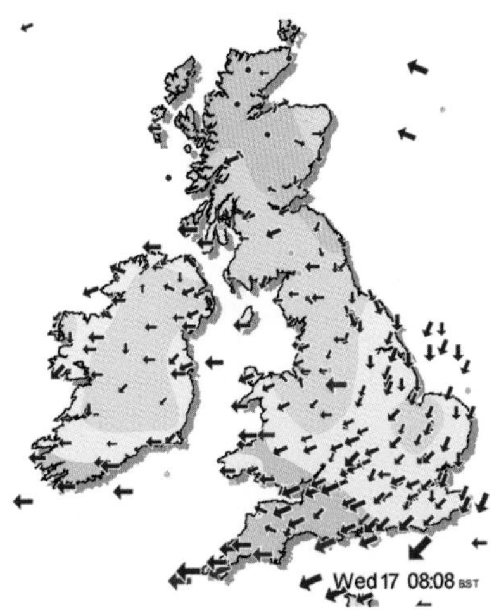


▲ Figure 3 *Wind velocities around the UK*

### Worked example: Velocity and displacement

Calculate the mean velocity of the woman in the previous worked example. She took 10 s to walk the 25 m east and a further 7.0 s to walk the 15 m north.

**Step 1:** Select the correct equation to use in this context.

$$\text{Mean velocity} = \frac{\text{displacement}}{\text{time taken}}$$

Do not be tempted to find the mean velocity of the walk east, then the mean velocity of the walk north and then add them with a vector triangle. This would be wrong!

**Step 2:** Since the overall displacement has been found through geometry or trigonometry, we can substitute this value into the equation.

$$\text{Mean velocity} = \frac{29.1...\,\text{m}}{10\,\text{s} + 7.0\,\text{s}} = \frac{29.1...\,\text{m}}{17\,\text{s}} = 1.7\,\text{m}\,\text{s}^{-1} \text{ (2 s.f.) in the}$$

direction of the displacement, which is E 31° N.

### Relative velocity and subtracting vectors

When measuring velocity you need to be clear what it is measured relative to. If you are flying an aircraft, you need to know your velocity relative to the ground in order to set the correct course, but if there is another aircraft flying nearby, you need to know if you will collide!

To find the velocity of another aircraft relative to yours, you subtract the velocity of your aircraft (relative to the ground) from that of the other one (relative to the ground). Because X − Y = X + (−Y), you subtract a vector by adding one of the same magnitude in the opposite direction.

## Worked example: The relative velocity of two aircraft

Figure 4 shows two aircraft on possible collision courses, drawn to scale. Calculate if these two aircraft will collide.

You need to find the velocity of the other aircraft relative to your aircraft to decide this.

**Step 1:** Add a velocity opposite to that of your aircraft to both aircraft, as shown in Figure 5.

**Step 2:** Calculate relative velocities by adding the two velocity vectors for each aircraft.

Relative velocity of your plane = 225 − 225 = 0 m s$^{-1}$ (since you are not moving relative to yourself).

For the relative velocity of the other aircraft, we cannot calculate this using trigonometric means as the angles are not given. This means that we cannot be sure if this is a right-angled triangle or not.

Measuring the red vector arrow in Figure 6 relative to the scale given gives us an approximation of 170 m s$^{-1}$.

Figure 6 shows the resultant velocity of the other aircraft relative to your aircraft. The magnitude of the velocity (the speed), by measurement on the vector addition triangle, is 170 m s$^{-1}$. This is not as important here as the direction, as the magnitude of the velocity tells you only how long it will be before any possible collision.

The direction of the relative velocity vector suggests the other aircraft will pass in front of you. However, it may be best to take evasive action!

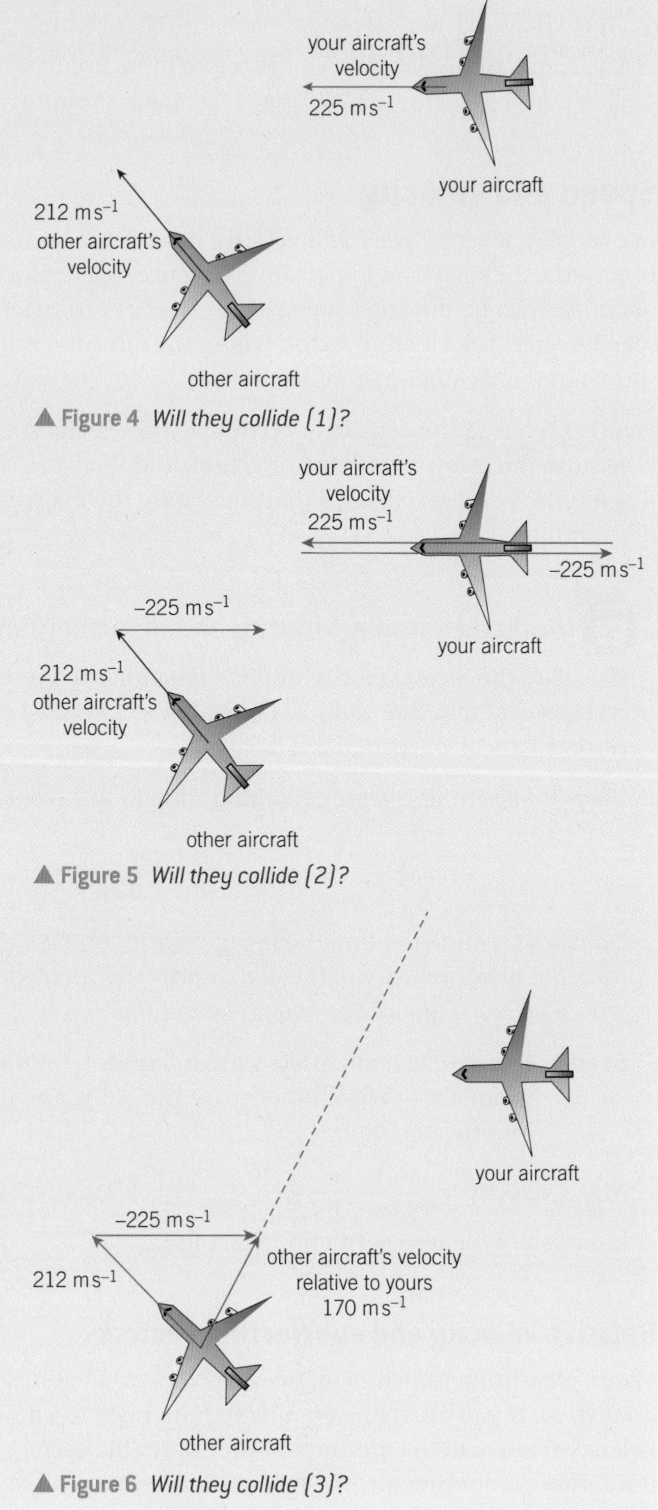


▲ Figure 4 *Will they collide (1)?*



▲ Figure 5 *Will they collide (2)?*



▲ Figure 6 *Will they collide (3)?*

## The father of relativity

If you ask most people who first had the idea of relativity, most will say Einstein. Certainly Einstein's theory of Special Relativity in 1905 was a ground-breaking development, but the principle of relativity had first been proposed 273 years earlier by Galileo.

Galileo argued that the natural motion of any object was not slowing down to a stop, as the ancient Greek thinker Aristotle stated, but moving in a straight line at a steady speed, that is, with a constant velocity. He also noted that if you were on a ship travelling at a constant velocity and not rocking from side to side you would be unable to tell that you were moving – a dropped object would seem to fall as if it were on stationary land rather than moving sideways at the same time.

What Galileo realised was that there was no fixed 'zero' of position from which measurements should be made, and that if two places of observation were moving at constant velocity relative to each other, the laws of physics would be the same. Making calculations of what is happening in one such place observed from the other one are called Galilean transformations.

**1** The surface of the Earth is not moving at a constant velocity, as it is spinning. Why does it seem that dropped objects fall vertically downwards?

▲ **Figure 7** *Galileo Galilei*

## Acceleration

Acceleration was previously defined as the rate at which velocity is changing, that is, $a = \frac{\Delta v}{\Delta t}$. In everyday terms, the acceleration of a car is often described as, for example, 0 to 60 mph in 11.3 s.

We do not normally measure velocity in mph, but in m s$^{-1}$, and 60 mph = 26.8 m s$^{-1}$, so the acceleration of this car,

$$a = \frac{26.8\,\text{m s}^{-1} - 0\,\text{m s}^{-1}}{11.3\,\text{s}} = 2.37\,(\text{m s}^{-1} \div \text{s}) = 2.37\,(\text{m s}^{-1} \times \text{s}^{-1}) = 2.37\,\text{m s}^{-2}$$

Acceleration is the rate of change of velocity and, like velocity, it is a vector. Because acceleration is a vector, it can also be represented using a vector arrow, just like with displacement and velocity.

**Synoptic link**

You have met acceleration in Topic 8.1, Graphs of motion.

## Acceleration and *g*

One well-known acceleration is that due to gravity, which is given the special symbol $g$. It has the same value, with small fluctuations, anywhere on the Earth's surface, that is, $g = 9.81\,\text{m s}^{-2}$ downwards.

High accelerations are often quoted in terms of $g$: a dragster car can accelerate from 0 to 44 m s$^{-1}$ in 0.86 s, giving an acceleration of 51 m s$^{-2}$.

As $\frac{51\,\text{m s}^{-2}}{9.8\,\text{m s}^{-2}} = 5.2$, this is equal to an acceleration of $5.2g$.

**Synoptic link**

You will see how to measure $g$ in Topic 8.4, Speeding up and slowing down.

## Worked example: A thrown ball

A ball is thrown horizontally at $5.0\,\text{m}\,\text{s}^{-1}$. Ignoring any air resistance, calculate its velocity 0.25 s later. The acceleration due to gravity, $g = 9.8\,\text{m}\,\text{s}^{-2}$.

**Step 1:** Use the definition of acceleration to calculate the change in velocity.

$$a = \frac{\Delta v}{\Delta t} \Rightarrow \Delta v = a\Delta t$$

**Step 2:** Substitute values in the equation.

$\Delta v = a\Delta t = g\Delta t = 9.8\,\text{m}\,\text{s}^{-2} \times 0.25\,\text{s} = 2.45\,\text{m}\,\text{s}^{-1}$ vertically downwards

**Step 3:** Add $v_{\text{vertical}}$ to the vector diagram.

Add this change in velocity to the initial velocity of $5.0\,\text{m}\,\text{s}^{-1}$ horizontally, as shown in Figure 8. On your vector diagram, label the resultant vector arrow and the angle it makes with the horizontal (or vertical).

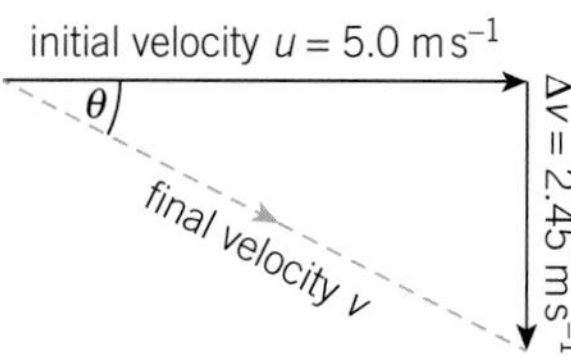


▲ Figure 8 *A thrown ball*

**Step 4:** Use measurement on Figure 8, or geometry and trigonometry, to find the magnitude and direction of the final velocity.

Using Pythagoras' theorem,

$$(\text{resultant magnitude})^2 = (5.0\,\text{m}\,\text{s}^{-1})^2 + (2.45\,\text{m}\,\text{s}^{-1})^2$$
$$= 31.0...\,\text{m}^2\,\text{s}^{-2}$$

Resultant velocity $= \sqrt{31.0...\,\text{m}^2\,\text{s}^{-2}} = 5.6\,\text{m}\,\text{s}^{-1}$ (2 s.f.) in the direction $\theta$

where $\tan\theta = \dfrac{2.45\,\text{m}\,\text{s}^{-1}}{5.0\,\text{m}\,\text{s}^{-1}} = 0.49 \Rightarrow \theta = 26°$ (2 s.f.) below the horizontal.

Note that the direction can be been seen from the labelled diagram.

**Hint**

Remember that acceleration $a$ is a vector. $a$ has the same direction as the velocity change $\Delta v$.

## Resolving a vector into components

Just as two vectors can be added to give a resultant vector, so a single vector can be resolved, or split, into two components at right angles to each other. The directions of these components are often horizontal and vertical.

In Figure 9, adding the components $s_H$ and $s_V$ tip-to-tail will give the displacement $s$. Note that either ADC or ABC could be the 'tip-to-tail' triangle for this addition. However, as $s_H$ and $s_V$ both transform the object originally at point A, the resolution into components is usually represented in the way shown in Figure 9.

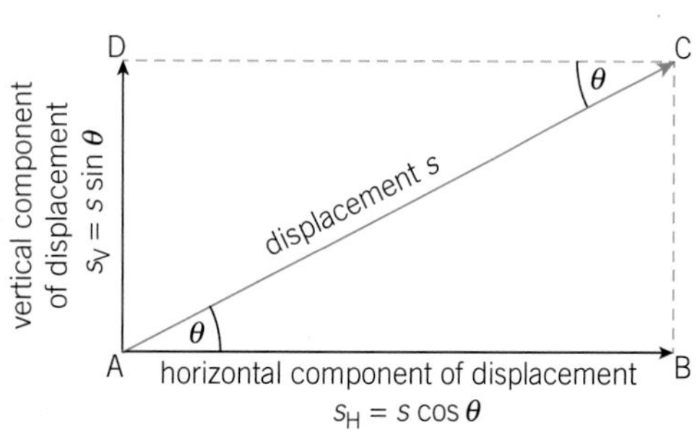


▲ Figure 9 *Components of displacement*

$$\cos\theta = \frac{AB}{AC} = \frac{\text{magnitude of } s_H}{\text{magnitude of } s} \Rightarrow s_H = s\cos\theta$$

$$\sin\theta = \frac{BC}{AC} = \frac{\text{magnitude of } s_V}{\text{magnitude of } s} \Rightarrow s_V = s\sin\theta$$

Horizontal and vertical motion can be calculated separately as the acceleration due to gravity acts vertically.

### Worked example: Calculating components of velocity

You are driving at a speed of $28\,\mathrm{m\,s^{-1}}$ in a direction N 25° W. Resolve this velocity into north-south and east-west components.

**Step 1:** Sketch the situation described in the question, adding rectangles to show the resolution of velocity (Figure 10).

Make sure you keep the directions the same as in the original sketch. You will often find it easiest to superimpose this resolution diagram on top of the sketch. Label the two components clearly.

**Step 2:** Use trigonometry to find the magnitudes of the two components. The opposite faces of the rectangle are equal, so you need only consider the triangle in which the angle is labelled.

$$\cos(25°) = \frac{\text{adjacent}}{\text{hypotenuse}} = \frac{v_N}{28\,\mathrm{m\,s^{-1}}} \Rightarrow v_N = \cos(25°) \times 28\,\mathrm{m\,s^{-1}}$$

$$= 0.906... \times 28\,\mathrm{m\,s^{-1}} = 25\,\mathrm{m\,s^{-1}}\ (2\text{ s.f.})$$

$$\sin(25°) = \frac{\text{opposite}}{\text{hypotenuse}} = \frac{v_W}{28\,\mathrm{m\,s^{-1}}} \Rightarrow v_W = \sin(25°) \times 28\,\mathrm{m\,s^{-1}}$$

$$= 0.422... \times 28\,\mathrm{m\,s^{-1}} = 12\,\mathrm{m\,s^{-1}}\ (2\text{ s.f.})$$

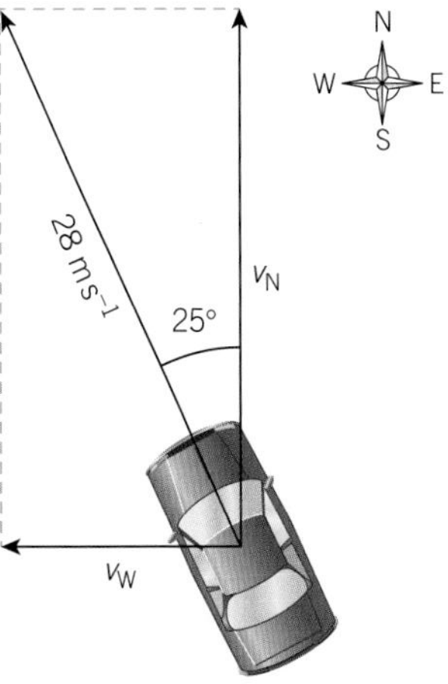


▲ **Figure 10** *Components of velocity*

## Summary questions

**1** On a treasure map, you must take 20 paces north from the big tree, then 6 paces east, to dig for the treasure. Draw an addition vector diagram for these displacements and use it to find the displacement of the treasure from the tree. (You can ignore the vertical displacement.) *(3 marks)*

**2** A rower can row at $2.0\,\mathrm{m\,s^{-1}}$ in still water. She needs to row straight across a river where the current is flowing at $0.8\,\mathrm{m\,s^{-1}}$.

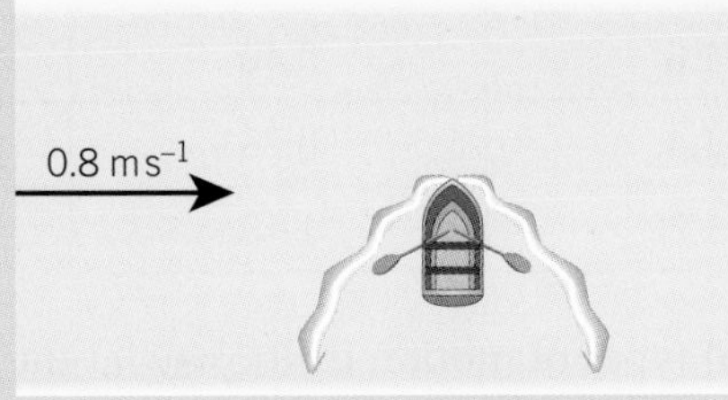


▲ **Figure 11** *Crossing the river*

Draw a vector addition diagram to show her resultant velocity relative to the river banks, and use it to calculate the direction in which she must row. *(4 marks)*

**3** An aircraft is flying relative to the air at $200\,\mathrm{m\,s^{-1}}$. The aircraft is heading north-west and there is a $50\,\mathrm{m\,s^{-1}}$ wind relative to the ground heading west. Use components of velocity to find the velocity of the aircraft relative to the ground. *(4 marks)*

# 8.3 Modelling motion

Specification reference: 4.2b(iii), 4.2c(xi), 4.2d(iii)

**Learning outcomes**

Describe, explain, and apply:

- graphical models in terms of vector additions to represent displacement and velocity
- computational models to represent changes of displacement and velocity in small time steps
- terminal velocity using an experiment with paper cases in air.

▲ **Figure 1** *Hurricane approaching Florida – a complex system*

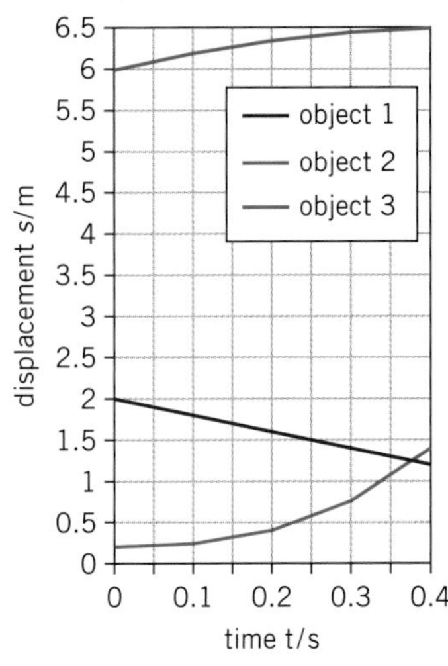


▲ **Figure 2** *Displacement–time graph of data in Table 1*

You have so far met descriptions of the behaviour of simple systems, and will meet ideas which explain these behaviours later on. However, real situations can be very complex, with many interacting variables to be considered. This is true of hurricanes.

One approach to analysing such systems is to use the large memory and fast processors of modern supercomputers to perform many simple calculations of tiny parts of the system. These can then be added together to describe the motion of the complete complex system.

These approaches are all based on dividing motion into tiny intervals or increments of time and defining what happens between one such increment and the next.

## Iterative computational models

**Iterative** means step-by-step. An iteration is a single step, usually a tiny time increment. Iterative models track what is happening as time progresses. Computers are normally used for these models but you can best see how they work by working through a few steps manually.

### Working through an iterative model

The table shows the displacements of three different objects 1, 2, and 3 after four time increments of 0.1 s. You can assume that all motions are along the same straight line, and all displacements are relative to the same fixed point.

▼ **Table 1** *Displacement data*

| Time $t$ / s | Displacement $s_1$ / m | Displacement $s_2$ / m | Displacement $s_3$ / m |
|---|---|---|---|
| 0.0 | 2.0 | 0.20 | 6.00 |
| 0.1 | 1.8 | 0.24 | 6.20 |
| 0.2 | 1.6 | 0.40 | 6.35 |
| 0.3 | 1.4 | 0.76 | 6.45 |
| 0.4 | 1.2 | 1.40 | 6.50 |

You can identify the types of motion by looking at the changes in displacement $\Delta s$ for successive time increments $\Delta t$. For object 1, $\Delta s$ is negative, so it is approaching the fixed point. It is also constant, so the velocity is constant. For object 2, $\Delta s$ is positive and increasing, so it is accelerating away from the fixed point. For object 3, $\Delta s$ is positive and decreasing, so it is decelerating away from the fixed point. The motion of these three objects is shown in Figure 2.

### Constructing an iterative computational model

In constructing a model, you write a set of instructions defining how $v$ and $s$ will change at regular time intervals $\Delta t$.

Velocity $v$ and displacement $s$ after each time interval $\Delta t$ are calculated using the rules:

$$\text{new } s = \text{previous } s + \Delta s$$

$$\text{new } v = \text{previous } v + \Delta v$$

Figure 3 shows how the extra $\Delta s$ and $\Delta v$ in the time increment $\Delta t$ between some time $t$ and one time interval later, $t + \Delta t$, are added to the previous values. A blue copy of the object with its velocity at time $t$ is duplicated below the object at time $t + \Delta t$ to show how the new velocity is obtained.

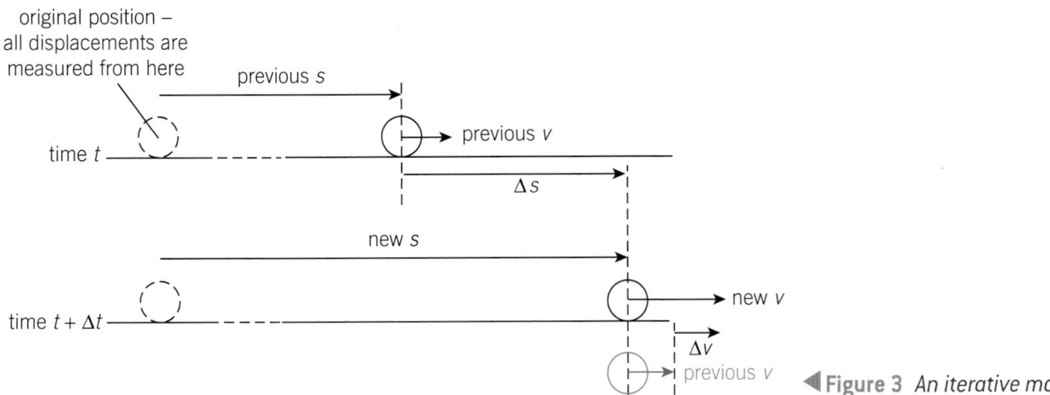


**Figure 3** *An iterative model*

In order to apply this model you need to calculate the change in $s(\Delta s)$, and the change in $v(\Delta v)$.

These come from the definitions of velocity and acceleration.

$$v = \frac{\Delta s}{\Delta t} \Rightarrow \Delta s = v\Delta t \quad \text{and} \quad a = \frac{\Delta v}{\Delta t} \Rightarrow \Delta v = a\Delta t$$

## Modelling free fall

To show how an iterative model can work, we will work though three iterations (steps) using a time interval $\Delta t$ of 0.1 s.

### Worked example: A crude model for free fall

Use an iterative model with time increment 0.1 s to investigate the changes in velocity and displacement for an object falling freely from rest.

In free fall, the acceleration $a$ is constant at $g = 9.8\,\text{m}\,\text{s}^{-2}$. This allows us to use the equation $\Delta v = a\Delta t$ to calculate the change in velocity $\Delta v$ in each time interval. Each $\Delta v$ will be the same. As we work through the model you will see how the *updated* values of $v$ affect the values of $\Delta s$ and therefore $s$.

▼ **Table 2**

| $t$/s | $a$/m s$^{-2}$ | $v$/m s$^{-1}$ | $s$/m | |
|---|---|---|---|---|
| 0.0 | 9.8 | 0 | 0 | ← **Step 1:** Put in the initial conditions of the model |
| 0.1 | 9.8 | 0 + 9.8 × 0.1 = 0.98 | 0 + 0 × 0.1 = 0 | ← **Step 2:** Use the previous values of $a$ and $v$ to calculate the new $v$ and $s$ |
| 0.2 | 9.8 | 0.98 + 9.8 × 0.1 = 1.96 | 0 + 0.98 × 0.1 = 0.098 | ← **Step 3:** Repeat Step 2 |
| 0.3 | 9.8 | 1.96 + 9.8 × 0.1 = 2.94 | 0.098 + 1.96 × 0.1 = 0.294 | ← **Step 4:** Repeat Step 2 |

Note that the velocity goes up in equal steps of $0.98\,\mathrm{m\,s^{-1}}$ every 0.1 s but that the increase in displacement gets larger and larger in each time interval.

In the model, the changes in $v$ and $s$ occur instantly at the end of each time interval, and then do not change until the end of the next time interval, as shown in the graphs of this model (Figure 4).

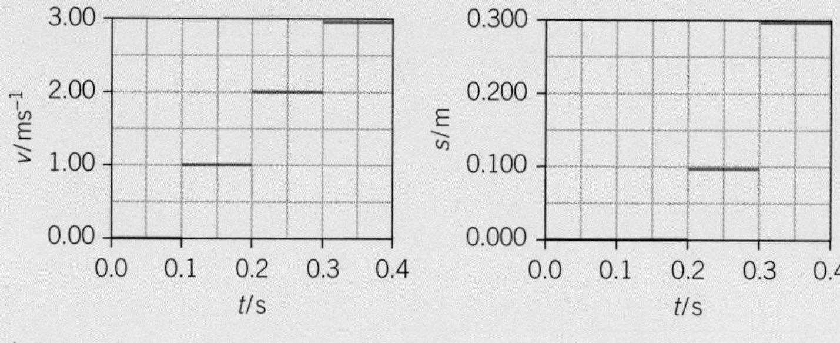


▲ **Figure 4** *Graphs for the free fall model*

In this worked example, you will have noticed how monotonous Step 2, Step 3, and Step 4 become. What would your response be if you were asked to do it for 30 intervals of 0.01 s instead of 3 intervals of 0.1 s? Or for 300 intervals of 0.001 s? This is one reason why computers are often used.

## Pros and cons of the iterative model

### Advantages to iterative models

If friction acts on a falling object, the acceleration is not constant but decreases. The faster an object is moving, the greater the upwards frictional force acting on it. Using algebra and calculus to tackle this problem is difficult, but an iterative model can be used just as easily as for the constant acceleration case above.

### Problems with iterative models

The graphs in Figure 4 do not look realistic. In reality, when an object is in free fall, the velocity and displacement change all the time, not in sudden jumps, and all the calculated values occur earlier than the graphs suggest as the model uses the previous values of $a$ and $v$ to make the calculation.

One way to improve the model is to use a time interval $\Delta t$ smaller than the 0.1 s used above. Figure 5 shows the effect of reducing $\Delta t$ five-fold, to 0.02 s.

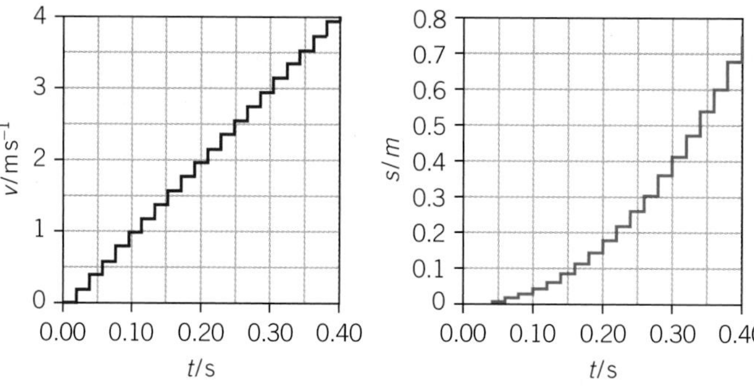


▲ **Figure 5** *An improved model*

### Graphical models

An incremental model can be based on vector addition rather than the spreadsheet approach above, and this can be useful when the direction of movement is changing. Although this can be analysed with a computational model using components, a graphical approach is visually clearer.

#### Worked example: Tracking the path of an object falling under gravity

For clarity, this will be applied to an object initially moving horizontally. If it were not, all the vectors would be in the same vertical line.

What path would be predicted for an object initially moving horizontally at $5.0\,\text{m s}^{-1}$, analysed with a time increment of 0.1 s? In this question, use $g = 10\,\text{m s}^{-2}$.

**Step 1:** Find the velocity change $\Delta v$ caused by gravity each 0.1 s.

As $g = 10\,\text{m s}^{-2}$, $\Delta v = a\Delta t = 10\,\text{m s}^{-2} \times 0.1\,\text{s} = 1.0\,\text{m s}^{-1}$ vertically downwards

**Step 2:** Choosing a suitable scale, draw a horizontal arrow near the top to represent $5.0\,\text{m s}^{-1}$ horizontally. Divide the horizontal axis of the graph paper into equal intervals each the same width as the initial velocity vector.

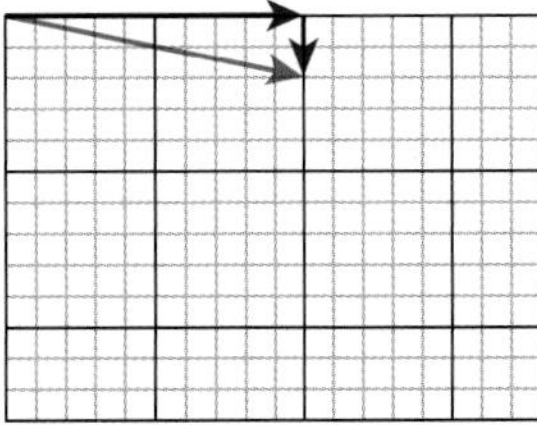

▲ **Figure 6** *Graphical model after Step 2*

At the tip of the initial velocity arrow, draw a $\Delta v = 1.0\,\text{m s}^{-1}$ arrow vertically downwards. Draw the resultant of these two velocities. In this example, the resultant velocities are shown in red whereas the two velocities being added will be in a different colour for each stage.

**Step 3:** Continue the resultant arrow on for the same horizontal distance as the length of the initial velocity vector. This is the same length and direction as the previous resultant. Add another velocity change $\Delta v$ at the tip, and add as before. This is the vector version of new $v$ = previous $v + \Delta v$.

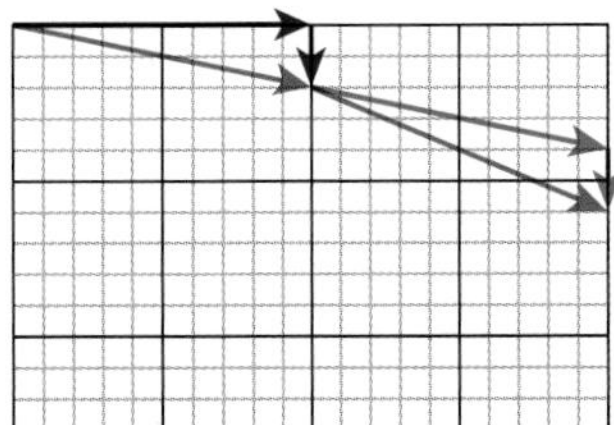

▲ **Figure 7** *Graphical model after Step 3*

**Step 4:** Repeat Step 3 again and again until you run out of graph paper.

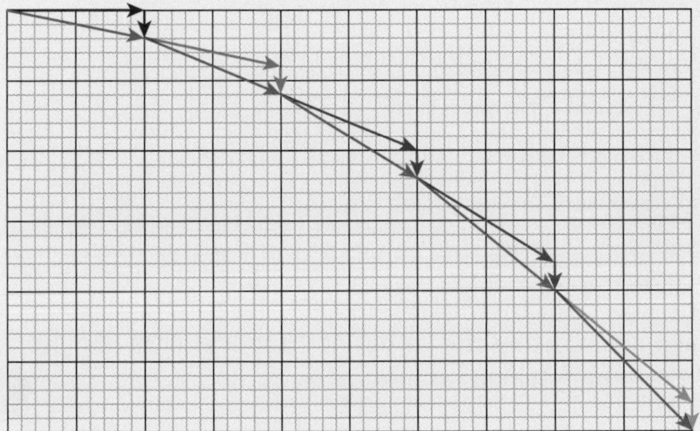

▲ **Figure 8** *Graphical model after Step 6*

**Synoptic link**

Projectiles will be studied in Topic 9.4, Projectiles.

As this worked example shows, the object thrown horizontally follows a curved path.

### Modelling more complex situations

In complex situations there are often simple ideas that can help with the modelling. When an object falls through air, it is common knowledge that the air acts to reduce the acceleration of gravity, and the faster the object falls, the greater the effect of air resistance. Engineering experiments suggest that a good model for this situation is that the downwards acceleration at any time is given by

$$a = 9.8\,\mathrm{m\,s^{-2}} - Kv^2$$

where $v$ is the velocity of the object through the air and $K$ is a constant which depends on the shape, size, and density of the object. For example, a thin needle of steel will have little air resistance and so a small value of $K$, but a balloon full of air will have considerable air resistance and a large value of $K$.

In this worked example you will see how the use of this engineering equation combines with the previous ideas to allow useful predictions of behaviour.

#### Worked example: Modelling an object falling through air

Use an iterative model with time increments of 0.1 s to investigate the changes in velocity and displacement over two seconds for an object falling from rest with a value of $K = 0.2\,\mathrm{m^{-1}}$ in the engineering equation $a = 9.8\,\mathrm{m\,s^{-2}} - Kv^2$. Display your results using a sketch graph.

**Step 1:** Set up the initial conditions in the first row of Table 3. The object has not started to fall, and there is no air resistance yet, so $a = g$.

Remember that each calculation uses the values of $a$ and $v$ from the previous step.

**Step 2:** Substitute values into $a = 9.8\,\mathrm{m\,s^{-2}} - Kv^2$. Use the initial value of $v$ to calculate $a$ at 0.1 s using $a_{new} = 9.8\,\mathrm{m\,s^{-2}} - (0.2\,\mathrm{m^{-1}}) \times (v_{previous})^2$.

Use the initial value of $a$ to calculate $v$ at 0.1 s using

$$v_{new} = v_{previous} + a_{previous} \times \Delta t.$$

Use the initial value of $v$ to calculate $s$ at 0.1 s using

$$s_{new} = s_{previous} + v_{previous} \times \Delta t.$$

**Step 3:** Repeat Step 2, using the values of $v$ and $a$ at 0.1 s to calculate the new values at 0.2 s.

**Step 4:** Repeat again, using the values of $v$ and $a$ at 0.2 s to calculate the new values at 0.3 s.

This completes the first four rows of the table.

▼ Table 3

| $t$/s | $a$/m s$^{-2}$ | $v$/m s$^{-1}$ | $s$/m |
|---|---|---|---|
| 0.0 | 9.8 | 0 | 0 |
| 0.1 | $9.8 - 0.2 \times 0^2 = 9.8$ | $0 + 9.8 \times 0.1 = 0.98$ | $0 + 0 \times 0.1 = 0$ |
| 0.2 | $9.8 - 0.2 \times (0.98)^2 = 9.61$ | $0.98 + 9.8 \times 0.1 = 1.96$ | $0 + 0.98 \times 0.1 = 0.098$ |
| 0.3 | $9.8 - 0.2 \times (1.96)^2 = 9.03$ | $1.96 + 9.61 \times 0.1 = 2.92$ | $0.098 + 1.96 \times 0.1 = 0.294$ |

Using a computer to repeat this several more times gives Figure 9, which shows how the values of acceleration, velocity, and displacement change over the first 2 seconds using this model.

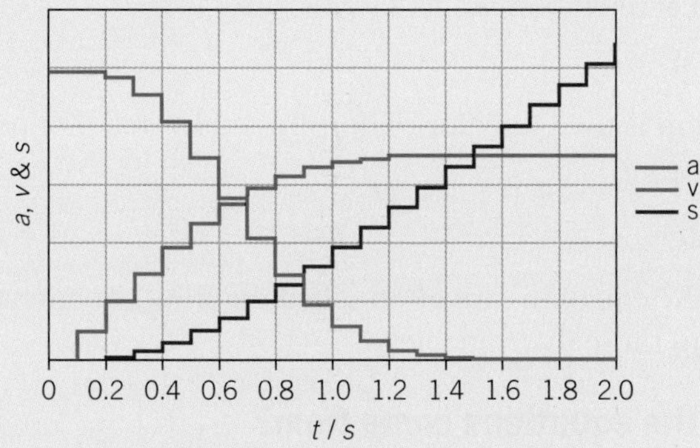


▲ **Figure 9** *Graphs for free fall with air resistance*

Even with the large value of $\Delta t = 0.1$ s, the essential features of free fall with air resistance are clear. The acceleration decreases to zero, the velocity becomes constant, and the displacement graph changes from a curve to a straight line. The constant velocity reached when falling against resistance is referred to as **terminal velocity**.

## Practical 4.2d(iii): Investigating falling cupcake cases

The paper cases used for baking cupcakes fall in quite a regular way without spinning round or capsizing, but they are greatly affected by air resistance and so reach terminal velocity within about a metre of fall.

With simple measurements of displacement and time, and an appropriate choice of experimental conditions, such as the height of the drop and over which part of its fall the cupcake case is being timed, it is possible to investigate how the terminal velocity reached depends on factors such as the mass of the falling object.

## Summary questions

1 A car accelerates from rest along a straight, horizontal road at a constant 3.4 m s$^{-2}$. Construct an iterative model with $\Delta t = 0.1$ s to find the distance travelled after 0.8 s. *(4 marks)*

2 In the Worked example *Modelling free fall with air resistance*, the object ends up moving at a constant velocity of 7.00 m s$^{-1}$. If the model is repeated with a much smaller time increment, it has the same constant velocity at the end. Explain this. *(2 marks)*

3 The rower in Figure 10 starts to row across the river at 2.0 m s$^{-1}$ relative to the water, without attempting to correct her course.

▲ **Figure 10** *Crossing the river*

*(4 marks)*

Use a graphical model to add the displacements at a time interval of $\Delta t = 1$ s to plot her path over the first four seconds.

# 8.4 Speeding up and slowing down

Specification reference: 4.2c(iii), 4.2d(ii)

**Learning outcomes**

Describe, explain, and apply:

- → the kinematic (*suvat*) equations for motion with constant acceleration in a straight line
- → how the acceleration of free fall can be determined using a trapdoor and an electromagnet.

You have seen that vectors and iterative computation can model the motion of objects. This topic deals with the use of algebra, which is often the most flexible approach to problem solving. Algebra is also particularly appropriate when dealing with the physical principles of force and energy, and relating them to motion. To do this, you need to use the kinematic equations that describe motion.

## Kinematic equations

There are four equations relating the following variables:

- displacement, $s$ (m)
- initial velocity, $u$ ($\mathrm{m\,s^{-1}}$)
- final velocity, $v$ ($\mathrm{m\,s^{-1}}$)
- uniform acceleration, $a$ ($\mathrm{m\,s^{-2}}$)
- time, $t$ (s).

Each equation is used only for uniformly accelerated motion in a straight line.

$$v = u + at \qquad s = \frac{u+v}{2}t \qquad s = ut + \frac{1}{2}at^2 \qquad v^2 = u^2 + 2as$$

The kinematic equations are often called the *suvat* equations, because they relate to those five variables.

**Synoptic link**

The use of algebraic functions for force and energy will be covered in Chapter 9, Modelling motion.

### Where do the equations come from?

The first two equations are simple re-arrangements of definitions you have already met.

$$a = \frac{\Delta v}{t} = \frac{v-u}{t} \Rightarrow at = v - u \Rightarrow v = u + at$$

Average velocity $= \frac{s}{t}$ and, if the acceleration or deceleration is uniform, average velocity $= \frac{(u+v)}{2}$.

$$\frac{u+v}{2} = \frac{s}{t} \Rightarrow s = \left(\frac{u+v}{2}\right)t$$

The third equation, often the most useful in practice, can be shown by referring to a graph of accelerated motion.

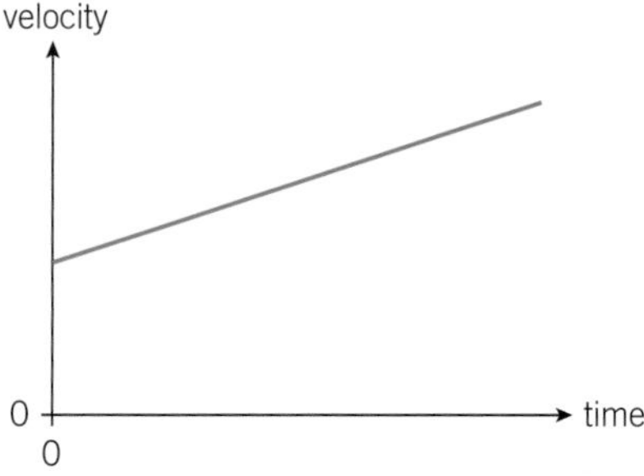


▲ **Figure 1** *Uniformly accelerated motion*

**Worked example: Relating an equation to a graph**

The graph in Figure 1 shows uniformly accelerated motion. Label $u$, $v$, and $t$, and explain how $a$ and $s$ may be found. Then derive an equation for $s$ in terms of $u$, $a$, and $t$.

**Step 1:** Draw horizontal and vertical construction lines from each end of the line of the graph.

**Step 2:** Label $u$, $v$, and $t$ on the graph and use this to label the sides.

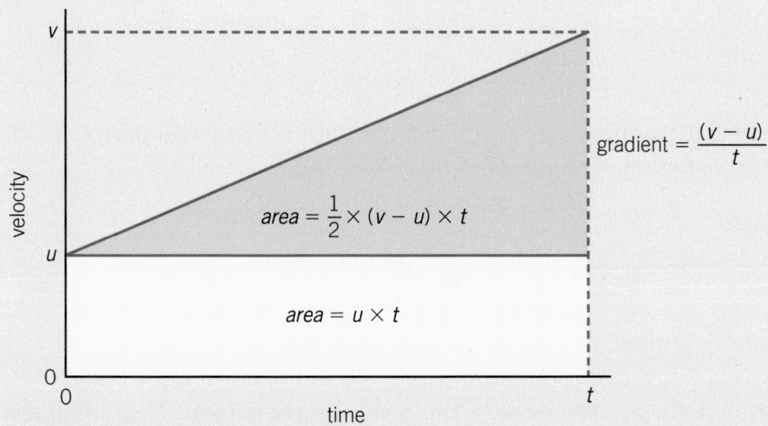


▲ **Figure 2** *Rectangle and triangle marked out*

**Step 3:** The displacement, $s$, is the area under the line.

$\Rightarrow s$ = area of rectangle + area of triangle

= base × height of rectangle + $\frac{1}{2}$ × base × height of triangle

$= u \times t + \frac{1}{2} \times t \times (v - u)$

This is almost there, but it has $v$ and not $a$.

But the first *suvat* equation is $v = u + at \Rightarrow (v - u) = u + at - u = at$

**Step 4:** Substitute $v - u = at$ into $s = u \times t + \frac{1}{2} \times t \times (v - u)$

$$s = ut + \frac{1}{2}t \times (at) = ut + \frac{1}{2}tat = ut + \frac{1}{2}at^2$$

Note that $ut$ = area of rectangle = the distance the object would have travelled if it had not accelerated, and $\frac{1}{2}at^2$ = area of triangle = the extra distance travelled by virtue of the gain in speed.

The fourth *suvat* equation can be obtained algebraically by substituting $s = \left(\frac{u+v}{2}\right)t$ into $v = u + at$ in order to eliminate $t$, but you will not need to do this. In Topic 9.3, Conservation of energy, you will see that it can be explained in energy terms more easily.

### Worked example: How deep is the well?

A stone is dropped down a deep well. A splash is heard 1.5 s after the stone is released.

How deep is this well?

**Step 1:** List the variables.

displacement $s$ = ?; initial velocity $u = 0$; final velocity $v$ = ?; acceleration $a = g = 9.8\,\text{m}\,\text{s}^{-2}$ and time $t = 1.5\,\text{s}$.

**Study tip**

**Be methodical in using equations**

Before any calculation, write down the values you have for $s$, $u$, $v$, $a$, and $t$. Once you have listed these you should be able to see which equation(s) to use. Remember to take account of direction for all vector quantities and that $g$ is a downward acceleration.

**Step 2:** Eliminate the unknown you do not wish to find.

We need $s$, so we can eliminate $v$. The equation we need is the one with $s$, $u$, $a$, and $t$.

$$s = ut + \frac{1}{2}at^2$$

$$s = 0 \times 1.5\,\text{s} + \frac{1}{2} \times 9.8\,\text{m s}^{-2} \times (1.5\,\text{s})^2$$

$$= 0 + 4.9\,\text{m s}^{-2} \times 2.25\,\text{s}^2 = 11\,\text{m}$$

The well is 11 m deep.

In this example, there was an obvious equation to use. In truth, however, it was not the only way to do it. Even though we had eliminated $v$ as unnecessary, there was no reason why you could not first use $v = u + at$ to calculate $v$, and then use $s = \left(\frac{u + v}{2}\right)t$ to find $s$. It just takes a little longer.

## Practical 4.2d(ii): Measuring the acceleration due to gravity, $g$

This can be done using the equation $s = ut + \frac{1}{2}at^2$.

For an object falling from rest, $u = 0$ and $a = g$, so the equation to be used is $s = \frac{1}{2}gt^2$.

Figure 3 shows one arrangement that can be used.

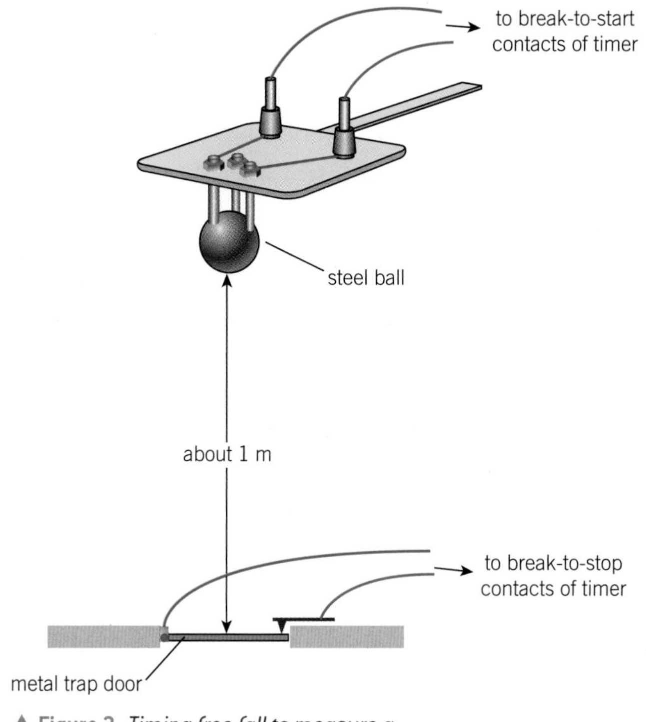


▲ Figure 3 *Timing free fall to measure g*

The steel ball is usually held in place at the top using an electromagnet. Switching off the electromagnet circuit releases the ball. This breaks the contacts at the top of the measured drop, and starts an accurate timer. Striking the trap door at the bottom of the measured drop breaks those contacts to the timer circuit, stopping the timing.

There are alternative ways of measuring $g$ – light gates can be set up to find the velocity at two different points along the fall of a dense object, attached to a timing card, or the fall of a compact object such as a golf ball can be video recorded and the recording analysed frame-by-frame.

## Car stopping distance and speed

In stopping a car, there are two factors to consider. One is the human reaction time, or *thinking time* before you press your foot hard on the brake pedal. The other is the time taken for the car to come to a halt. For each of these times, there is a distance the car will travel – the *thinking distance* is the distance the car travels before your press the pedal, and the *braking distance* is the distance the car travels during deceleration. The stopping distance is the sum of these two distances, and this must be less than the distance to the hazard you have just seen if an accident is not to happen.

Two extracts on car stopping distances from the Highway Code are shown in Figure 4.

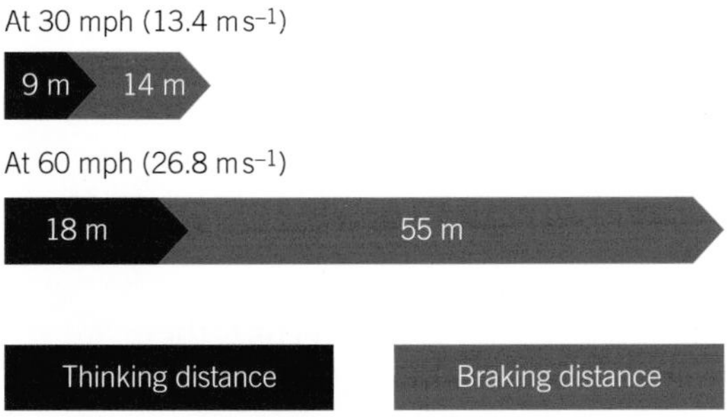

At 30 mph (13.4 m s$^{-1}$)

9 m | 14 m

At 60 mph (26.8 m s$^{-1}$)

18 m | 55 m

Thinking distance | Braking distance

▲ **Figure 4** *Thinking distance and braking distance*

As the speed of the car increases, the thinking distance increases in proportion – this is because the speed has doubled, from 30 mph to 60 mph in this case, so it travels twice as far in the same thinking time.

However, the braking distance goes up about four times because there are two factors at play:

- the average speed has doubled, so it travels twice as far each second
- $\Delta v$ has doubled from 30 mph to 60 mph, so the time taken for the same deceleration to reduce the speed to zero becomes twice as long.

The thinking and stopping distances in the Highway Code are based on assumed values of human reaction time and a car's deceleration. These factors can be affected by a number of factors such as tiredness or icy road conditions.

### Worked example: Car stopping distance

The Highway Code suggests safe braking distances for cars going at different speeds. Calculate the braking deceleration that the Highway Code assumes. At 30 mph, which is $13.4\,\mathrm{m\,s^{-1}}$, the Highway Code indicates the breaking distance to be at least 14 m.

**Step 1:** List the variables.

distance $s = 14\,\mathrm{m}$; initial velocity $u = 13.4\,\mathrm{m\,s^{-1}}$;
final velocity $v = 0\,\mathrm{m\,s^{-1}}$; acceleration $a = ?$; and $t = ?$

**Step 2:** Eliminate the unknown you do not need to find.

This is $t$.

**Step 3:** Identify the equation required.

The equation we need is the one with $s$, $u$, $v$, and $a$.

$$v^2 = u^2 + 2as$$

**Step 4:** Substitute values into the equation.

$$(0\,\mathrm{m\,s^{-1}})^2 = (13.4\,\mathrm{m\,s^{-1}})^2 + 2 \times a \times 14\,\mathrm{m}$$

$$0 = 179.56\,\mathrm{m^2\,s^{-2}} + 28\,\mathrm{m} \times a$$

$$-179.56\,\mathrm{m^2\,s^{-2}} = 28\,\mathrm{m} \times a$$

$$a = \frac{-179.56\,\mathrm{m^2\,s^{-2}}}{28\,\mathrm{m}} = -6.4\,\mathrm{m\,s^{-2}}\ (2\ \text{s.f.})$$

The minus sign shows that it is decelerating, that is, $\Delta v = v - u$ is in the opposite direction to $s$ and $u$.

## Applying the kinematic equations to two-dimensional motion

As displacement, velocity, and acceleration are all vectors, there are often cases where they do not all lie on the same line. To use the equations in these cases, it is necessary to resolve one or more of these values into appropriate directions.

When resolving, you often have a choice as to which perpendicular directions you choose to resolve along, as shown in this worked example.

### Worked example: Trolley on a ramp

A trolley is released from rest to roll a distance of 2.5 m along a ramp inclined at 15° to the horizontal. Calculate the speed it is going when it reaches the 2.5 m mark. You can ignore friction.

**Step 1:** Choose two directions at right angles to resolve components along. You could resolve vertically and horizontally, but here it is easier to choose one of the directions to be along the ramp, as that is the way the trolley is moving.

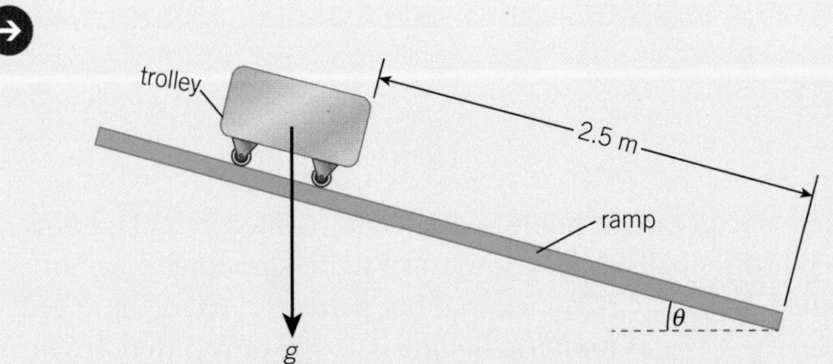


▲ **Figure 5** *Trolley on an inclined ramp 1*

The components of the acceleration due to gravity, $g$, are $g_{perpendicular}$, which has no effect, and $g_{parallel}$, which accelerates the trolley along the ramp.

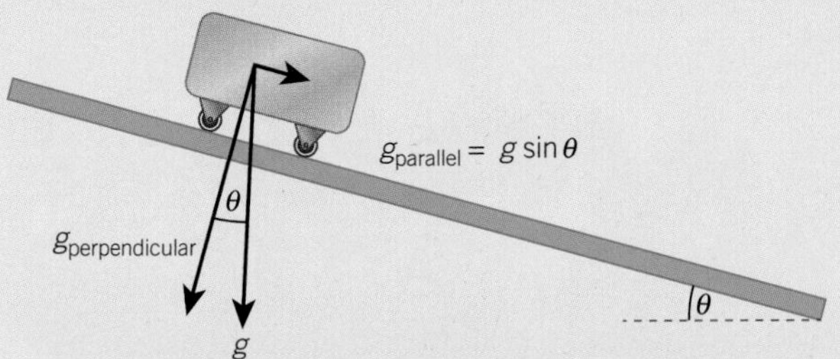


▲ **Figure 6** *Trolley on an inclined ramp 2*

**Step 2:** List the variables.

Along the ramp, $s = 2.5\,\text{m}$; $u = 0$; $v = ?$; $a = g_{parallel}$; $t = ?$

**Step 3:** Find $g_{parallel}$.

$$g_{parallel} = g\sin\theta = 9.8\,\text{m}\,\text{s}^{-2}\sin(15°) = 2.53...\,\text{m}\,\text{s}^{-2}$$

The rest of this worked example follows the example on for car stopping distance.

**Step 4:** Eliminate the unwanted variable.

This is $t$.

**Step 5:** Identify the correct equation.

This is the one with $s$, $u$, $v$, and $a$.

$$v^2 = u^2 + 2as$$

**Step 6:** Substitute values into the equation.

$v^2 = u^2 + 2as = 0^2 + 2 \times 2.53...\,\text{m}\,\text{s}^{-2} \times 2.5\,\text{m} = 12.6...\,\text{m}^2\,\text{s}^{-2}$

$v = \sqrt{(12.6...\,\text{m}^2\,\text{s}^{-2})} = 3.6\,\text{m}\,\text{s}^{-1}$ (2 s.f.)

Note here that $v = -3.6\,\text{m}\,\text{s}^{-1}$ is also a solution to $v = \sqrt{(12.7\,\text{m}^2\,\text{s}^{-2})}$. In this case the question asked for the speed, so the direction is not required.

## Summary questions

1 Use the braking distance data in Figure 4 to show that the deceleration to a final velocity $v = 0$ is the same at 60 mph as it was at 30 mph. *(3 marks)*

2 A car is travelling in a straight line. It has an initial velocity of $12\,\text{m}\,\text{s}^{-1}$. It accelerates at $3.5\,\text{m}\,\text{s}^{-2}$ for 4 s, then decelerates at $1.5\,\text{m}\,\text{s}^{-2}$ for 3.5 s. By considering the acceleration and deceleration stages separately, calculate the distance travelled since the start of the acceleration. *(4 marks)*

3 A ball is thrown upwards at a velocity of $11\,\text{m}\,\text{s}^{-1}$ at an angle of 50° to the horizontal. By resolving the velocity into appropriate components, find the time taken to reach its highest point. *(3 marks)*

4 A ball is thrown vertically upwards at $12\,\text{m}\,\text{s}^{-1}$ to dislodge a kite stuck in a tree 5 m above the thrower. Calculate the time before the ball reaches the kite. Explain why this question has two solutions. *(5 marks)*

# Physics in perspective

## Galileo takes on gravity

In a famous experiment published in 1638, Galileo described how he timed bronze balls rolling down a straight, smooth groove in an inclined plank. He reasoned that the pattern of changing speed would be the same as for a free-falling ball, the only difference being that its slower rate of change would make it easier to measure. He had already deduced that uniform acceleration would produce displacements proportional to the square of the time taken, and he wished to verify this.

▲ **Figure 1** *A 19th century painting showing Galileo (the tallest figure, just left of the centre) demonstrating his revolutionary concepts to Don Giovanni de Medici and members of his court.*

Galileo wrote, in *Dialogues Concerning Two New Sciences*, translated by Crew & de Salvio

*"We rolled the ball along the channel, noting, in a manner presently to be described, the time required to make the descent. We repeated this experiment more than once in order to measure the time with an accuracy such that the deviation between two observations never exceeded one-tenth of a pulse beat. Having performed this operation and having assured ourselves of its reliability, we now rolled the ball only one-quarter the length of the channel; and having measured the time of its descent, we found it precisely one-half of the former. Next we tried other distances, comparing the time for the whole length with that for the half, or with that for two-thirds, or three-fourths, or indeed for any fraction; in such experiments, repeated a full hundred times, we always found that the spaces traversed were to each other as the squares of the times, and this was true for all inclinations of the place, i.e. of the channel, along which we rolled the ball.*

*For the measurement of time, we employed a large vessel of water placed in an elevated position; to the bottom of this vessel was soldered a pipe of small diameter giving a thin jet of water, which we collected in a small glass during the time of each descent, whether for the whole length of the channel or for a part of its length; the water thus collected was weighed, after each descent, on a very accurate balance; the differences and ratios of these weights gave us the differences and ratios of the times, and this with such accuracy that, although the operation was repeated many, many times, there was no appreciable discrepancy in the results"*

## Summary questions

1 Galileo described in detail how he lined the groove in the plank with parchment and polished it, and used a smooth, very round bronze ball. Explain why he needed to take these steps.

2 Galileo had earlier reasoned that the displacement is proportional to the square of the time for constant acceleration. Give the equation that describes this kind of motion.

3 Figure 2 illustrates Galileo's apparatus. The end of the 12-cubit plank was lifted by different heights between 1 and 2 cubits, and various fractions of the measured length are marked.

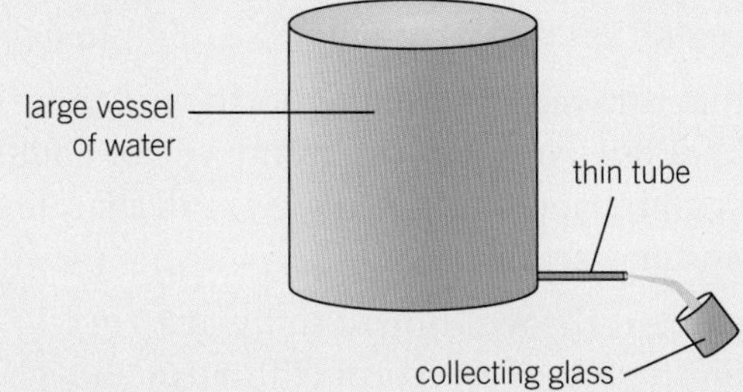


Galileo's 'stop clock'

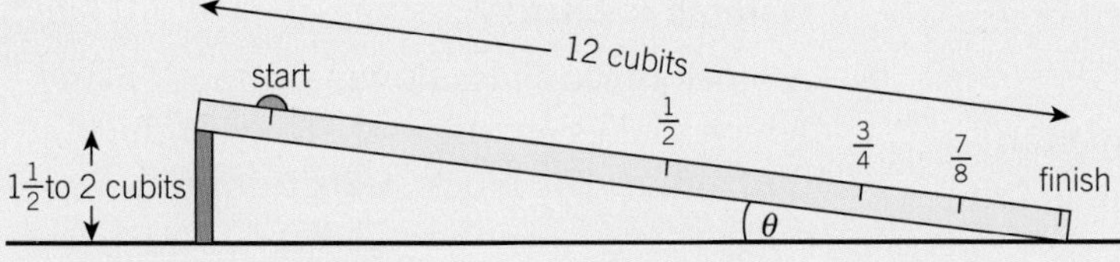


The inclined plank

▲ Figure 2

### Study tip

Cubits were an old unit used for measuring distance. One cubit is equal to roughly half a metre.

a Calculate the maximum and minimum values of the angle $\theta$ between the plank and the horizontal that he would have used.

b Galileo would have done this experiment with assistants. Describe a procedure that they might have followed.

c Accurate timing was not possible in the 17th Century, but weighing could be done to great precision. Galileo writes 'We repeated this experiment more than once in order to measure the time with an accuracy such that the deviation between two observations never exceeded one-tenth of a pulse beat'. Suggest how he might have calibrated his 'stop clock' in order to convert his weights into 'pulse beats'.

4 Using a value for $\theta$ of 8°,

a calculate the component of the acceleration of gravity down the slope $g = 9.8\ \mathrm{m\,s^{-2}}$

b use this value of acceleration to calculate the time taken to travel a distance of 12 cubits along the plank 1 cubit = distance from the elbow to the tip of the middle finger

c given that Galileo wrote that 'the deviation between two observations never exceeded one-tenth of a pulse beat', write down the expected time from **(b)** together with its uncertainty.

5 Galileo is often described as the person who did most to establish scientific method. Use examples from the extract of his book given above to explain how his approach to this experiment supports this claim.

## Practice questions

1 A car travelling in a straight line accelerates at a constant rate from rest to a velocity of $20\,\mathrm{m\,s^{-1}}$ over a time of 4.0 s, continues at this velocity for 2.0 s, and then decelerates at $2.5\,\mathrm{m\,s^{-2}}$ for 3.0 s.

Which of following shows a velocity-time graph for this motion?

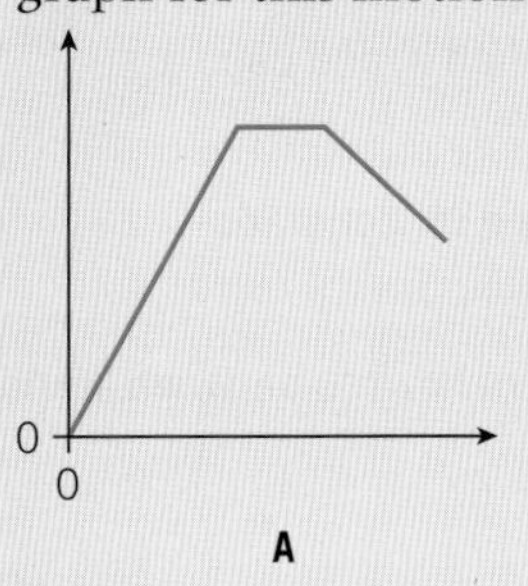


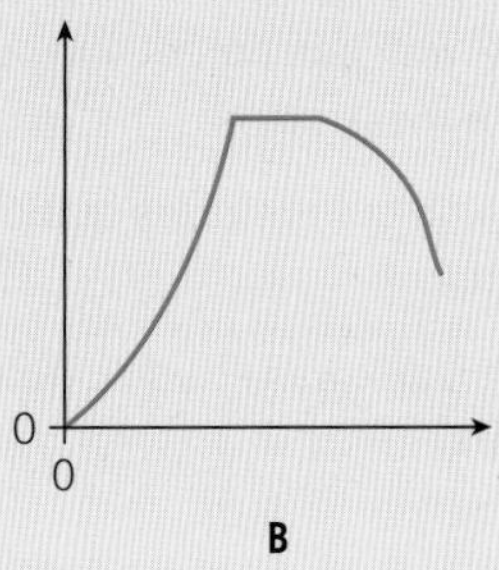


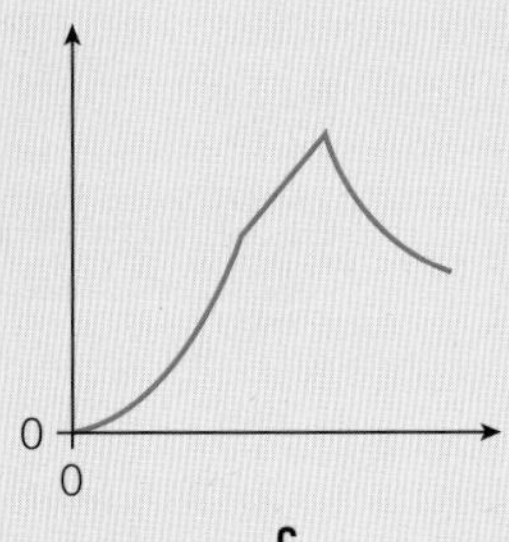


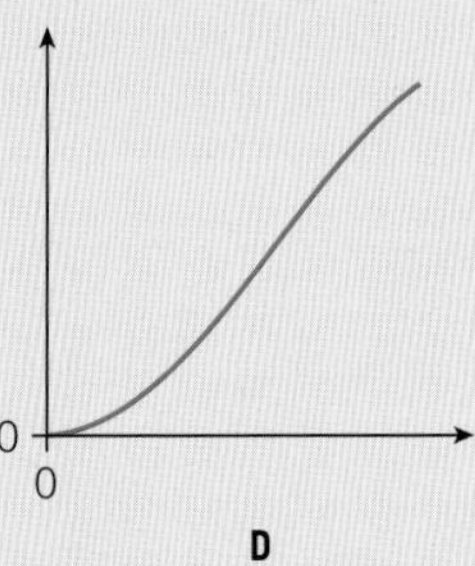


▲ Figure 1

2 Which of the graphs, **A**, **B**, **C** or **D**, given in question 1, shows a displacement-time graph for the motion described?

3 A stone is thrown with a velocity of $8.0\,\mathrm{m\,s^{-1}}$ down a well 20 m deep. What is its speed to 2 s.f. when it hits the bottom? ($g = 9.8\,\mathrm{m\,s^{-2}}$)

**a** $20\,\mathrm{m\,s^{-1}}$

**b** $21\,\mathrm{m\,s^{-1}}$

**c** $390\,\mathrm{m\,s^{-1}}$

**d** $460\,\mathrm{m\,s^{-1}}$

4 A woman dives from a high diving board into a swimming pool. She starts the dive by leaping upwards. Sketch a velocity-time graph for the dive from the moment she leaves the board to the moment that she enters the water. Label the graph fully. *(3 marks)*

5 Table 1 gives data values for an object falling freely from rest under gravity.

▼ Table 1

| t/s | s/m |
|---|---|
| 0 | 0 |
| 0.5 | 1.2 |
| 1.0 | 5.0 |
| 1.5 | 11 |
| 2.0 | 20 |
| 2.5 | 31 |

Plot a displacement-time graph from these data. By finding the velocity at two appropriate places, determine the acceleration due to gravity. *(6 marks)*

6 A hiker travels 8.4 km due north, and then 5.2 km due west. Represent the vector addition on graph paper and calculate the displacement from the starting point. *(4 marks)*

7 A stone is thrown horizontally at $15\,\mathrm{m\,s^{-1}}$ from the top of a vertical cliff, 50 m above the sea. Calculate the distance from the bottom of the cliff to the place where the stone hits the water. $g = 9.8\,\mathrm{m\,s^{-2}}$. *(3 marks)*

8 A block of ice slides down a smooth plane which makes an angle of 18° with the horizontal. The block starts from rest.

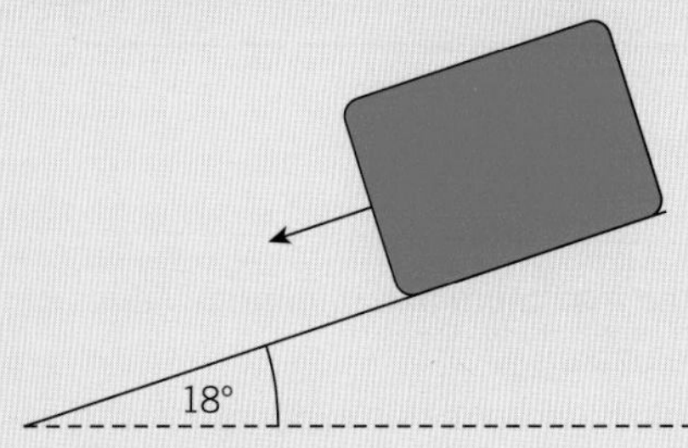


▲ Figure 3

**a** Calculate the component of $g$ parallel to the plane. *(2 marks)*

**b** Calculate the time taken for the block to slide 48 cm from rest along the plane, assuming there are no frictional forces. *(2 marks)*

**9** Figure 1 is the velocity-time graph for a short train journey between two stations joined by a straight track.

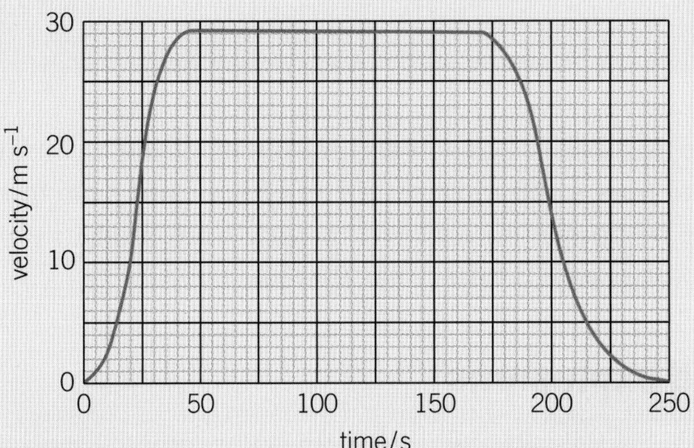


▲ Figure 2

**a** Describe, in words, the journey shown in Figure 2. *(3 marks)*

**b** The graph is not symmetrical. Suggest and explain a reason why the graph for the last 20 s of travel is not a mirror-image for the first 20 s of travel. *(2 marks)*

**c** Calculate the acceleration at time $t = 215$ s. Show your working clearly. *(3 marks)*

**d** Use the graph to find the distance between the two stations. *(3 marks)*

**10** An aircraft is scheduled to fly from London to Belfast, a distance of 510 km in a direction N 40° W (a bearing of 320°). The aircraft has a cruising speed of 240 m s$^{-1}$ in still air.

**a** Calculate the northwards and westward component of the aircraft's velocity when travelling on a still day. *(2 marks)*

**b** On the day of the flight, there is a wind of velocity 15 m s$^{-1}$ towards the east.

Using a clearly-labelled scale drawing, find the direction in which the aircraft must fly to reach Belfast without any change of course, and the magnitude of the velocity of the aircraft relative to the ground. *(5 marks)*

**c** Find the extra time the aircraft will take to fly to Belfast in this wind, compared with the time it would take on a still day. *(2 marks)*

**11** Two students are measuring the acceleration due to gravity, $g$, by timing the fall of a ball bearing.

**a** They time the ball falling five times for a height of 0.65 m and obtain the results in Table 2.

▼ Table 2

| time / s | 0.361 | 0.372 | 0.354 | 0.378 | 0.367 |
|---|---|---|---|---|---|

Use the data to find the mean time taken for the ball to fall this height. Give the uncertainty. *(2 marks)*

**b** The students repeat the experiment for a range of heights, obtaining the results shown in Table 3.

▼ Table 3

| $s$ / m | $t$ / s | $\Delta t$ / s |
|---|---|---|
| 0.65 | | |
| 0.70 | 0.38 | 0.03 |
| 0.75 | 0.40 | 0.03 |
| 0.80 | 0.42 | 0.04 |
| 0.85 | 0.43 | 0.04 |
| 0.90 | 0.44 | 0.05 |

Copy Table 3, including your results from part **(a)**, and add an extra column headed $t^2 / s^2$. Fill in the values for this column. *(2 marks)*

**c** Explain why a graph of $s$ (on the $y$-axis) against $t^2$ (on the $x$-axis) would be expected to give a straight line through the origin, and explain how the gradient is related to $g$, the acceleration due to gravity. *(3 marks)*

**d** Use the data in the table from part (b) to plot a graph of $s$ against $t^2$, including uncertainty bars for $t$, and use this to obtain a value for $g$ and its uncertainty. You can assume all the uncertainty bars are of the same size. *(6 marks)*

**e** Identify any systematic error in the experiment, and suggest and explain one way in which the two students could improve their experiment. *(3 marks)*

# 9 MOMENTUM, FORCE, AND ENERGY

## 9.1 Conservation of momentum

Specification references: 4.2a(ix), 4.2b(i), 4.2c(iv), 4.2c(vi)

**Learning outcomes**

Describe, explain, and apply:

- → the principle of conservation of momentum
- → the term: momentum
- → the equation $p = mv$.

What does a game of snooker have in common with firing a cannon? They both illustrate one of the most useful and important principles of physics, which can be applied in contexts ranging from nuclear physics to sending space probes around the Solar System.

### Collisions and explosions

The collision of two snooker balls and the firing of a cannon both illustrate the principle of conservation of momentum. Each of these interactions is best analysed with 'before' and 'after' diagrams.

▲ Figure 1 *A collision*

▲ Figure 2 *A cannon being fired*

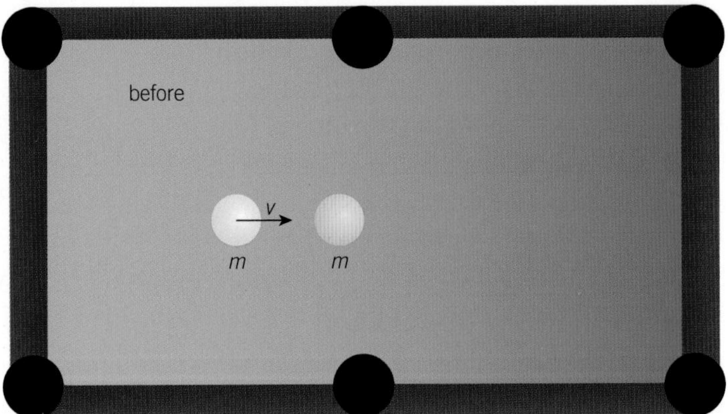


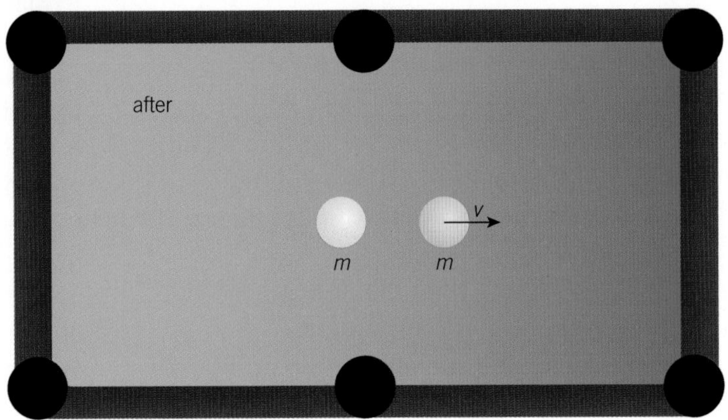


▲ Figure 3 *Collision in snooker*

The two snooker balls have identical mass, $m$. In a 'stun' shot the white ball, initially with velocity $v$, stops on impact and the yellow ball moves off with the same velocity $v$. If the balls had different masses, this would not have happened – a very light white ball hitting a massive yellow one would rebound, whilst the yellow one would move only very slowly as a consequence.

In the cannon, the cannon and ball are both stationary before firing. When the gunpowder explodes, the lighter ball moves off at a great speed $v_{ball}$, whilst the heavier cannon recoils with a smaller velocity $v_{cannon}$.

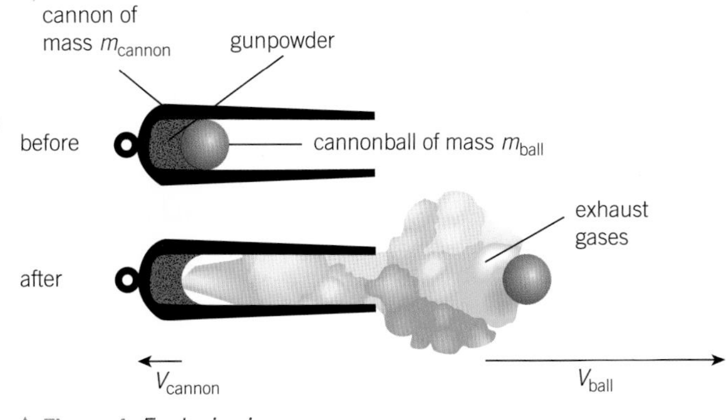


▲ **Figure 4** *Explosion in a cannon*

## Conservation of momentum

In the examples above, both the mass of the interacting objects and their velocities need to be considered when describing the interaction. The linking factor in these two events is **momentum**, $p$, which is defined as

$$\text{momentum } p\ (\text{kg m s}^{-1}) = \text{mass } m\ (\text{kg}) \times \text{velocity } v\ (\text{m s}^{-1})$$

### Worked example: Momentum of a bullet

Calculate the momentum of a 3.0 g bullet travelling at $180\,\text{m s}^{-1}$.

**Step 1:** Identify the equation.

$$p = mv$$

**Step 2:** Substitute values into the equation, making sure to use SI base units.

$$p = 3.0 \times 10^{-3}\,\text{kg} \times 180\,\text{m s}^{-1} = 0.54\,\text{kg m s}^{-1}$$

Note the units of momentum are $\text{kg} \times \text{m s}^{-1} = \text{kg m s}^{-1}$.

**Hint**

Like velocity, momentum is a vector.

For the two snooker balls, the total momentum before the collision = $mv + 0 = mv$, where $v$ is taken to be positive in the direction left→right. The total momentum after the collision = $0 + mv = mv$, the same as the total momentum before the collision.

For the cannon firing, the masses are different, but the velocities after firing are also different, and in opposite directions – if the cannon is 200 times more massive than the cannonball ($m_{cannon} = 200\,m_{ball}$), then its recoil velocity will be 200 times smaller ($v_{ball} = -200\,v_{cannon}$). The total momentum before the explosion = 0, as nothing is moving. After the explosion, ignoring the momentum of the exhaust gases,

$$\begin{aligned}\text{Total momentum} &= m_{cannon} \times v_{cannon} + m_{ball} \times v_{ball} \\ &= 200\,m_{ball} \times v_{cannon} + m_{ball} \times (-200\,v_{cannon}) \\ &= 200\,m_{ball} \times v_{cannon} - 200\,m_{ball} \times v_{cannon} = 0\end{aligned}$$

These examples illustrate an important law in physics – *the principle of conservation of momentum*. This principle states that, for *any* interaction, total momentum before = total momentum after.

The two examples above were simplified to make the issues clear. In reality, there are other interactions taking place in both cases. The principle holds true even in those cases – if all of the momentum changes in any interaction are included, then momentum is conserved.

### Worked example: Recoil of a cannon

One type of 17th century cannon, a demi-culverin, had a mass of 1500 kg and could fire a 3.6 kg cannonball at a speed of $120\,\text{m}\,\text{s}^{-1}$. Calculate the recoil velocity of the cannon, and explain why it would be larger than this in practice.

**Step 1:** Calculate the momentum $p_{\text{ball}}$ of the cannonball after the explosion.

$$p_{\text{ball}} = m_{\text{ball}} \times v_{\text{ball}} = 3.6\,\text{kg} \times 120\,\text{m}\,\text{s}^{-1} = 432\,\text{kg}\,\text{m}\,\text{s}^{-1}$$

**Step 2:** Apply the principle of conservation of momentum to the total momentum before and after firing.

Total momentum before = 0, so total momentum after also must = 0

**Step 3:** Calculate the momentum of the cannon after explosion, $p_{\text{cannon}}$.

From Step 2, $p_{\text{cannon}} + p_{\text{ball}} = 0$

$p_{\text{cannon}} + 432\,\text{kg}\,\text{m}\,\text{s}^{-1} = 0$ so $p_{\text{cannon}} = -432\,\text{kg}\,\text{m}\,\text{s}^{-1}$

**Step 4:** Use the definition of momentum to find the recoil velocity $v_{\text{cannon}}$.

$$p_{\text{cannon}} = m_{\text{cannon}} \times v_{\text{cannon}}$$

$$-432\,\text{kg}\,\text{m}\,\text{s}^{-1} = 1500\,\text{kg} \times v_{\text{cannon}}$$

$$v_{\text{cannon}} = \frac{-432\,\text{kg}\,\text{m}\,\text{s}^{-1}}{1500\,\text{kg}} = -0.29\,\text{m}\,\text{s}^{-1} \text{ (2 s.f.)}$$

In practice, the gases produced in the explosion also have momentum. As the total forward momentum of cannonball + exhaust gases must be larger than $432\,\text{kg}\,\text{m}\,\text{s}^{-1}$, the momentum of the cannon must be of the same size, so it will recoil faster than $0.29\,\text{m}\,\text{s}^{-1}$.

## Other types of collisions

You have now seen the principle of conservation of momentum applied to two situations. This principle can be applied for every type of collision, even when two bodies move off in the same direction afterwards, when two bodies collide head-on, or when two bodies make a glancing collision and move off at different angles to their path of approach.

### Worked example: A glancing collision

A white snooker ball of mass 0.16 kg moving at $6.0\,\text{m}\,\text{s}^{-1}$ hits a yellow ball of the same mass. After the collision, the two balls move off as shown in Figure 5. Calculate the final speed $v$ of the yellow ball.

**Synoptic link**

You have met how vectors can be resolved in Topic 8.2, Vectors.

This can be solved by resolving the velocities in two directions, $x$ and $y$.

Take $x$-direction to be the initial path of the white ball (where positive is left to right).

Take $y$-direction to be at right-angles to the $x$-direction (where positive is up).

By applying the principle of conservation of momentum to the $y$-direction, the momentum component for the white ball will be negative.

**Step 1:** Find the component of momentum for each ball after the collision in the $y$-direction.

White: $p_y = mv\cos(30°)$
$= 0.16\,\text{kg} \times -3.0\,\text{m}\,\text{s}^{-1} \times 0.866$
$= -0.41568\,\text{kg}\,\text{m}\,\text{s}^{-1}$

Yellow: $p_y = mv\cos(60°)$
$= 0.16\,\text{kg} \times v \times 0.5 = 0.08\,\text{kg} \times v$

**Step 2:** Apply the principle of conservation of momentum in the $y$-direction.

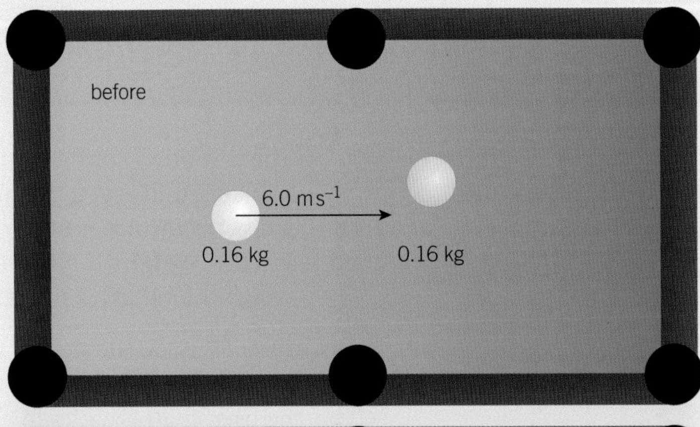


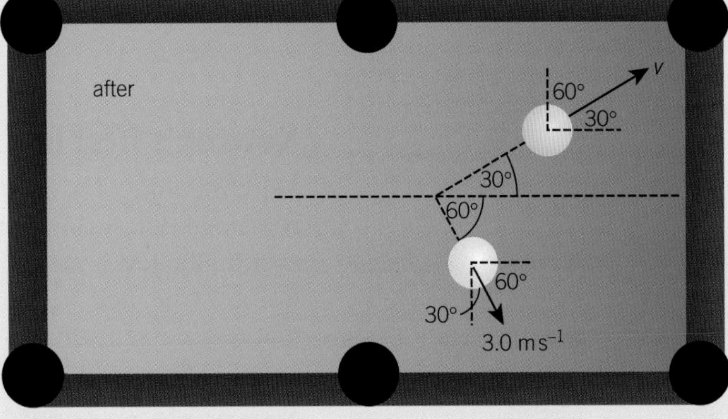


▲ **Figure 5** *Glancing collision*

$$p_y \text{ before} = 0 = p_y \text{ after}$$

$$0 = (-0.41568\,\text{kg}\,\text{m}\,\text{s}^{-1}) + (0.08\,\text{kg} \times v)$$

$$0.08\,\text{kg} \times v = 0.41568\,\text{kg}\,\text{m}\,\text{s}^{-1} \Rightarrow v = \frac{0.41568\,\text{kg}\,\text{m}\,\text{s}^{-1}}{0.08\,\text{kg}} = 5.2\,\text{m}\,\text{s}^{-1} \text{ (2 s.f.)}$$

This question can also be solved if you apply the principle of conservation of momentum in the $x$-direction. You should reach the same answer from the following calculation.

$$0.16\,\text{kg} \times 6.0\,\text{m}\,\text{s}^{-1} = (0.16\,\text{kg} \times 3.0\,\text{m}\,\text{s}^{-1} \times \cos(60°)) + (0.16\,\text{kg} \times v \times \cos(30°))$$

## Summary questions

**1** Calculate the momentum of **a** a 1000 kg car travelling at 60 mph (27 m s$^{-1}$); *(1 mark)*

**b** a person walking at a typical walking pace. (You will need to estimate values.) *(3 marks)*

**2** In a nuclear reaction, an atom of mass $4.0 \times 10^{-25}$ kg changes by emitting a particle of mass $6.6 \times 10^{-27}$ kg at a speed of $1.6 \times 10^{7}$ m s$^{-1}$. Calculate the speed at which the new atom recoils. State any assumptions that you make. *(3 marks)*

**3** In a head-on collision between a car of mass 900 kg and a van of mass 1300 kg, the two vehicles, locked together, skid in the direction the car was travelling. The van was travelling at 11 m s$^{-1}$ before impact. Calculate whether the car was exceeding the speed limit (13.4 m s$^{-1}$) before the two vehicles crashed. *(3 marks)*

# 9.2 Newton's laws of motion and momentum

Specification references: 4.2a(viii), 4.2a(ix), 4.2b(i), 4.2c(v)

**Learning outcomes**

Describe, explain, and apply:

- → Newton's laws of motion
- → Newton's third law as a consequence of the principle of conservation of momentum
- → the terms: force, mass, impulse
- → the equation $F = ma = \frac{\Delta(mv)}{\Delta t}$ where the mass is constant.

▲ **Figure 1** *Statue of Sir Isaac Newton in Trinity College, Cambridge. Of Newton, Wordsworth wrote, "Newton with his prism and silent face, the marble index of a mind forever voyaging through strange seas of thought, alone."*

**Study tip**

Newton's first law can be summarised as:

*A body will have a constant velocity, or stay at rest, unless an external force acts upon it.*

There is often confusion between mass and force, which is made worse by using units such as kg, stones, and pounds, which are measures of **mass**, or the amount of matter in a body, and calling these units of weight. **Weight** is a **force**, and all forces are pushes and pulls – the weight of a mass is the pull of gravity acting on it.

In 1687 Isaac Newton published his *Principles of Natural Philosophy*, arguably the most important scientific book ever written, in which he used mathematics to explain the relationship between force and motion and the movements of planets and moons. He started this work with three assumptions. These he called his *three laws of motion*.

## Newton's first law – inertia

Since the time of Aristotle, most people believed that things will not move unless they are pushed or pulled, and that moving bodies eventually 'lost impetus' and slowed down. Galileo showed that this was not true, and Newton's first law summarises this idea of **inertia** – that the momentum of a body will not change unless a force acts on it.

## Newton's second law – force

A weaker force and a stronger force will both change the momentum of an object, but the stronger force will make it happen faster. This means that force is proportional to the rate at which momentum changes. Since $p = mv$, we can say

$$F \propto \frac{\Delta p}{\Delta t} \propto \frac{\Delta(mv)}{\Delta t}$$

This is Newton's second law. As momentum change $\Delta p$ is a vector quantity, force $F$ must also be a vector quantity in the same direction.

### The newton

Although the equation $F \propto \frac{\Delta(mv)}{\Delta t}$ implies the need for a constant of proportionality, providing that we use SI base units (kg, m s$^{-1}$, and s) then $F = \frac{\Delta(mv)}{\Delta t}$ where $F$ is in newtons (N).

#### Force and acceleration

Providing that the mass of a body stays constant then we can write Newton's second law in a different form. As $p = mv$, if $m$ stays constant then $\Delta p = m\Delta v$ . The equation now becomes

$$F = \frac{\Delta p}{\Delta t} = \frac{m\Delta v}{\Delta t} = m\frac{\Delta v}{\Delta t} = ma$$

$F = ma$ is often the most convenient form of Newton's second law.

## Worked example: Force accelerating a car

A luxury car of mass 2400 kg accelerates from 0 to 60 mph ($26.8\,m\,s^{-1}$) in 4.7 s. Calculate the mean resultant force on the car during that time.

There are two ways of doing this.

**Method 1: Using momentum change**

**Step 1:** Calculate the change in momentum.

Initial momentum = 0 (it's not moving)

$$\text{Final momentum } p = mv = 2400\,kg \times 26.8\,m\,s^{-1} = 64\,300\,kg\,m\,s^{-1}$$

$$\text{Momentum change } \Delta p = 64\,300\,kg\,m\,s^{-1} \;\; 0 = 64\,300\,kg\,m\,s^{-1}$$

**Step 2:** Divide by the time it took for that momentum change.

$$F = \frac{\Delta p}{\Delta t} = \frac{64\,300\,kg\,m\,s^{-1}}{4.7\,s} = 14\,000\,N \text{ (2 s.f.)}$$

**Method 2: Using acceleration**

**Step 1:** Use the *suvat* equations to find the acceleration, $a$.

$$s = ?,\ u = 0,\ v = 26.8\,m\,s^{-1},\ a = ?,\ \text{and } t = 4.7\,s$$

**Step 2:** Discard the variable which is not needed ($s$) and use the equation with the other four variables.

$$v = u + at \Rightarrow a = \frac{v - u}{t} = \frac{26.8\,m\,s^{-1} - 0}{4.7\,s} = 5.70...\,m\,s^{-2}$$

**Step 3:** Substitute values into $F = ma$.

$$F = ma = 2400\,kg \times 5.70...\,m\,s^{-2} = 14\,000\,N \text{ (2 s.f.)}$$

## Impulse

The statement of Newton's second law $F = \frac{\Delta p}{\Delta t}$ can be rearranged to give $\Delta p = F\Delta t$. $F\Delta t$ is called the **impulse** of the force. If the force is larger, or the time it acts for is larger, then the impulse is greater. This means there is a greater momentum change produced.

Crumple zones in cars are designed to increase the time $\Delta t$ taken in a collision. Since $\Delta p = F\Delta t$, for the same change in momentum the force $F$ on the car is reduced as the time $\Delta t$ is increased.

Seat belts and air bags have exactly the same role in protecting the driver and passengers – seat belts gradually stretch and air bags gradually deflate, increasing the time taken for a person to come to a halt if the car stops very suddenly.

The relationship between impulse and momentum change can be applied in cases where $F$ is not constant, as in kicking a football (Figure 2).

### Study tip

Newton's second law can be summarised as:

*The momentum change per second produced by an external force is in the same direction as the force, and proportional to it.* $F = \frac{\Delta p}{\Delta t} = ma$

### Hint

Remember that acceleration, $a = \frac{\text{change in velocity } \Delta v}{\text{change in time } \Delta t}$

### Synoptic link

You have met acceleration in Topic 8.1, Graphs of motion.

### Synoptic link

You first met *suvat* equations in Topic 8.4, Speeding up and slowing down.

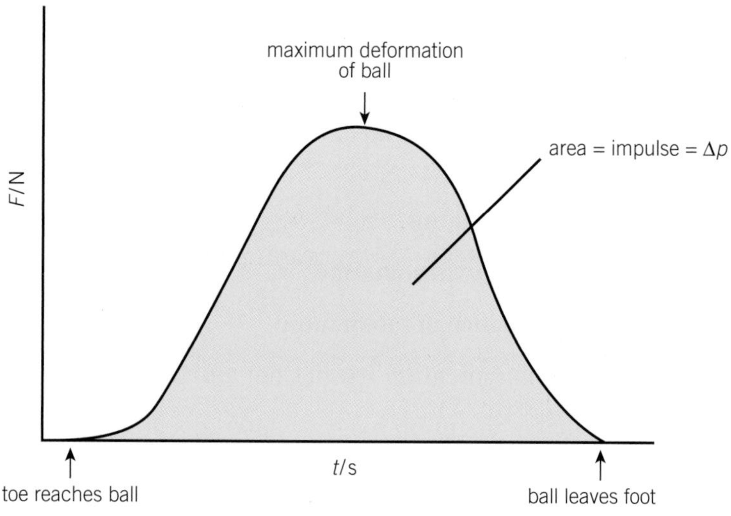


▲ **Figure 2** *Kicking a football*

## Synoptic link

Using the total area under a $v$–$t$ graph to find distance travelled was covered in Topic 8.1, Graphs of motion.

When the toe first meets the ball, the ball just starts to deform, so the force is not great. The force rises to a maximum as the ball accelerates away and drops to zero as the ball leaves the foot. For a constant force, the impulse $F\Delta t$ would be the area under a line $\Delta t$ long and $F$ high. In this case, the impulse is the total area under the graph, just as the distance travelled is the total area under a $v$–$t$ graph.

## Newton's third law

This is exactly equivalent to the principle of conservation of momentum. Figure 3 shows an interaction between two bodies **A** and **B** where **A** gains momentum and **B** loses momentum. The nature of the force between **A** and **B** is not important – it could be a gravitational or electromagnetic force, or simply from contact.

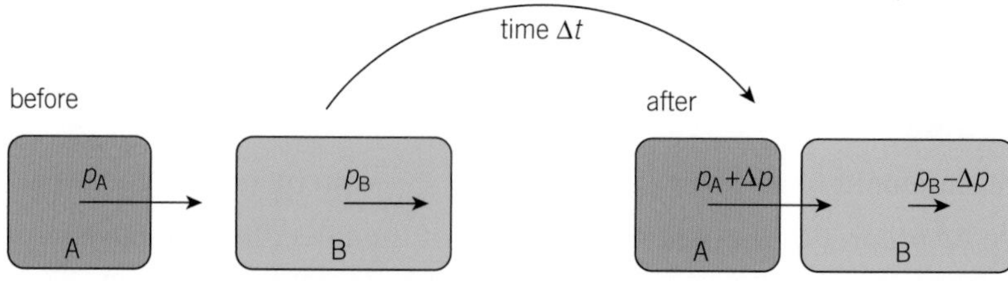


▲ **Figure 3** *An interaction between two bodies*

Momentum is always conserved, so the momentum lost by B is that gained by A. A's momentum change is + $\Delta p$ and B's momentum change is – $\Delta p$. The momentum changes occur over the same time $\Delta t$, so from $F_A = \frac{\Delta p}{\Delta t}$ and $F_B = \frac{-\Delta p}{\Delta t}$, $F_A = -F_B$.

This is true for *all* interactions. Forces always come in pairs, equal in size and opposite in direction. If **A** pulls **B** in one direction, then **B** pulls **A** with an identical force in the opposite direction.

## Study tip : Draw a diagram

To avoid confusion, always draw clear diagrams showing all the forces present. Remember that the two forces in a Newton's third law pair act on different objects.

### Mass and weight

A free-falling object of mass $m$ near the Earth's surface accelerates downwards at $g = 9.8\,\text{m}\,\text{s}^{-2}$. From Newton's second law, the force needed to produce this acceleration is $F = ma = mg$. This force is called the *weight* of the object. A supported object, such as a book on a shelf, is not accelerating and so by Newton's first law has zero resultant force – the shelf must be exerting an equal upwards force.

The weight $mg$ is caused by the gravitational pull of the Earth. By Newton's third law,

downwards pull of the Earth on the book = upwards pull of the book on the Earth

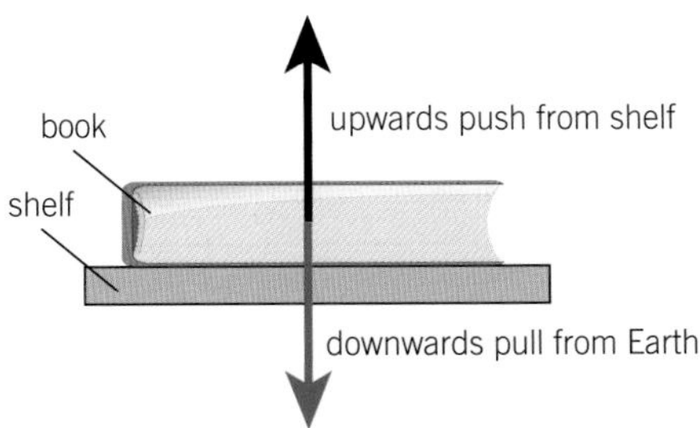


▲ **Figure 4** *Forces on a book on a shelf*

**Study tip**

Forces always occur in equal and opposite pairs between the two interacting bodies, $F_A = -F_B$

### Worked example: Does the Earth move?

A 65 kg diver dives from a 10 m diving board into a pool. He takes 1.4 s to cover the 10 m. How far does the Earth move upwards? Mass of Earth = $6.0 \times 10^{24}$ kg.

**Step 1:** Find the force on the diver.

The force is the diver's weight = $mg = 65\,\text{kg} \times 9.8\,\text{m s}^{-2}$
$= 637\,\text{N}$ downwards

**Step 2:** Identify Newton's third law pair.

Downward force on diver = upward force on the Earth
= 637 N upwards

**Step 3:** Find the upwards acceleration of the Earth using Newton's second law.

$$F = ma \Rightarrow a = \frac{F}{m} = \frac{637\,\text{N}}{6.0 \times 10^{24}\,\text{kg}} = 1.06... \times 10^{-22}\,\text{m s}^{-2}\text{ upwards}$$

**Step 4:** Use the *suvat* equations to find the displacement $s$.

$s = ?$, $u = 0$, $v = ?$, $a = 1.1 \times 10^{-22}\,\text{m s}^{-2}$, $t = 1.4\,\text{s}$. Eliminate $v$ which is not needed.

$$s = ut + \tfrac{1}{2}at^2 = 0 + \tfrac{1}{2}at^2 = 0.5 \times 1.1 \times 10^{-22}\,\text{m s}^{-2} \times (1.4\,\text{s})^2$$
$$= 1.0 \times 10^{-22}\,\text{m (2 s.f.)}$$

Yes, the Earth does move, but only by a distance of $1.0 \times 10^{-22}$ m, which is approximately $10^{12}$ times smaller than the size of a hydrogen atom.

## Summary questions

1 A car of mass 1100 kg travelling at 15 m $s^{-1}$ collides with a wall and stops in a time of 0.1 s. Calculate the deceleration of the car and use this value to find the average force acting on it during the collision. *(3 marks)*

2 When a gun fires a bullet, it recoils. Explain this using:

- **a** the principle of conservation of momentum; *(1 mark)*
- **b** Newton's third law. *(2 marks)*

3 When a car stops suddenly, you feel yourself being thrown forward. Explain in terms of Newton's laws what is actually happening. *(2 marks)*

4 During the collision in question 1, a passenger of mass 75 kg is decelerated uniformly to rest by a seatbelt. The belt stretches so much that the passenger travels a total distance of 2.1 m between the front of the car hitting the wall and the passenger coming to rest. Calculate the impulse of the force acting on the passenger and use this to calculate the mean force acting on him. *(4 marks)*

5 A rocket engine ejects 500 g of exhaust gases each second for 3 s. The velocity of the gases, relative to the rocket, is 250 m $s^{-1}$. The mass of the rocket is 2.5 kg.

- **a** Calculate the force acting on the rocket. *(2 marks)*
- **b** Explain why the acceleration will not remain constant for the 3 s of the rocket's flight. *(2 marks)*

# 9.3 Conservation of energy

## Specification references: 4.2a(v), 4.2b(i), 4.2c(vii), 4.2c(viii), 4.2c(ix)

In a lecture, the Nobel-winning physicist Richard Feynman said, 'There is a fact, or if you wish a law, governing all natural phenomena that are known to date. There is no exception to this law – it is exact so far as is known. The law is called the *conservation of energy*. Energy is a mathematical principle, a numerical quantity which does not change when something happens. It is not a description of a mechanism, or anything concrete; it is just a strange fact that we can calculate some number and when we finish watching nature go through her tricks and calculate the number again, it is the same'.

If you look in a dictionary it will tell you that energy 'is a measure of the ability to do work'. But what is meant by *work*?

### Learning outcomes

Describe, explain, and apply:

- → the principle of conservation of energy
- → the terms: work, energy
- → the equation work done $\Delta E = F\Delta s$
- → the equation $E_k = \frac{1}{2}mv^2$
- → the equation for gravitational energy, $\Delta E_{grav} = mg\Delta h$.

## Force and work

In the previous topic, you saw that a force $F$ acting on an object over a time $\Delta t$ had an effect called *impulse*, where impulse = $F \times \Delta t$, which was the change in the momentum of the object. Another important measure of what a force does is obtained when you multiply the force $F$ by the displacement $\Delta s$ of the object it acts on. This effect is given the name **work** $\Delta E$ and is measured in joules (J), when $F$ is in N and $s$ in m, giving the equation

$$\text{Work } \Delta E \text{ (J)} = \text{force } F \text{ (N)} \times \text{displacement } \Delta s \text{ (m)}$$

▲ **Figure 1** *Richard Feynman*

### Worked example: Work done in accelerating a car

In the previous topic you saw that the resultant force accelerating a car from 0 to 60 mph (26.8 m s$^{-1}$) in 4.7 s was 13 700 N. Calculate the work done during this acceleration.

**Step 1:** Write down the *suvat* variables for this situation.

$$s = ?,\ u = 0,\ v = 26.8\,\text{m s}^{-1},\ a = ?,\ t = 4.7\,\text{s}$$

$a$ can be eliminated as it is not required here.

**Step 2:** Select the correct *suvat* equation and substitute in values.

$$s = \frac{u + v}{2}t = \frac{(0 + 26.8\,\text{m s}^{-1})}{2} \times 4.7\,\text{s} = 13.4\,\text{m s}^{-1} \times 4.7\,\text{s} = 62.98\,\text{m}$$

**Step 3:** Calculate the work done.

$$E = F \times s = 13\,700\,\text{N} \times 62.98\,\text{m} = 862\,826\ \text{J}$$
$$= 860\,\text{kJ (2 s.f.)}$$

## Conservation of energy

**Energy** is the capacity to do work. The fuel in the car, burning in air, does 860 kJ of work on the car to accelerate it, so the fuel has less capacity to do work than it had. If the car decelerates to a halt, the brakes exert a force on the car and do work on the brake linings, heating them. The moving car has the capacity to do work (heating the brake linings) which it did not have when it was still. The energy of any moving object is called its **kinetic energy**.

In this example, we can say that the fuel is acting as an energy store, and that doing 860 kJ of work in speeding up the car reduces the fuel and air energy store by 860 kJ. The moving car is now an energy store. Would it now have 860 kJ? In reality, the kinetic energy store of the moving car will be less than 860 kJ as there are numerous other forces present, all doing work.

## Kinetic energy and work done

To obtain a useful equation for the kinetic energy store of a moving object, consider a uniformly accelerating car.

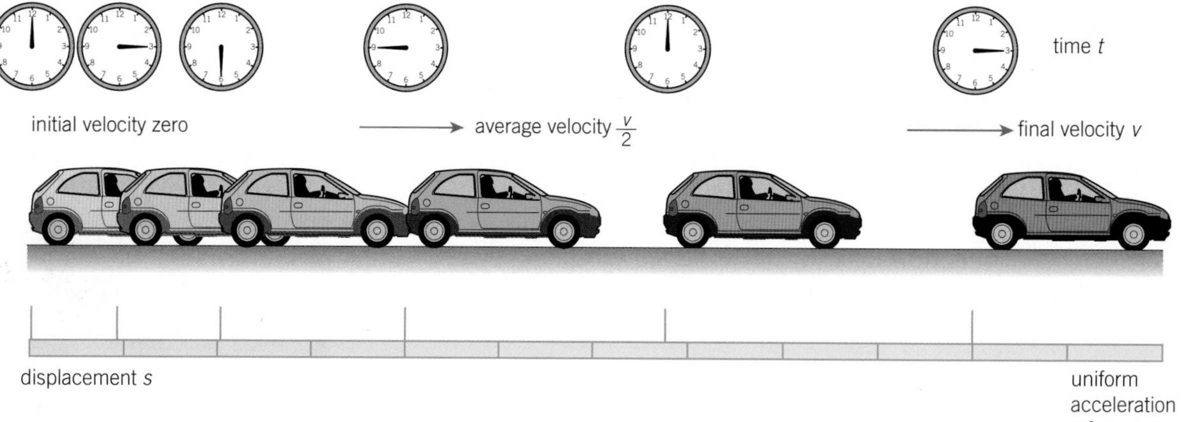


▲ **Figure 2** *Calculating kinetic energy increase*

The accelerating force is given by Newton's second law, $F = ma$.

The kinetic energy gained by the car is the work done in accelerating it, as it had no kinetic energy beforehand, as it was stationary.

$$\text{Kinetic energy } E_k = Fs = mas$$

We can get $a\,s$ from the *suvat* equation $v^2 = u^2 + 2as$.

$$\text{as } u = 0,\ v^2 = 0 + 2as = 2as \Rightarrow \tfrac{1}{2}v^2 = a\,s$$

$$\text{Kinetic energy, } E_k = m(as) = m(\tfrac{1}{2}v^2) = \tfrac{1}{2}mv^2$$

### Worked example: Calculating the kinetic energy of a bee

A bumble bee has a mass of 0.50 g, and flies at 80 cm s$^{-1}$. Calculate its kinetic energy.

**Step 1:** Convert the data to SI base units.

$$m = 0.5 \times 10^{-3}\,\text{kg},\ v = 80 \times 10^{-2}\,\text{m s}^{-1}$$

**Step 2:** Substitute values into the kinetic energy equation.

$$E_k = \tfrac{1}{2}mv^2 = 0.5 \times (0.5 \times 10^{-3}\,\text{kg}) \times (80 \times 10^{-2}\,\text{m s}^{-1})^2 = 0.000\,16\,\text{J}$$
$$= 0.16\,\text{mJ}$$

## Gravitational potential energy and work done

Another way in which a force can do work is to lift something. Gravity gives masses weight, which near the Earth's surface will make them accelerate downwards at $a = g = 9.8\,\text{m s}^{-2}$ in the absence of any other forces. As $F = ma$, the force with which an object of mass $m$ is pulled downwards (its weight) $= mg$.

The force that is needed to support an object in the Earth's gravitational field is its weight $mg$ and that is the force you need to apply to lift it at a steady speed. When it is lifted through a displacement $h$ – the symbol '$h$' is used rather than '$s$' to remind us it is a height, measured vertically – then

$$\text{Work done} = \text{force} \times \text{displacement} = (mg) \times h = mgh$$

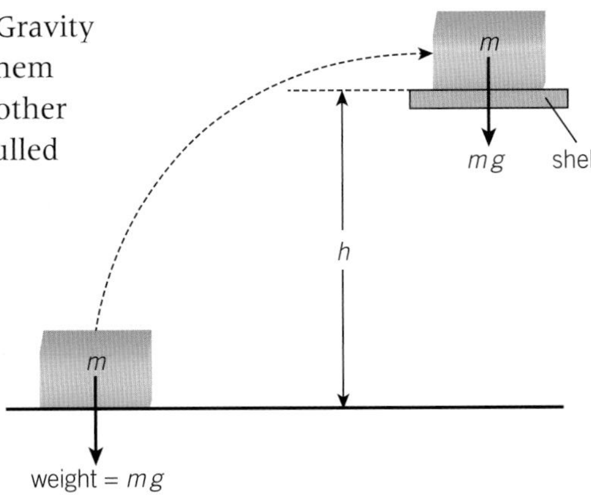


▲ **Figure 3** *Calculating gravitational potential energy increase*

The increase in gravitational potential energy, $\Delta E_{\text{grav}} = mgh$

Gravitational potential energy acts as a store of energy – the work done in lifting it can be transferred from this store if it is allowed to fall.

### Worked example: Energy exchange as an apple falls

A 0.10 kg apple falls 3.5 m from a tree. Calculate the speed of the apple when it reaches the ground.

**Step 1:** State the law of conservation of energy in this situation.

Kinetic energy gained = change in gravitational potential energy

**Step 2:** Using the appropriate equations, calculate each term in this equation.

$$E_k = \tfrac{1}{2}mv^2 = 0.5 \times 0.10\,\text{kg} \times v^2 = (0.050\,\text{kg}) \times v^2$$

$$\Delta E_{\text{grav}} = mgh = 0.10\,\text{kg} \times 9.8\,\text{m s}^{-2} \times 3.5\,\text{m} = 3.43\,\text{J}$$

**Step 3:** Substitute the terms into the conservation of energy equation and solve.

$$(0.050\,\text{kg}) \times v^2 = 3.43\,\text{J} \Rightarrow v^2 = \frac{3.43\,\text{J}}{0.05\,\text{kg}} = 68.6\,\text{m}^2\,\text{s}^{-2}$$

$$v = \sqrt{(68.6\,\text{m}^2\,\text{s}^{-2})} = 8.3\,\text{m s}^{-1}\ (2\text{ s.f.})$$

The question above could equally be solved using *suvat* equations (in particular $v^2 = u^2 + 2as$) as multiplying this equation throughout by $\frac{1}{2}m$, putting $a = g$, and putting $s = h$ results in

$$\tfrac{1}{2}mv^2 = \tfrac{1}{2}mu^2 + mgh$$

Final kinetic energy = initial kinetic energy + change in gravitational potential energy

## Summary questions

1 Using a reasonable estimate of the velocity, calculate the kinetic energy of a 1100 kg car speeding down a motorway. *(2 marks)*

2 A catapult fires a 0.05 kg stone vertically into the air. It is accelerating the stone over a distance of 20 cm with an average force of 1.6 N. Use the principle of conservation of energy to find the distance the stone rises into the air. *(4 marks)*

3 Show that kinetic energy is related to momentum by the equation $E_k = \frac{P^2}{2m}$. *(3 marks)*

4 A trolley of mass 0.8 kg is accelerated from rest along a horizontal surface as shown in Figure 4.

▲ **Figure 4**

The accelerating weight has a mass of 600 g. Ignoring friction, calculate the velocity of the trolley when the weight has fallen 55 cm. *(5 marks)*

# 9.4 Projectiles

## Specification references: 4.2a(ii), 4.2a(iii)

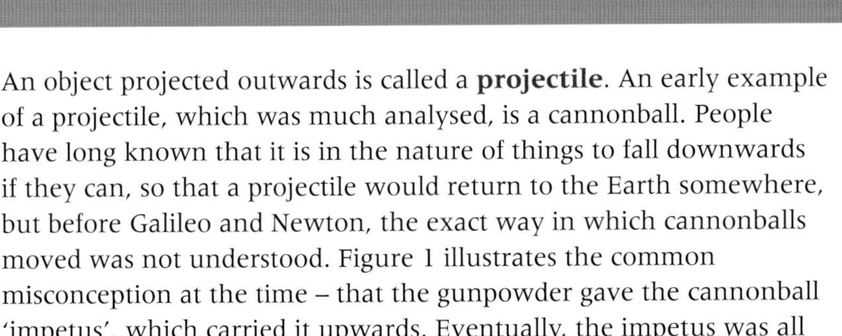

An object projected outwards is called a **projectile**. An early example of a projectile, which was much analysed, is a cannonball. People have long known that it is in the nature of things to fall downwards if they can, so that a projectile would return to the Earth somewhere, but before Galileo and Newton, the exact way in which cannonballs moved was not understood. Figure 1 illustrates the common misconception at the time – that the gunpowder gave the cannonball 'impetus', which carried it upwards. Eventually, the impetus was all used up, and the cannonball fell. Gunnery was not a precise science in the 16th century.

### Learning outcomes

Describe, explain, and apply:

- → the trajectory of a body moving under constant acceleration, in one or two dimensions
- → the independent effects of perpendicular components of a force.

## Horizontal and vertical components

The weight $mg$ of a body can be resolved into perpendicular components at an angle $\theta$ as shown in Figure 2.

As the angle $\theta$ becomes larger, the component $mg\cos\theta$ becomes smaller and smaller. In the limit, when $\theta = 90°$

$$mg\cos\theta = mg\cos(90°) = mg \times 0 = 0$$

Weight is a vertical force – it has no horizontal component and cannot change the horizontal velocity component of a moving object. Air resistance will affect this component, but in this section we are assuming that air resistance is negligible. For any projectile, the initial velocity $u$ can be resolved into vertical and horizontal components as shown in Figure 3.

The vertical component $u\sin\theta$ is accelerated downwards by the weight of the object – this vertical acceleration = $g$ downwards.

The horizontal component $u\cos\theta$ is not affected by gravity, and continues at this value as long as the projectile is in motion. This means that forces perpendicular to one another act *independently*.

A serve in tennis is hit from a height of approximately 3 m. If simply dropped from that height, you would estimate that the ball would reach the ground in around 0.8 s. If the server hits the ball horizontally at that height, the ball still takes just the same amount of time to reach the ground. Its horizontal (component of) velocity simply carries it towards you as it falls.

In reality, the server usually hits the ball at a downward angle, giving it quite a large downward component of velocity. So the ball reaches the ground near you even sooner, just as it would if it had been thrown downwards.

Figure 4 shows the independent vertical and horizontal motion of a tennis ball struck horizontally.

▲ **Figure 1** *Medieval projectile analysis*

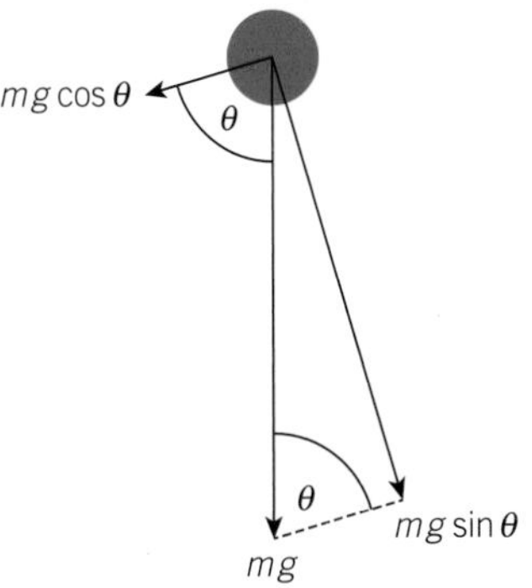


▲ **Figure 2** *Components of weight*

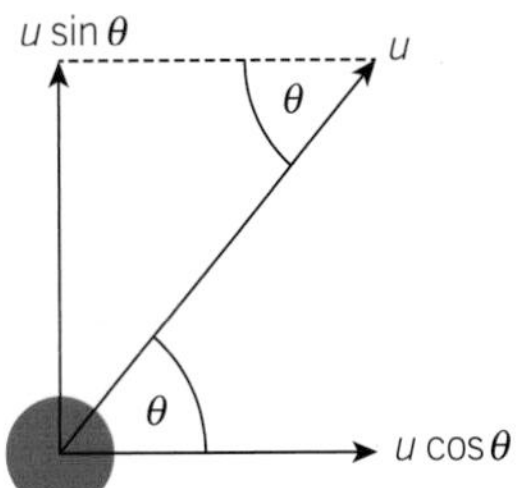


▲ **Figure 3** *Resolving the initial velocity of a projectile*

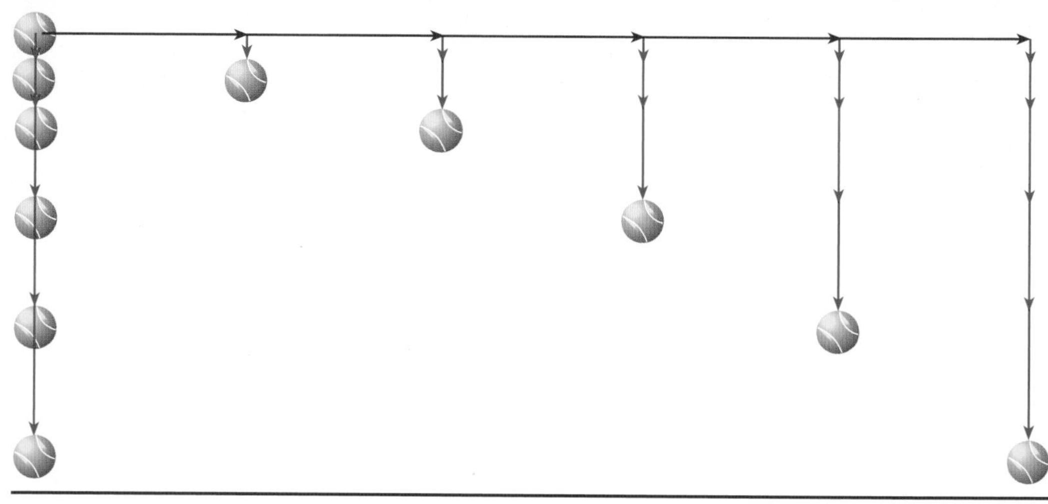


▲ **Figure 4** *Falling and going sideways*

**Synoptic link**

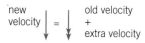

Resolving forces into components follows the same method as resolving vectors. You met this in Topic 8.2, Vectors.

**Synoptic link**

You have met the iterative model in Topic 8.3, Modelling motion.

▲ **Figure 5** *Fountains send water along parabolic paths*

Detailed mathematical analysis using the equation $s = ut + \frac{1}{2}at^2$ applied to the horizontal and vertical components of displacement gives the result that the shape of the trajectory of a projectile is a parabola. Or, rather, it has this shape if the forces that are the result of the ball moving through and spinning in the air, for example, air resistance, are unimportant. This means that a parabola is a good approximation of the path of a moving cricket ball, a fair approximation for a moving football or tennis ball, and a very bad approximation for a moving badminton shuttlecock, since air resistance plays a huge part in the movement of a shuttle.

The path of a tennis ball during a lob can be analysed using iterative modelling. In a short interval of time, the downward acceleration will have added a vertical downward component to the velocity. The resultant velocity is the vector sum of the two. So the ball's motion is tilted down a little, and it slows down a little. In the next moment, the same thing happens, and again, and again, so that the ball's path always curves downwards. Sooner or later, the downward acceleration will have removed all of the upward component of velocity, and the ball will begin travelling downwards.

In Figure 6 (next page), in each time interval $\Delta t$ the blue arrow is the displacement vector assuming that the ball had the same velocity as in the previous $\Delta t$, and the red arrow is the extra vertical displacement $\frac{1}{2}g(\Delta t)^2$ produced by gravity in that time. The vector sum gives the resultant displacement shown in purple.

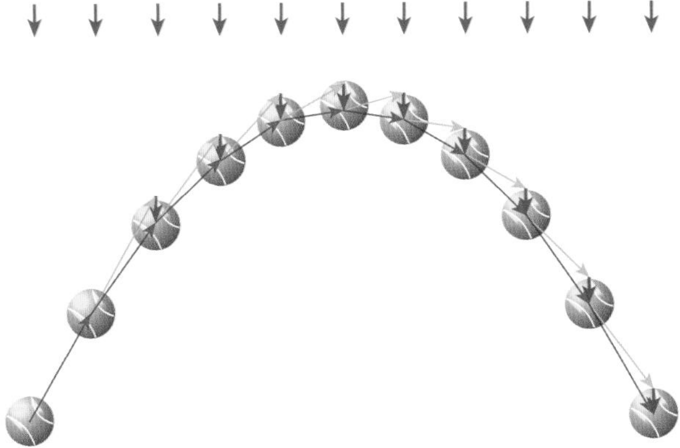


▲ **Figure 6** *Lobbing a ball in the air*

## Worked example: Range of a projectile

A ball is thrown with an initial velocity of $18\,\mathrm{m\,s^{-1}}$ at an angle of 60° to the horizontal. Calculate how far it travels before hitting the ground. You can assume it was thrown from a height of 1.5 m above the ground.

**Step 1:** Decide on the components required to answer this question.

Use vertical motion to find the time taken to reach the ground, then horizontal motion to find how far it has travelled.

**Step 2:** Identify the equations required to resolve the components of $u$.

$$\text{Vertical component} = u\sin\theta$$

$$\text{Horizontal component} = u\cos\theta$$

**Step 3:** Substitute values and evaluate components.

Vertical component $= u\sin\theta = (18\,\mathrm{m\,s^{-1}}) \times \sin(60°) = 15.5...\,\mathrm{m\,s^{-1}}$

Horizontal component $= u\cos\theta = (18\,\mathrm{m\,s^{-1}}) \times \cos(60°) = 9.0\,\mathrm{m\,s^{-1}}$

**Step 4:** Use values from Step 3 to calculate vertical motion using *suvat*.

Taking downwards as positive, $s_v = 1.5\,\mathrm{m}$, $u_v = -15.5...\,\mathrm{m\,s^{-1}}$, $a = g = 9.8\,\mathrm{m\,s^{-2}}$, $t = ?$

$$s_v = u_v t + \tfrac{1}{2}at^2$$

$$\begin{aligned}1.5\,\mathrm{m} &= (-15.5...\,\mathrm{m\,s^{-1}})t + 0.5 \times (9.8\,\mathrm{m\,s^{-2}})t^2 \\ &= (4.9\,\mathrm{m\,s^{-2}})\,t^2 - (15.5...\,\mathrm{m\,s^{-1}})t\end{aligned}$$

$(4.9\,\mathrm{m\,s^{-2}})t^2 - (15.5...\,\mathrm{m\,s^{-1}})t - 1.5\,\mathrm{m} = 0$ is a quadratic equation $ax^2 + bx + c = 0$ where $a = 4.9\,\mathrm{m\,s^{-2}}$, $b = -15.5...\,\mathrm{m\,s^{-1}}$, and $c = -1.5\,\mathrm{m}$.

The solution to $ax^2 + bx + c = 0$ is $x = \dfrac{-b \pm \sqrt{b^2 - 4ac}}{2a}$

In this case

$$t = \frac{-(-15.5\ldots\,\mathrm{m\,s^{-1}}) \pm \sqrt{(-15.5\ldots\,\mathrm{m\,s^{-1}})^2 - 4(4.9\,\mathrm{m\,s^{-2}})(-1.5\,\mathrm{m})}}{2(4.9\,\mathrm{m\,s^{-2}})}$$

$$= \frac{-15.5\ldots\,\mathrm{m\,s^{-1}} \pm \sqrt{273\,\mathrm{m^2\,s^{-2}}}}{9.8\,\mathrm{m\,s^{-2}}}$$

$$t = \frac{15.5\ldots\,\mathrm{m\,s^{-1}} + 16.5\ldots\,\mathrm{m\,s^{-1}}}{9.8\,\mathrm{m\,s^{-2}}} = -\frac{0.9\ldots\,\mathrm{m\,s^{-1}}}{9.8\,\mathrm{m\,s^{-2}}} \text{ or } \frac{32.09\ldots\,\mathrm{m\,s^{-1}}}{9.8\,\mathrm{m\,s^{-2}}}$$

$$= -0.093\ldots\,\mathrm{s} \text{ or } 3.27\ldots\,\mathrm{s}$$

The negative solution means that, if the ball had been thrown from ground level, it would have followed this path had it been thrown (slightly faster) 0.093... s earlier.

**Step 5:** Using $t$ from Step 4, find the range for horizontal motion.

The speed is constant, so $s_{\mathrm{h}} = u_{\mathrm{h}}t = 9.0\,\mathrm{m\,s^{-1}} \times 3.27\ldots\,\mathrm{s} = 29\,\mathrm{m}$ (2 s.f.)

## Summary questions

1 Explain how the imagined path of the cannonball shown in Figure 1 illustrates the old 'impetus' theory of motion. Explain how the path of a real projectile differs from this. *(3 marks)*

2 A projectile is launched from ground level at 25 m s$^{-1}$ at an angle of 45° to the horizontal. Calculate the range of the projectile and show that a 1° change in angle in either direction will reduce the range. *(6 marks)*

3 A basketball player launches a ball from a height of 2.0 m towards a hoop as shown in Figure 7. Show, using calculations, that the ball will go through the hoop. *(4 marks)*

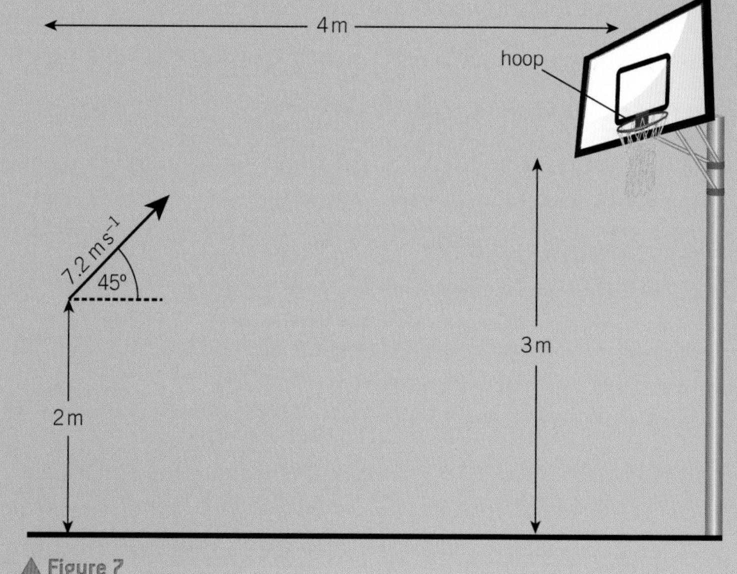


Figure 7

# 9.5 Work and power

## Specification references: 4.2a(iv), 4.2a(vi), 4.2 b(i), 4.2c(x)

### Learning outcomes

Describe, explain, and apply:

→ calculations of work done, including cases where the force is not parallel to the displacement

→ power as the rate of transfer of energy

→ the term: power

→ the equations for power, where $P = \frac{\Delta E}{t}$ and $P = Fv$.

If the weights in Figure 1 were being held up by a stand instead of by a woman, no work would be done. This is because work is done only when a force moves its point of application.

No movement = no work

Work is done in lifting the weights, but not in just holding them up. The weightlifter in Figure 1 would probably disagree – she would say she is working hard to hold the weights in this position.

This is because biological systems are different. If you stand for a long time holding a heavy object – this could actually be your own weight – you do expend energy due to muscle fibres continually relaxing and contracting. Forces are moving their points of application – inside your muscles – and so work is being done. You can tell that your muscles are working, because their low efficiency means that you get hot, and you also get fatigued, which indicates your body's need for more energy.

▲ Figure 1 *Doing work*

## Work done by a force at an angle

When a constant force $F$ moves an object by a displacement $s$, then the work done = $Fs$ only if the force is in the direction of displacement. You first met work done in the direction of displacement in Topic 9.3, Conservation of energy. Figure 2 shows a constant force $F$ acting on an object which can move only in the direction shown in orange, which is at an angle $\theta$ to the direction of the force. The force moves the object from $P_1$ to $P_2$.

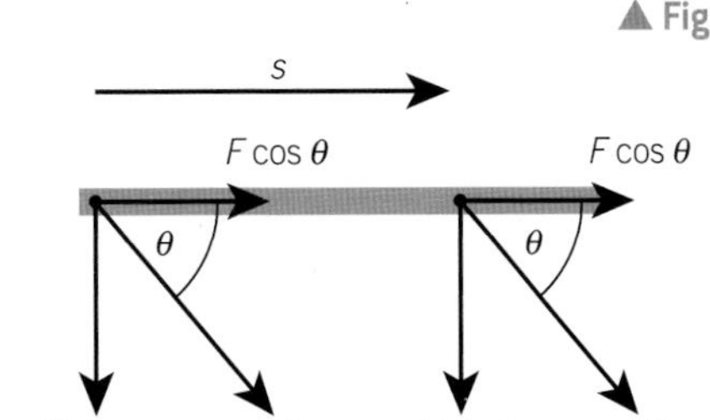

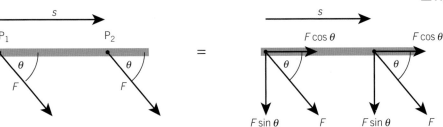


▲ Figure 2 *Force and work*

When $F$ is resolved into components perpendicular to $s$ and parallel to $s$, as in Figure 2, you can see that the object is not moving in the direction of the perpendicular component $F\sin\theta$ at all, so this component does no work. Work is being done only by the parallel component $F\cos\theta$ so

$$\text{Work done} = (F\cos\theta) \times s = Fs\cos\theta$$

Although both force and displacement are vector quantities, work, like energy, is a scalar quantity and has no direction. The way in which two vector quantities $F$ and $s$ separated by an angle $\theta$ multiply together to

**Synoptic link**

You first met vector and scalar quantities in Topic 8.2, Vectors.

give a scalar quantity is known as a *scalar product*, and is common in physics.

A force often produces a displacement in a different direction. All that is required is that the displacement is constrained to be in a particular direction. One example is sailing, where the shape of the boat, and the presence of a keel, define the direction in which it must travel.

### Worked example: Work done in sailing

Figure 3 shows a boat sailing forwards under the action of wind acting at 56° to the direction of travel. The wind pushes on the sail with a force of 500 N. Calculate the work done in sailing the boat forwards a distance of 25 m.

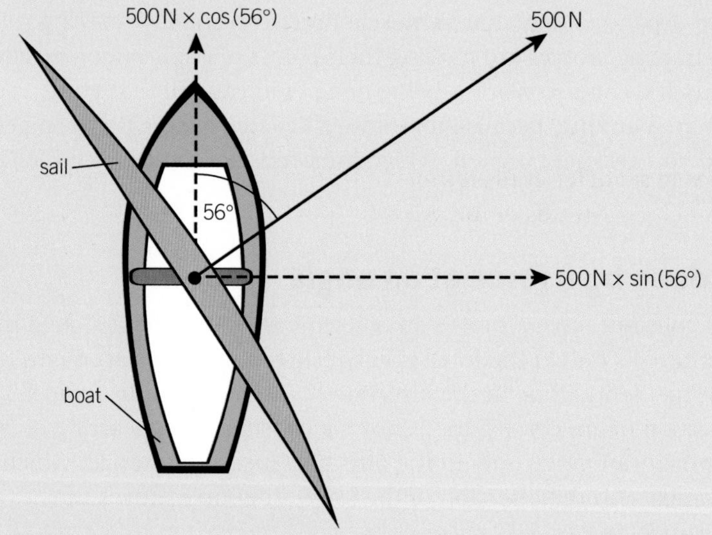


▲ **Figure 3** *Work done in sailing*

**Step 1:** Resolve the force into components to find the component in the direction of movement.

Forwards component = $F\cos\theta$ = 500 N × cos (56°)
= 500 N × 0.559... = 279.5... N

**Step 2:** Multiply this component by the displacement in this direction.

Work done = forwards component of force × displacement

= 279.6 N × 25 m = 7000 J (2 s.f.)

Note that this two-stage process is exactly the same as using the equation work done = $Fs\cos\theta$.

Although the component of the force perpendicular to the direction of the motion does no work, this does not mean it has no effect. In sailing, it makes the boat tilt over, so the crew need to lean out to prevent it overturning.

## Work done and the principle of conservation of energy

The principle of conservation of energy is such a powerful tool in physics that whenever energy seems to go missing, it raises the question – to where must the energy have been transferred? This energy transfer can often be explained as work done against resistive forces, and usually results in heating.

A force is frequently moving its point of application, doing work in circumstances where it is not clear where the energy is transferred. In Figure 4 the skydiver is falling at his terminal velocity (about $60\,\mathrm{m\,s^{-1}}$ or 130 m.p.h.) and is clearly losing gravitational potential energy. As he's not accelerating, his kinetic energy remains constant.

As the skydiver falls, he can feel the resistive force exerted on him by the air through which he is falling, and by Newton's third law he must be exerting an equal force on the air.

▲ Figure 4 *Skydiver at terminal velocity*

### Synoptic link

You met the relationship between gravitational potential energy and kinetic energy in Topic 9.3, Conservation of energy, and Newton's third law in Topic 9.2, Newton's laws of motion and momentum.

### Worked example: Work done on the air

Calculate the amount of energy transferred to the air each second by a 70 kg skydiver at his terminal velocity of $60\,\mathrm{m\,s^{-1}}$. Assume that about two-thirds of the original gravitational potential energy is transferred to the air, with the rest heating the skydiver's suit.

**Step 1:** Calculate using *s u v a t* equations how far the skydiver falls in this time.

He is not accelerating, so $v = u$ and $a = 0$, so $s = \frac{u+v}{2}t$ and $s = ut + \frac{1}{2}at^2$ both become $s = ut$.

$$\text{height } h = s = ?,\ u = 60\,\mathrm{m\,s^{-1}},\ t = 1\,\mathrm{s}$$

$$h = s = ut = 60\,\mathrm{m\,s^{-1}} \times 1\,\mathrm{s} = 60\,\mathrm{m}$$

**Step 2:** Calculate the change in gravitational potential energy in this time.

$$\Delta E_{grav} = mgh = 70\,\mathrm{kg} \times 9.8\,\mathrm{m\,s^{-2}} \times 60\,\mathrm{m} = 41\,160\,\mathrm{J}$$

**Step 3:** Use the information in the question to find how much energy is transferred to the air.

$$\text{Work done on the air} \approx \frac{2}{3} \times 41160\,\mathrm{J} = 30\,000\,\mathrm{J}\ (1\text{ s.f.})$$

### Study tip

For calculations involving rough estimates, answers should be rounded to one significant figure.

## Power

**Power** is the rate at which work is done, that is, the rate at which energy is transferred. It was the focus of the Worked example above – the energy transferred to the air in one second is the power *dissipated* by the gravitational potential energy change of the skydiver heating the air. The word 'dissipated' is frequently used to refer to energy transferred to the surroundings, usually heating them.

Assuming that the displacement resulting from a force is in the direction of the force itself, we can simply relate the power dissipated to the force.

$$\text{Power} = \frac{\text{energy transfer}}{\text{time}} = \frac{\text{work done}}{\text{time}} = \frac{Fs}{t} = F\frac{s}{t} = Fv$$

## Worked example: Drag force on the Eurostar train

A Eurostar train is driven by two engines, each providing an output power of 5.6 MW. Calculate the drag force on the Eurostar when it is travelling at a constant speed of 186 m.p.h. ($83\,\mathrm{m\,s^{-1}}$).

**Step 1:** Calculate the force provided by the engines at this velocity.

$$P = Fv \Rightarrow F = \frac{P}{v} = \frac{2 \times 5.6 \times 10^6\,\mathrm{W}}{83\,\mathrm{m\,s^{-1}}} = 134\,939.7\ldots\,\mathrm{N}$$

**Step 2:** Apply Newton's first law.

The train is not accelerating, so the resultant force on it must be 0.

Forward force provided by engines = drag force in the opposite direction

so drag force = 134 939.7... N = 130 kN (2 s.f.)

## Dinorwig Power Station

For many of the UK power stations, particularly the coal-burning and nuclear power stations, changing the output power takes many hours. Sudden changes in demand for electricity – this can be when everyone switches on an electric kettle to make a cup of tea at the end of a popular television programme – result in demands which cannot be met rapidly. To help with situations like this, the pumped-storage power station in Dinorwig in North Wales was constructed.

The turbines which generate electricity are inside Elidir mountain, as are 16 km of underground tunnels leading up to the storage reservoir near the top of the mountain. During times of low electricity demand, the electricity generators are used as motors to pump water from the lower lake through a height of 500 m to the storage reservoir, which has a capacity of $7.0 \times 10^6\,\mathrm{m^3}$. Water has a density of $1000\,\mathrm{kg\,m^{-3}}$, so the total mass of water in the storage reservoir = $7.0 \times 10^9$ kg when it is full.

The water in the storage reservoir may be rapidly run back down to the lower lake through the turbine tunnels to cope with surges in electricity demand, and can deliver a power of 1.7 GW within 16 s of being turned on.

**1** Calculate the total gravitational potential energy difference when all the water runs from the storage reservoir to the lower lake, and use this value to calculate the time for which Dinorwig can deliver a power of 1.7 GW. The generators operate on average with an efficiency of 75%.

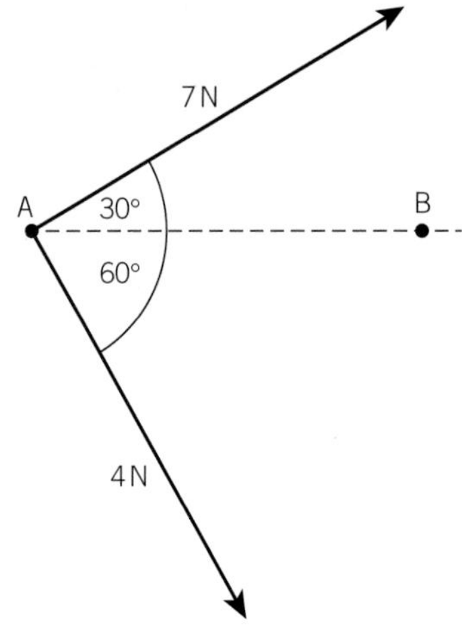


▲ **Figure 5** *An object under the action of two forces*

## Summary questions

**1** A car drives at a steady speed of $27\,\mathrm{m\,s^{-1}}$ along a motorway against drag forces of 3000 N. Calculate the power dissipated against the drag forces. *(2 marks)*

**2** Figure 5 shows two forces acting on an object which moves in the direction shown by the dotted line.

Calculate the work done in moving the object from A to B, a distance of 60 cm. *(3 marks)*

**3** A car of mass 900 kg accelerates from 0 to $26.8\,\mathrm{m\,s^{-1}}$ in 9.2 s. The thrust from the engine is constant at 3500 N during this period. By comparing the kinetic energy gained by the car with the total work done by the engine, calculate the work done against resistive forces. *(5 marks)*

# Module 4.2 Summary

## Motion

### Graphs of motion

- measuring: speed, velocity, acceleration, and displacement
- graphs of accelerated motion, including gradient and area under graph
- investigating the motion of a trolley using light gates or other techniques

### Vectors

- the terms: vector, scalar, resultant, component
- the representation of displacement, velocity, and acceleration as vectors
- the addition of two vectors by drawing a 'tip to tail' triangle
- finding relative velocity
- the resolution of a vector into two components at right angles to each other

### Modelling motion

- computational models to represent changes of displacement and velocity
- graphical models in terms of vector additions to represent motion
- using time intervals in iterative models with relation to accuracy of prediction
- investigating falling cupcake cases

### Speeding up and slowing down

- kinematic (*suvat*) equations for straight-line motion with uniform acceleration: $v = u + at$, $s = ut + \frac{1}{2}at^2$, $v^2 = u^2 + 2as$
- measuring $g$

## Momentum, force and energy

### Newton's laws of motion and momentum

- momentum $p = mv$
- conservation of momentum
- the terms force, mass, and impulse
- Newton's laws of motion
- Newton's third law as a consequence of the principle of conservation of momentum
- the equation: $F = \frac{\Delta p}{\Delta t} = ma$ where the mass is constant

### Conservation of energy

- the principle of conservation of energy
- the terms: work, energy
- the equation: $E_k = \frac{1}{2}mv^2$, $\Delta E_{grav} = mg\Delta h$, work done $\Delta E = F\Delta s$

### Projectiles

- the trajectory of a body moving under constant acceleration
- the independent effects of perpendicular components of a force

### Work and power

- the terms: work and power
- the equation: work = (component of force in the direction of the displacement) × displacement
- power is the rate of transfer of energy
- the equations: $P = \frac{\Delta E}{\Delta t}$ and $P = Fv$

# Physics in perspective

## Racing car design

▲ **Figure 1** *Ralph de Palma winning the 1914 Vanderbilt Cup in his 115 brake horsepower Mercedes Grand Prix race car*

When motor racing started, the aim of car designers was to use ever more powerful engines. Figure 1 shows a 1914 Mercedes, which had a mass of about a tonne. It was fast, with a top speed of 115.0 mph and an engine power of 115 brake horsepower. It won many races at that time, but it was hardly what you would call streamlined.

### Drag and supercharging

As speeds increased in the 1920s and 1930s, designers realised that they needed to tackle drag force. This force is generally described by the equation

$$F_{\mathrm{D}} = \frac{1}{2}\rho C_{\mathrm{D}} A v^2$$

where $\rho$ is the density of fluid through which the object is moving (in this case, air), and $v$ and $A$ are the velocity and cross-sectional area of the moving object. $C_{\mathrm{D}}$, the drag coefficient, is a dimensionless number, usually less than 1. Increasing $v$ greatly increases the drag, but increasing $v$ is what car designers wanted — thankfully, both $C_{\mathrm{D}}$ and $A$ can be reduced. $A$ is reduced by lowering the profile of the car, and $C_{\mathrm{D}}$ is reduced by finding ways to streamline the vehicles so that the air flows smoothly over them.

**Study tip**

Brake horsepower is a unit of power, often used in motoring. 1 BHP ≈ 0.75 kW.

Automobile designers also introduced supercharging, which involved pumping extra air into the engine so that it could burn fuel faster and generate greater power, up to 180 kW. A modern development of this, which uses waste heat and so takes less energy from the engine, is called turbocharging.

### Acceleration and stability

Acceleration is change in velocity. In motor racing, it comes up in 3 main situations.

- Speeding up — the easiest way to improve this is increasing the engine power. This increases the forces involved in driving the car, and so the driver must be protected against the force with which the seat pushes him or her forwards during acceleration.
- Slowing down — the braking system must be able to safely bring the car, along with the driver, to a stop in a reasonable time.
- Cornering may not involve a change of speed, but it does involve a change in direction and so also a change in velocity. The force needed to push the car into a new direction has to be provided by friction with the ground. So the wheel size and tyres used must maximise the contact between the vehicle and the ground.

▲ **Figure 2** *Using a wind tunnel to investigate air flow around a car*

As cars became more and more streamlined they became less stable. Just as aircraft reaching a certain speed take off, so too do a car's wheels leave the ground if it is travelling fast enough. Furthermore, race tracks became smaller in size, resulting in tighter bends and more laps to be driven, so more corners had to be negotiated at speed. In the 1960s and

1970s, car design began integrating technology developed in the aircraft industry, particularly from the use of wind tunnels to investigate air flow around vehicles. As a result, cars started to sprout wings, fins and skirts.

The combined effect of the changes introduced was to increase the drag coefficient $C_D$ considerably, but in such a way that the resultant drag force was downwards — this is usually referred to as 'downforce'. The overall behaviour of the car became rather like that of an aircraft wing, but upside down. This increased the 'grip' of the tyres on the ground, an effect further increased by the use of wide wheels with smooth, soft tyres called 'slicks'.

▲ **Figure 3** *A modern Formula 1 car, with wide, smooth 'slicks'*

## Safety and Formula 1

Formula 1 refers to the highest class of single-seat car racing, and the 'formula' is the set of regulations which must apply to all cars. By the 1970s, speeds had increased dramatically but safety had not, and deaths were common. As a result, drivers campaigned for improved safety, which gave rise to modern Formula 1 Regulations.

Besides the use of protective clothing and regular inspections of vehicles and track, a number of safety measures were integrated into the cars. These included crumple zones, automatic cut-off of fuel lines and a 'survival cell' in which the driver's cockpit was built. Public safety was also considered, with the introduction of double guard rails and wide verges to the track.

Other regulations restrict the performance of vehicles — these include increasing the minimum mass of vehicles, restricting the engine capacity (and so keeping the power down), and banning some aerodynamic developments which increased the downforce. These restrictions make the race fairer and also less risky for the drivers. As car designers continually come up with technological improvements, the Formula 1 regulations have to be refined each year.

## Summary questions

**1** Ralph de Palma won the 1914 Vanderbilt Cup in a car of mass 1080 kg with a top speed of $51.5\ \mathrm{m\,s^{-1}}$ and an engine output power of 85.8 kW. The race consisted of 35 laps on a circuit of length 13.6 km.

- **a** Calculate the shortest time that this race could have taken and explain why the race would have taken longer than this.
- **b** After the race, de Palma demonstrated the car's maximum speed by driving in a straight line on horizontal ground. Draw a labelled diagram showing all the forces acting on the car at this steady speed. You can assume that the rear wheels are providing the driving force.
- **c** Assuming that 80% of the car's output power is delivered to work done against resistive forces, calculate the value of the resultant resistive force when driving at a constant $51.5\ \mathrm{m\,s^{-1}}$.

**2** The drag force acting on a car is given by $F_D = \frac{1}{2}\rho C_D A v^2$. Assuming that a family car has a drag coefficient of 0.32 and a maximum cross-sectional area of $2.1\ \mathrm{m^2}$, calculate the drag force while driving down the motorway at 70 mph ($31.3\ \mathrm{m\,s^{-1}}$). The density of air, $\rho = 1.2\ \mathrm{kg\,m^{-3}}$.

## Practice questions

1 Which of the following combinations of SI base units is equivalent to the joule, J?

A $kg\,m\,s^{-1}$

B $kg\,m\,s^{-2}$

C $kg\,m^2\,s^{-2}$

D $kg\,m^2\,s^{-3}$

2 An object of mass 2.5 kg travelling at $6.4\,m\,s^{-1}$ collides with and sticks to a stationary object of mass 5.5 kg. The combined masses move forwards.

After the collision, which of the following statements are correct?

1 The kinetic energy of the combined masses is 51.2 J.

2 The momentum of the combined masses is 16 N s.

3 The velocity of the combined masses is $2\,m\,s^{-1}$.

A 1, 2, and 3 are correct

B Only 1 and 2 are correct

C Only 2 and 3 are correct

D Only 1 is correct

3 A rifle of mass 3.2 kg fires a 3.0 g bullet horizontally with a velocity of $370\,m\,s^{-1}$. Ignoring the effect of any exhaust gases produced in the firing, calculate the speed at which the rifle recoils. *(3 marks)*

4 Two identical 4.0 kg masses, heading directly towards each other, each with a speed of $0.8\,m\,s^{-1}$, collide and take a time of 0.15 s to come to rest.

a Calculate the force acting on each mass during the collision. *(2 marks)*

b Calculate the energy dissipated in the collision. *(2 marks)*

5 A roller-coaster car of mass 900 kg is stationary at the top of the track, and then falls a height $h$ to the lowest point, as in Figure 1.

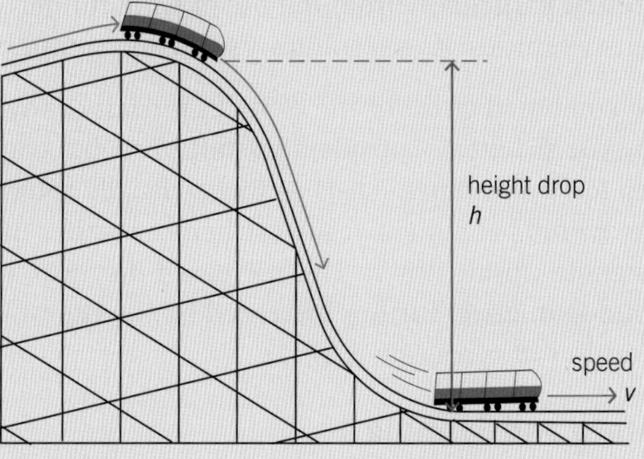


▲ Figure 1

a Calculate the maximum velocity of the roller-coaster car at the bottom of the track if $h$ = 60 m.
Acceleration due to gravity, $g = 9.8\,m\,s^{-2}$. *(3 marks)*

b Explain why the velocity calculated in **(a)** is likely to be less than this value. *(1 mark)*

6 A car of mass 1100 kg accelerates from rest to $27\,m\,s^{-1}$ in 5.6 s.

a Calculate the mean resultant force acting on the car. *(2 marks)*

b Calculate the mean power output of the car. *(2 marks)*

7 In a nuclear decay reaction, a nucleus of mass $3.6 \times 10^{-25}$ kg emits a particle of mass $6.6 \times 10^{-27}$ kg with kinetic energy $8.0 \times 10^{-13}$ J.

a Calculate the kinetic energy of the nucleus after the reaction. *(4 marks)*

b The same nucleus can also decay by emitting a particle of mass $9.1 \times 10^{-31}$ kg with kinetic energy $1.4 \times 10^{-15}$ J. Explain how the kinetic energy of the nucleus after the reaction would compare with the value obtained in **(a)**. *(4 marks)*

**c** In a nuclear reaction, the kinetic energy of any particle with kinetic energy $E_k$ given by the equation $E_k > (10^{15}\,m^2\,s^{-2}) \times$ (mass of particle in kg) cannot be obtained from the standard $E_k = \frac{1}{2}mv^2$: a relativistic equation must be used. Check whether either of the emitted particles in **(a)** and **(b)** is travelling fast enough for this condition to apply. *(2 marks)*

**8** In an experiment, a cylindrical wooden rod was dropped vertically from different heights into a container of water, and the depth to which it sunk before floating back upwards was measured.

**a** For a drop height of 12 cm, the following results were obtained.

▼ Table 1

| depth / cm | 5.2 | 5.8 | 3.4 | 5.7 | 5.4 | 5.5 |
|---|---|---|---|---|---|---|

Use the data to obtain a mean value for depth at this drop height together with its uncertainty. Justify any decision you make in selecting data. *(3 marks)*

**b** Add your result from **(a)** to a copy of Table 2, and use this data to plot a graph on a copy of Figure 2. Add a line of best fit to the graph. *(2 marks)*

▼ Table 2

| height / cm | 2 | 4 | 6 | 8 | 10 | 12 |
|---|---|---|---|---|---|---|
| depth / cm | 0.4 | 1.5 | 2.6 | 3.6 | 4.6 | |

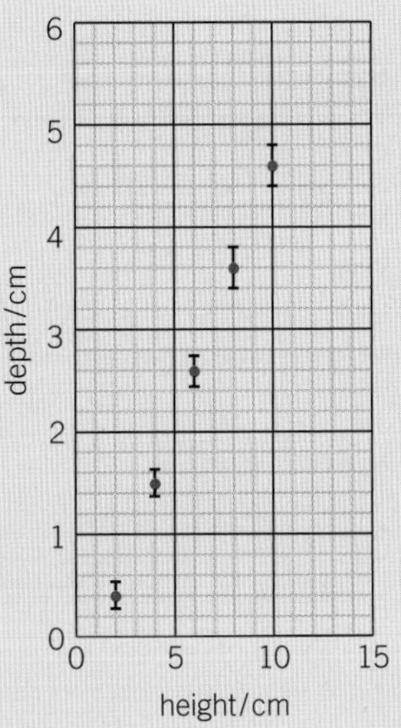


▲ Figure 2

**c** It is suggested that the depth reached by the rod is directly proportional to the height from which the road was dropped. Explain why this is not supported by the data. *(2 marks)*

**d** Suggest and explain reasons why the data do not support the expected relationship. *(4 marks)*

**9** A tennis ball of mass 59 g, moving at 50 m s$^{-1}$, is struck by a racket and travels straight back at 37 m s$^{-1}$. The racket is in contact with the ball for a time of 3.5 ms.

**a** Use the momentum change of the ball to calculate the mean force exerted by the racket in returning the ball. *(4 marks)*

**b** The ball leaves the racket horizontally at a height of 1.2 m above the ground. The playing area of the tennis court is 23 m long, and the player receiving the ball is standing 3.0 m from the back of this area. By calculating the distance the ball travels before striking the ground, find whether the ball will be 'in', reaching the ground inside the playing area. Assume that air resistance is negligible. ($g = 9.8$ m s$^{-2}$) *(4 marks)*

**c** Explain in terms of the components of velocity of the ball why hitting the ball slightly upwards at 37 m s$^{-1}$ may have resulted in the ball being 'out' for a certain range of angles. No calculation is required. *(3 marks)*

# Paper 1-style questions

## Section A — Multiple Choice

1 The upper frequency limit of human hearing is about 20 kHz.

Which value gives the minimum rate needed to accurately sample sound of this frequency?

A > 10 kHz B > 20 kHz

C > 40 kHz D > 80 kHz

2 'Conductive putty' is a malleable material which conducts electricity. A piece of this putty is rolled into the cylinder shown in Figure 1. The conductance between the ends of the cylinder is measured as 0.18 S.

▲ Figure 1

This piece of putty is then rolled out further to make a uniform cylinder 12 cm long. What will be the conductance between the ends of this new cylinder?

A 0.08 S B 0.12 S

C 0.27 S D 0.41 S

3 The circuit in Figure 2 contains three identical lamps, each marked *6.0 V, 0.2 A*.

'Normal brightness' is the brightness of one bulb with a current of 0.2 A.

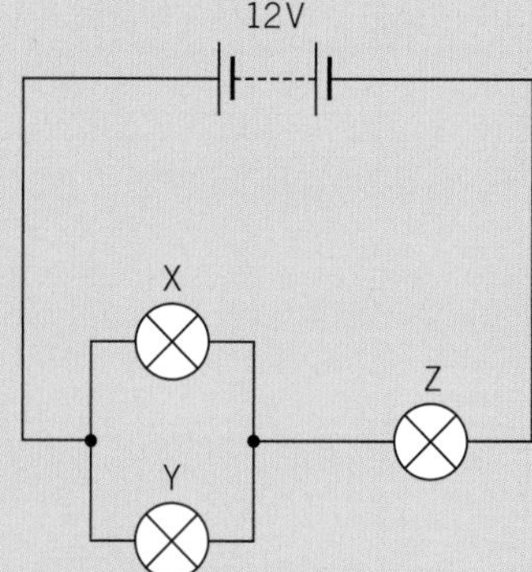


▲ Figure 2

Which of the following statements about the bulbs in Figure 2 are correct?

1 Bulbs X and Y have the same brightness

2 The p.d. across bulb Y is more than 6 V

3 Bulb Z is dimmer than normal

A 1, 2, and 3 are correct

B Only 1 and 2 are correct

C Only 2 and 3 are correct

D Only 1 is correct

4 Light passing through a narrow gap diffracts. Which of the following changes would decrease the amount by which the light diffracts?

A Decreasing the intensity of the light

B Decreasing the width of the gap

C Decreasing the wavelength of the light

D Decreasing the amplitude of the light

5 A converging lens of power +7.0 D focuses an image at a distance of 15 cm from the lens. Which of the values below is the object distance?

A −0.14 m B −3.0 m

C −0.33 m D −0.08 m

6 Which of the following units are the correct units for stress?

A $\text{N m}^{-3}$ B $\text{J m}^{-3}$

C $\text{J m}^{-1}$ D $\text{N m}^{-1}$

7 The work function of potassium is $3.5 \times 10^{-19}$ J. What is the maximum speed of the emitted photoelectrons when the potassium is illuminated by light of wavelength $3.9 \times 10^{-7}$ m?

Electron mass = $9.1 \times 10^{-31}$ kg

Planck constant = $6.63 \times 10^{-34}$ J s

A $3.5 \times 10^{11}\ \text{m s}^{-1}$ B $5.9 \times 10^{5}\ \text{m s}^{-1}$

C $4.2 \times 10^{5}\ \text{m s}^{-1}$ D $2.9 \times 10^{5}\ \text{m s}^{-1}$

8 Which of the following values is the best estimate of the de Broglie wavelength of an electron travelling at ten percent of the speed of light?

Electron mass = $9.1 \times 10^{-31}$ kg

Planck constant = $6.63 \times 10^{-34}$ J s

A $2.4 \times 10^{-12}$ m B $4.8 \times 10^{-12}$ m

C $2.4 \times 10^{-11}$ m D $4.8 \times 10^{-11}$ m

9 A metal spring extends elastically by 0.17 m. The energy stored by the spring at this extension is 0.87 J. What is the force constant $k$ of the spring?

**A** $60\,\mathrm{N\,m^{-1}}$ **B** $30\,\mathrm{N\,m^{-1}}$

**C** $10\,\mathrm{N\,m^{-1}}$ **D** $5\,\mathrm{N\,m^{-1}}$

10 Graphs **A** to **D** describe four different straight-line trips, each in two parts. In each graph, the $x$-axis variable is time.

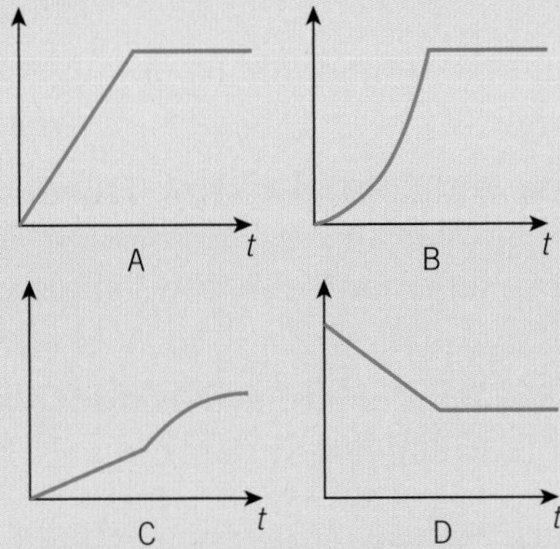


Assuming the $y$-axis variable is displacement, which graph shows deceleration over one part of the trip?

11 Figure 3 shows two soft balls of equal mass rolling towards each other on a horizontal, frictionless surface.

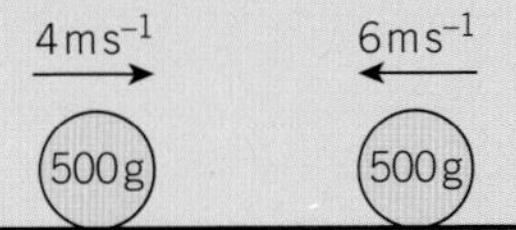


▲ **Figure 3**

What is the magnitude of the total momentum of the two balls after the collision?

**A** 1 N s **B** 5 N s

**C** 1000 N s **D** 5000 N s

12 What is the maximum possible total kinetic energy of the two balls in Figure 3 after the collision, assuming they stick together?

**A** 0.5 J **B** 1 J

**C** 9 J **D** 25 J

## Section B

1 Figure 5 shows an oscilloscope waveform of a pure sound wave.

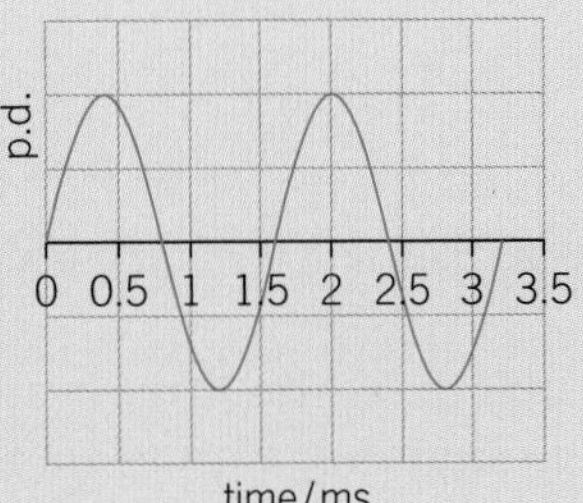


▲ **Figure 4**

Calculate the frequency of the note. Give the units. *(3 marks)*

2 A data projector has an illuminated display that is 48 mm wide. A lens projects an image of the display on to a screen. The width of the image is 1.7 m.

a Calculate the magnification of the image. *(1 mark)*

b The distance between the illuminated display and the lens in 8 cm. Calculate the distance from the lens to the screen. *(1 mark)*

c Calculate the power of the lens. *(2 marks)*

3 Light is partially polarised when it reflects from a sheet of glass. A student investigates this phenomenon using a polarising filter, as shown in Figure 5.

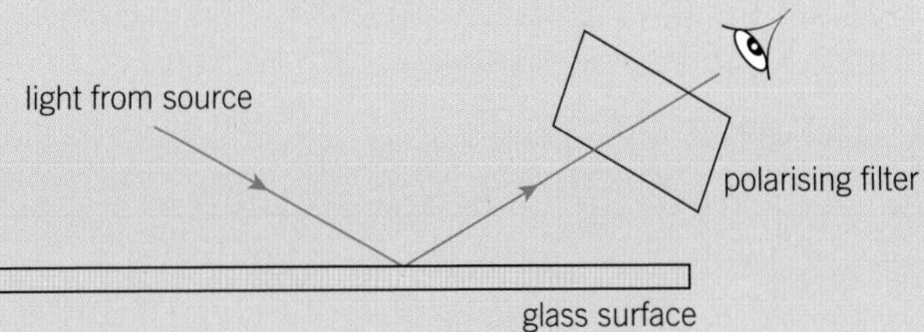


▲ **Figure 5**

a Describe how the student could use the filter to observe the polarisation of light. *(2 marks)*

b State the observation that would make the student conclude that the light was only partially polarised. *(1 mark)*

4 Figure 5 represents light waves incident on a lens. The lens focuses the waves at point *F*.

▲ Figure 6

a Copy Figure 4 and draw the three waves between the lens and the point of focus. *(2 marks)*

b What does Figure 6 indicate about the distance between the light source and the lens? Explain your answer. *(2 marks)*

5 A lens has power +15 D.

The curvature of light waves incident on the lens is −3 D.

a Calculate the curvature of the waves as they pass out of the lens. *(1 mark)*

b Calculate the distance from the lens at which the light is focused. *(1 mark)*

6 Figure 7 shows the result of plane wavefronts passing through a converging lens.

The lens is replaced with one of the same shape and dimensions, made from a material with a higher refractive index.

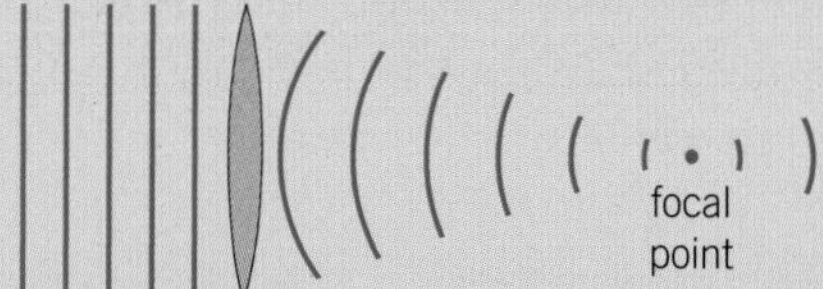


▲ Figure 7

a Copy Figure 7, then draw another diagram to show the effect of using a lens with *higher* refractive index on the wavefronts to the right of the lens. *(2 marks)*

b State with a reason whether the power of the new lens in part **(a)** is larger, smaller, or the same as that of the original lens. *(1 mark)*

*OCR Physics B Paper 2860 January 2009*

7 A memory cell has an area of $1 \times 10^{-14}$ m$^2$. It stores 1 bit of information.

The total area of memory cells on a microchip is $6 \times 10^{-5}$ m$^2$.

Calculate the number of bytes of information the chip can store. *(2 marks)*

8 An analogue signal has a voltage variation of 6.7 V. It is sampled with 1024 voltage levels.

a Calculate the resolution of each sample. *(1 mark)*

b Calculate the number of bits required for 1024 levels. *(1 mark)*

c The noise in the signal gives a random voltage variation of ± 0.1 V. Calculate the number of bits worth using to sample the signal. *(3 marks)*

9 Figure 8 shows light entering glass from air. The angles of incidence and refraction are given.

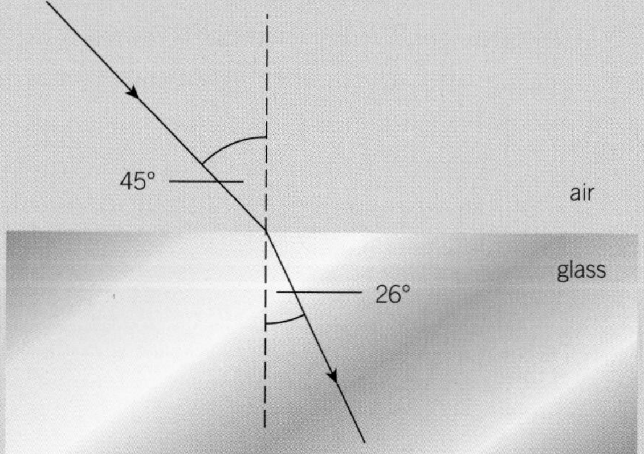


▲ Figure 8

a Use Figure 8 to calculate the refractive index of the glass. *(2 marks)*

b The velocity of light in air $= 3.00 \times 10^8$ m s$^{-1}$. Calculate the velocity in the glass. *(2 marks)*

c Light passes from the glass into air. The angle of incidence in the glass is 35°. Calculate the angle of refraction in the air. *(2 marks)*

**10** A variable resistor is connected to a battery of e.m.f. $\varepsilon$ and and internal resistance $r$ via an ammeter. The p.d. $V$ across the terminals of the battery is measured for different values of current $I$ to give the graph in Figure 9.

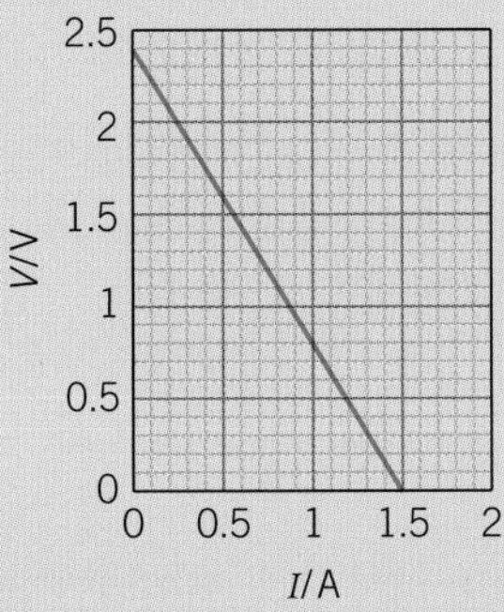


▲ **Figure 9**

**a** Explain why the graph shows that the e.m.f. of the battery is 2.4 V. *(1 mark)*

**b** Use the graph to calculate the internal resistance $r$ of the battery. *(2 marks)*

**11** Table 1 displays the electrical conductivity of four different materials at room temperature.

▼ **Table 1**

| material | gold | lithium | pure silicon | doped silicon |
|---|---|---|---|---|
| conductivity / S m$^{-1}$ | $4.9 \times 10^7$ | $1.2 \times 10^7$ | $4.7 \times 10^{-4}$ | 22 |

**a** Gold and silicon are both metals. In metals, each atom provides one free electron to the cloud responsible for conduction. Suggest and explain two reasons for the difference in resistivity between gold and silicon. *(2 marks)*

**b** The doped silicon has one atom in every ten million replaced by a boron atom. Suggest why this tiny proportion of foreign atoms increases the conductivity so significantly. *(2 marks)*

**12** A student makes the following measurements to find the tensile stress in a wire:

tension in wire = 147 ± 1 N

cross-sectional area = $(0.86 \pm 0.10) \times 10^{-6}\,\text{m}^2$

**a** Calculate the value of the tensile stress in the wire. *(1 mark)*

**b** Calculate the largest possible value of the stress, given the uncertainties in the data. *(2 marks)*

**c** The student wishes to reduce the uncertainty in the value of the stress.

State which measurement you would choose to improve to achieve this.

Explain your choice. *(1 mark)*

*OCR Physics B Paper G491 June 2010*

**13** Metals can show plastic behaviour under stress.

**a** Explain what is meant by the terms plastic behaviour and stress. *(2 marks)*

**b** Use what you know about the microscopic structure of metals to describe how they can show plastic behaviour. *(2 marks)*

**14** A light-emitting diode emits photons of energy $3.7 \times 10^{-19}$ J.

**a** Calculate the wavelength of the radiation emitted. *(2 marks)*

**b** The output power of the diode is 40 mW. Calculate the number of photons emitted each second. *(2 marks)*

**15** Figure 10 shows some energy levels in an atom. Calculate the frequency of the photon emitted when an electron falls from level A to level C. *(2 marks)*

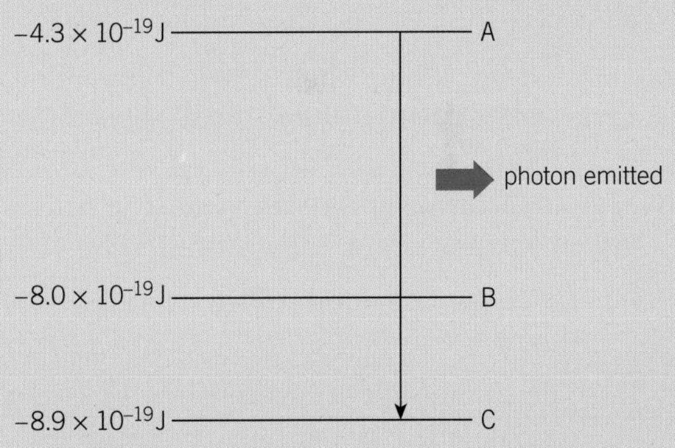


▲ **Figure 10**

**16** A photon travels from X to Y. Three possible paths for the photon are shown in Figure 11.

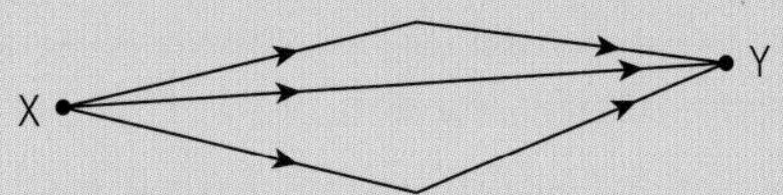


▲ **Figure 11**

The phasors at Y for the three paths are shown in Figure 12. Each has the same amplitude $A$.

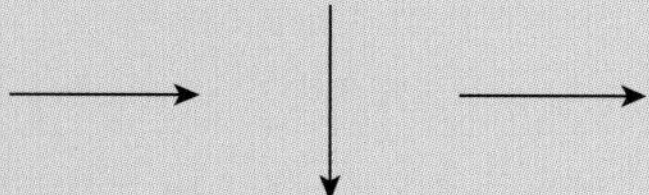

▲ Figure 12

a Draw a diagram of the phasor arrows adding tip-to-tail and show the resultant phasor. *(2 marks)*

b Show by calculation that the amplitude of the resultant phasor is between $2A$ and $2.5A$. *(2 marks)*

**17** A standing wave is formed on a stretched wire, as shown in Figure 13.

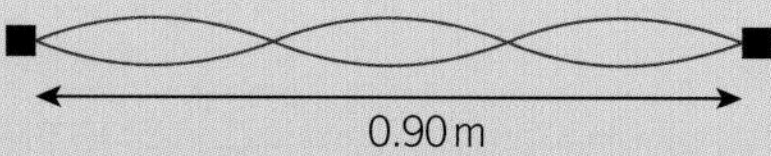


▲ Figure 13

a State the wavelength of the standing wave on the wire. *(1 mark)*

b The frequency of vibration of the wire is 360 Hz. Calculate the speed of transverse waves on the wire. *(2 marks)*

c On a copy of the diagram, label the positions of a displacement node and a displacement antinode with the letters $N$ and $A$. *(1 mark)*

**18** A stone is dropped down a well. The time for it to reach the water is measured to be 1.6 s.

a Calculate the depth of the well, assuming that $g = 9.8\,\text{m}\,\text{s}^{-2}$. *(2 marks)*

b The timing has an uncertainty of ±0.1 s. Calculate the uncertainty in your calculated depth in part **(a)**. *(2 marks)*

c The speed of sound in air is $340\,\text{m}\,\text{s}^{-1}$. Use your answer to **(a)** to calculate the time taken for the sound of the splash to reach the person timing the fall, and comment on how the systematic error will affect the estimate of the depth obtained in **(a)**. *(2 marks)*

**19** A car travels a distance $s$ in a time $t$ with constant acceleration $a$. In this time, the velocity of the car increases from an initial velocity $u$ to a final velocity $v$.

The equations below model the motion.

$$s = \frac{(u + v)t}{2} \qquad \text{equation 1}$$

$$v = u + at \qquad \text{equation 2}$$

a Rearrange each of these equations to make $t$ the subject of the equation. *(2 marks)*

b Equate the two expressions for $t$ and hence show that $v^2 = u^2 + 2as$. *(1 mark)*

**20** On a test drive on a straight, horizontal track, a high-performance car accelerates from 0 to 60 mph ($27\,\text{m}\,\text{s}^{-1}$) in 6.2 s. The mass of the car and driver is 1400 kg.

a Calculate the mean resultant accelerating force over the 6.2 s. *(2 marks)*

b Draw a labelled sectional diagram showing all forces acting on the car during this acceleration. Label the forces with appropriate descriptions, but without values. *(3 marks)*

c The car continues to accelerate to its top speed of 162 mph ($72\,\text{m}\,\text{s}^{-1}$). Assuming that the accelerating force between the wheels and the road is the same as in part **(a)**, calculate the power dissipated when travelling at this speed. *(3 marks)*

**21** This question is about stretching polythene.

a A long narrow sample strip of polythene is cut from a shopping bag. It stretches elastically up to a strain of 0.082 at a stress of 14 MPa. This is the elastic limit of the material.

(i) Calculate the Young modulus of the polythene and state the unit. *(3 marks)*

(ii) The cross-sectional area of the sample is $1.9 \times 10^{-7}\,\text{m}^2$.

Calculate the force applied to the sample to produce a stress of 14 MPa. *(2 marks)*

**b** Figure 14 shows the stress against strain graph for the sample to its breaking point.

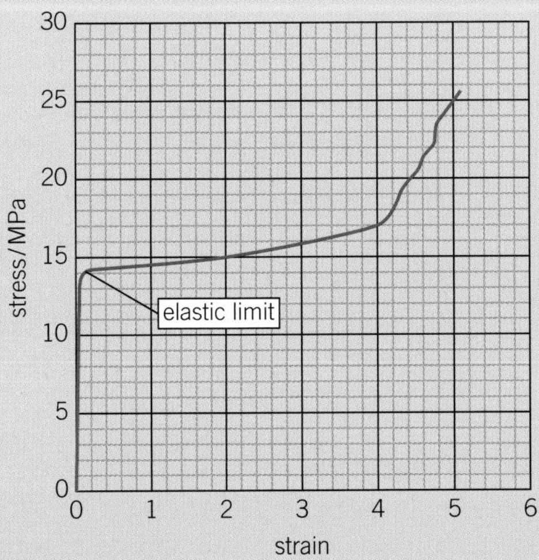


▲ **Figure 14**

(i) Describe the behaviour of the sample as it is stretched from the elastic limit to its breaking point. *(2 marks)*

(ii) Use Figure 14 to calculate the extension of the sample at the breaking point. The original length of the sample is 15 cm. *(3 marks)*

**c** Suggest and explain what is happening to the **long chain molecules** in the sample between the elastic limit and the break point as stress is increased slowly.

You may wish to use labelled diagrams. *(4 marks)*

*OCR Physics B Paper G491 June 2012*

**22** This question is about electron diffraction.

The equation for single slit diffraction is $\lambda = b \sin \theta$, where $b$ is the width of the slit and $\theta$ is the angle to the first minimum.

**a** (i) Show that an electron accelerated through a potential difference of 900 V will gain kinetic energy of about $1.4 \times 10^{-16}$ J. *(1 mark)*

(ii) The kinetic energy of an electron is related to its momentum in the equation

$$\text{kinetic energy} = \frac{\text{momentum}^2}{2m}$$

where $m$ is the mass of an electron $= 9.11 \times 10^{-31}$ kg.

Calculate the momentum $mv$ of the electron. *(2 marks)*

(iii) Use your answer from (a)(ii) to calculate the angle to the first minimum when electrons of energy $1.4 \times 10^{-16}$ J pass through a gap of width $4 \times 10^{-9}$ m *(3 marks)*

**b** State how the position of the first minimum will change when the electrons are accelerated through a greater potential difference. Explain your answer. *(3 marks)*

**23** A photocell generates electric current by the photoelectric effect.

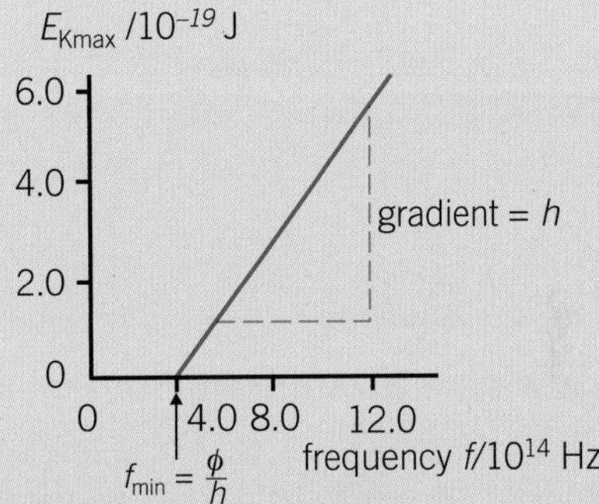


▲ **Figure 15**

Light of a wavelength $4.6 \times 10^{-7}$ m is incident on a metal plate called the photocathode.

**a** Show that the energy of a photon of this light is about $4 \times 10^{-19}$ J. *(2 marks)*

About $3 \times 10^{17}$ photons strike the photocathode every second. A current of 1.3 mA is detected.

**b** (i) Calculate the number of electrons released from the photocathode each second. *(2 marks)*

(ii) Suggest and explain why the number of electrons released each second is smaller than the number of photons incident on the photocathode each second. *(2 marks)*

**c** When red light of wavelength $6 \times 10^{-7}$ m is incident on the photocathode there is no current detected, even though the number of photons striking the photocathode each second is greater than $3 \times 10^{17}$. Explain this observation. *(2 marks)*

# Paper 2-style questions

## Section A

**1** This question is about the operation of a gas-filled pixel in a plasma TV screen. A plasma is a conducting ionised gas. It is formed by a high voltage pulse across a pair of electrodes in the pixel as shown in Figure 1.

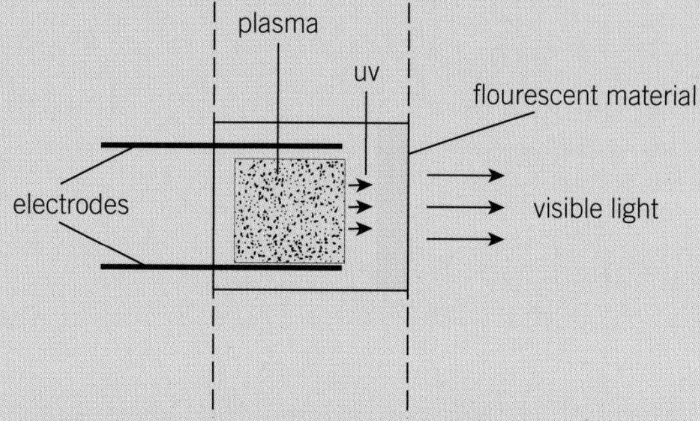


▲ **Figure 1** *Schematic diagram of a pixel in plasma display*

**a** Describe what is meant by an ionised gas. (*2 marks*)

**b** Plasma emits UV radiation at a frequency of $2.9 \times 10^{15}$ Hz.

Calculate the wavelength of this radiation.

Speed of light = $3.0 \times 10^{8}\,\mathrm{m\,s^{-1}}$ (*1 mark*)

**c** Gas atoms can be ionised by collision with fast-moving electrons. The p.d. between the electrodes provides energy for these electrons.

Calculate the energy gained by an electron of charge $1.6 \times 10^{-19}$ C when it passes through a p.d. of 240 V. (*2 marks*)

**d** Once started by a high voltage pulse the plasma in a pixel can be maintained at a lower voltage. The plasma can be ended by switching off the voltage. Figure 2 shows how the current in the gas in a pixel changes as the p.d. is raised to 290 V and lowered back to 0 V.

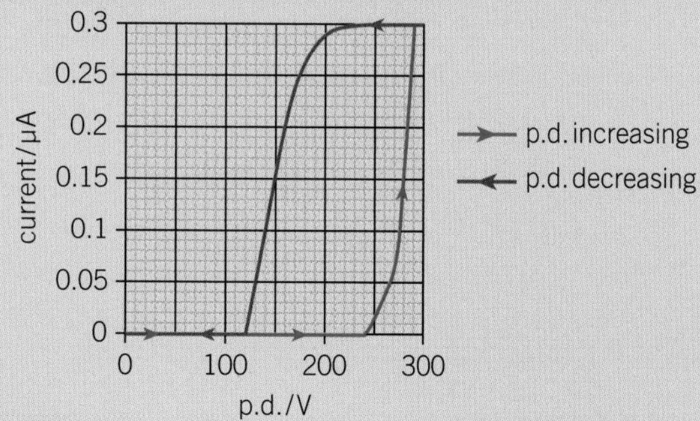


▲ **Figure 2**

(i) **1** State the voltage at which ionisation starts.

**2** State the voltage at which ionisation stops. (*2 marks*)

(ii) There are $6.2 \times 10^{6}$ pixels in the display. When emitting visible light pixels operate at 180 V.

Use data from Figure 2 to calculate the total operating power of the display with all the pixels on.

(*3 marks*)

*OCR Physics B Paper G491 Jan 2011*

**2** This question is about the materials from which cutting tools such as drill bits are made.

**a** (i) Metals have a polycrystalline structure.

Explain the term *polycrystalline* as applied to the structure of a metal.

You may wish to use labelled diagrams in your answer. (*2 marks*)

(ii) Drill bits can be made from steel alloy. Figure 3 shows the microstructures of pure iron metal and a steel alloy.

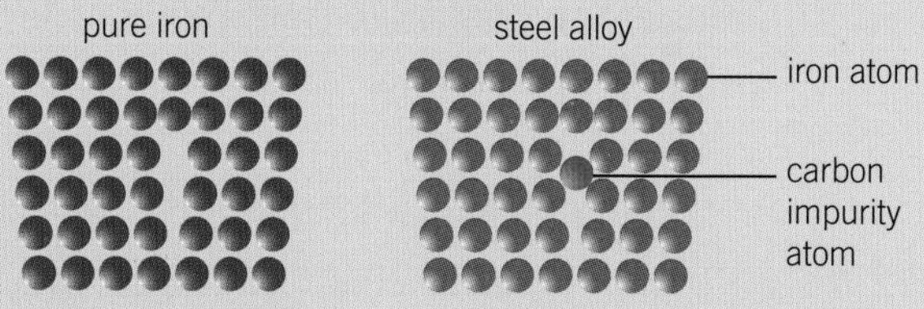


▲ **Figure 3**

Steel alloy containing carbon is less ductile than pure iron.

State the meaning of the term *ductile* and describe how Figure 3 can be used to help explain why steel is less ductile than iron. (*3 marks*)

**b** (i) Diamond is much harder than steel. This gives a diamond-coated steel drill bit an advantage over a steel one.

**1** State what is meant by *hardness*.

**2** Explain the advantage. (*2 marks*)

(ii) The atoms in steel have metallic bonding and in diamond the atoms have covalent bonding. Describe these types of bonding. Use your description to explain the difference in hardness between steel and diamond.
You may wish to use labelled diagrams in your answer. *(4 marks)*

*OCR Physics B Paper G491 Jan 2013*

**3** This question is about *relative* and *resultant* velocities.

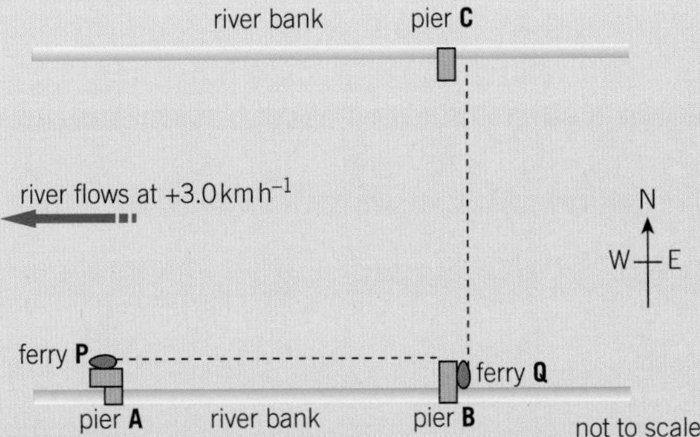


▲ **Figure 4**

Figure 4 shows part of a wide river on which there are three piers. The river flows from east to west at a constant velocity of +3.0 km h$^{-1}$ as shown.

**a** Ferry **P** travels from pier **A** to pier **B**, and then back again. The ferry travels at a speed of 5.0 km h$^{-1}$ through still water.

(i) Calculate the velocity of the ferry relative to the river bank as it sails.

**1** From **A** to **B**

**2** From **B** to **A** *(2 marks)*

(ii) Piers **A** and **B** are 2.0 km apart.

Show that the total sailing time for a return journey for ferry **P**, sailing from pier **A** to **B** and back again to **A**, is 1.25 hours.
Ignore the time taken for the boat to turn around at pier **B**. *(2 marks)*

**b** There is another pier **C** directly across the river from pier **B**, as shown in Figure 4.

A second ferry **Q** travels between piers **B** and **C** which are 2.0 km apart. This ferry also travels at a speed of 5.0 km h$^{-1}$ through still water.

(i) By scale drawing, or some other method of your choosing, show that the ferry **Q** must sail in a direction 37 degrees east of north in order to travel due north across the river, from pier **B** to pier **C**. *(2 marks)*

(ii) Show that the resultant velocity of this ferry relative to the river bank is 4.0 km h$^{-1}$ due north. *(2 marks)*

**c** Ferry **Q** sets off from pier **B** on an outward bound journey to **C** at the same time as ferry **P** sets off from pier **A** towards pier **B**.

Show that the bearing of ferry **Q** from ferry **P** is about 27 degrees east of north, when **Q** just reaches pier **C**. *(2 marks)*

*OCR Physics B Paper 2861 June 2004*

**4** Figure 7 shows the graph of force against extension for a metal wire **A**.

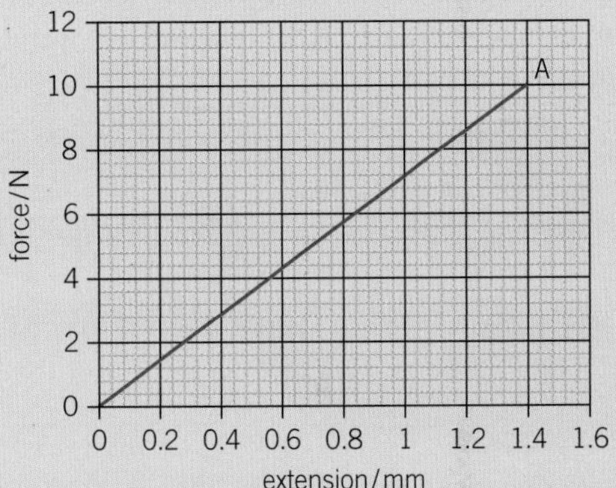


▲ **Figure 7**

**a** (i) Draw on a copy of Figure 5 the graph you would expect for a wire of the same material and diameter as **A**, but of **twice** the original length. Label this graph **B**. *(1 mark)*

(ii) Draw on a copy of Figure 5 the graph you would expect for a wire of the same material and length as **A**, but of **double** the original diameter. Label this graph **C**. *(1 mark)*

**b** (i) State **one** piece of evidence from the graph which suggests that the stretching of the wire (by a force of 10 N) is elastic. *(1 mark)*

(ii) Wire **A** has a cross-sectional area of $7.8 \times 10^{-8}\,m^2$ and an original length of 2.00 m.

Calculate the Young modulus of the material of the wire. *(3 marks)*

**c** Describe metallic bonding on the atomic scale. Include in your description an explanation of how metals such as wire **A** can show elastic behaviour.

In your explanation, you should make clear how the bonding between atoms can account for the large-scale elastic behaviour of the material. *(4 marks)*

*OCR Physics B Paper G491 Jan 2011*

## Section B

**1** This question is about taking a self-portrait with a mobile phone camera.

A camera on a mobile phone has a lens of focal length 4.5 mm. It is held 0.5 m from the photographer's face.

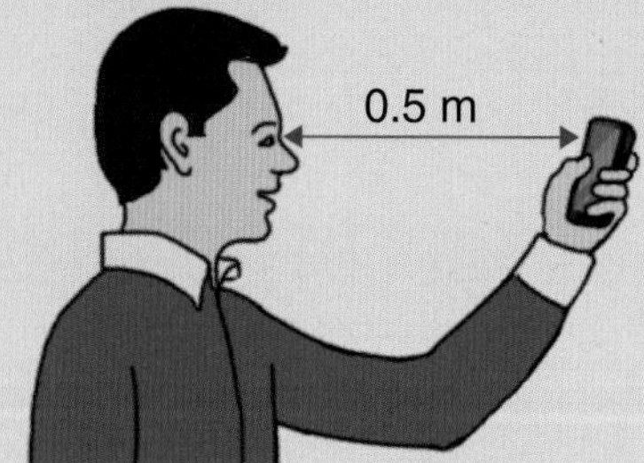


▲ Figure 6

**a** (i) Show that the power of the lens is about 220 D. *(1 mark)*

(ii) Calculate the distance behind the lens that the image of the photographer will be focused. *(2 marks)*

(iii) Calculate the magnification of the image. *(2 marks)*

The photographer's face has approximate dimensions of 270 × 225 mm. This fills the picture area of the light-sensitive chip.

**b** Use your answer to **(a)**(iii) to calculate the dimensions of the light-sensitive chip. *(2 marks)*

**c** The light-sensitive chip has 1200 × 1000 pixels. Calculate the resolution of the image of the face and comment on whether the image could resolve an individual eyelash of diameter 0.1 mm. *(3 marks)*

The camera stores three colours for each pixel, each with 256 levels of intensity. The camera memory has 0.9 GB available to store image files.

**d** (i) Calculate how many images can be stored in the memory. *(3 marks)*

(ii) Suggest how this number could be increased without increasing the size of the memory. Describe the advantages and disadvantages of your suggestion. *(3 marks)*

**2** One type of component, called a PTC thermistor, has a resistance which varies with temperature, as shown by the solid line in Figure 7. An 'ordinary' NTC thermistor, as used in sensing circuits, has a resistance which varies as shown by the dashed line.

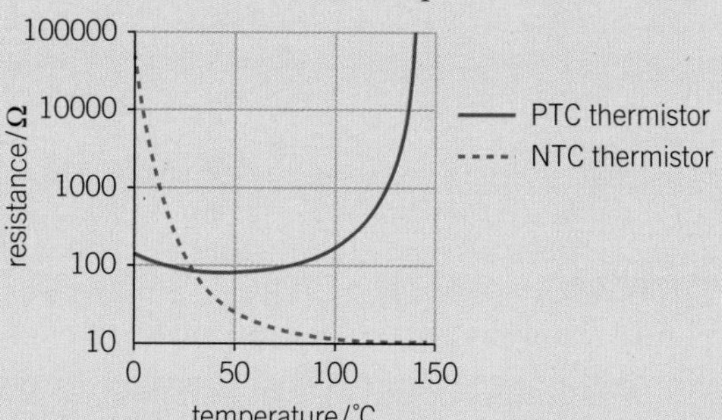


▲ Figure 7

**a** Compare the behaviour of these two components at different temperatures and suggest, in each case, reasons for the variation of resistance shown by the graph.

This question tests your ability to construct and develop a sustained and coherent line of reasoning. *(6 marks)*

**b** In one sensor application, a chemical reaction vessel which needs to be kept at 50 ± 2 °C needs to be monitored. Explain why the PTC thermistor is a poorer temperature sensor than the NTC thermistor for this application. *(2 marks)*

**c** The PTC thermistor is connected to a 6 V battery of negligible internal resistance, as shown in Figure 8

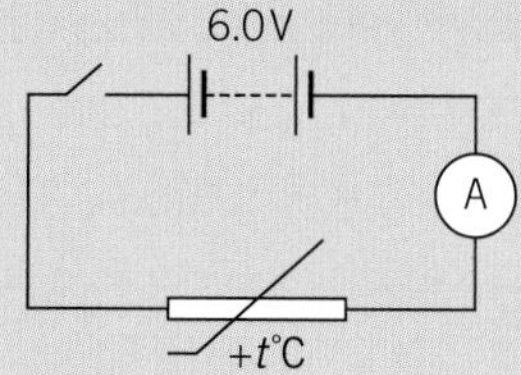


▲ Figure 8

The temperature is 15 °C when the circuit is set up.

(i) Calculate the power dissipated in the PTC thermistor when the switch is closed. *(3 marks)*

(ii) After closing the switch, it is noted that there is an initial change in the current, but that it eventually settles on a fixed value. Explain this observation. *(4 marks)*

## Section C

**1** This question is about an experiment to determine the focal length of a converging lens.

A student uses a converging lens, a 12 V filament lamp and a ground glass screen. Distances are measured using a pair of metre rules. The basic set up is shown in Figure 9.

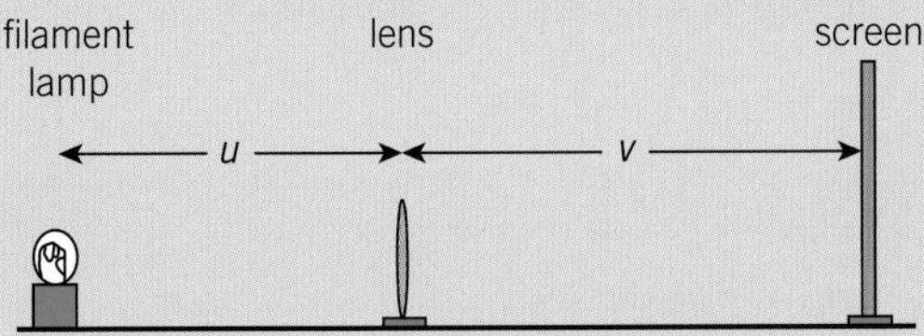


▲ Figure 9

The object distance is varied between −0.3 m and −1.4 m.

**a** (i) Suggest a possible cause of systematic error in measuring the object distance $u$. *(1 mark)*

(ii) Suggest a possible cause of uncertainty in the measurement of image distance $v$. *(1 mark)*

**b** Here are two comments students made about the uncertainty in measuring the image distance $v$.

- The actual value of the uncertainty increases as $v$ increases.
- The percentage uncertainty in the measurement of $v$ remains approximately constant over the range of measurements.

(i) Suggest and explain why the value of the uncertainty in the reading may increase with $v$. *(2 marks)*

(ii) Explain why the percentage uncertainty can remain the same even though the actual value changes. *(2 marks)*

**c** The student recorded this pair of values $u = 1.000\,\text{m}$ and $v = 0.260\,\text{m}$

Show that this pair of values leads to a value to a value of the focal length of the lens of about 0.21 m. *(2 marks)*

The uncertainty in the $v$ value was estimated at ±10 mm. The uncertainty in the $u$ value was estimated at ±1 mm. It was thought that the uncertainty in $u$ could be ignored in considering the uncertainty in the final result.

**d** Using the uncertainty in $v$, calculate the highest and lowest values for the focal length using the data pair in **(c)** to show that the uncertainty in the focal length result is more than ±5 mm. *(4 marks)*

**e** The student took a range of $u$ and $v$ readings.

She recalled the equation $\frac{1}{v} = \frac{1}{u} + \frac{1}{f}$ and decided to calculate $\frac{1}{u}$ and $\frac{1}{v}$ values and plot the graph in Figure 10.

Explain why she assumed this would give a straight-line graph. *(2 marks)*

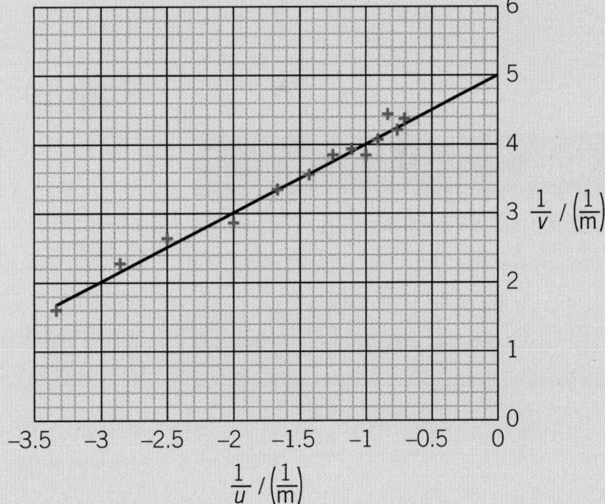


▲ Figure 10

**f** Use the graph to find the focal length of the lens. *(2 marks)*

**g** Suggest why the result obtained from the best-fit line in Figure 10 is better than the mean of individual calculations of focal length from $v$ and $u$ data pairs. *(1 mark)*

# Maths Appendix

## Changing the subject of an equation

An equation shows the relationship between variables. You can change an equation to make any of the variables into the subject of the equation. To do this you do the inverse operation to both sides on an equation to isolate the quantity that you want. For example, dividing is the inverse operation of multiplying. You always add or subtract terms first, and then divide, multiply, or perform other operations, for example, square root.

### Worked example: Make *d* the subject

Change the equation $a = b + \frac{c}{d}$ to make $d$ the subject.

**Step 1:** Add or subtract terms to isolate the term involving $d$.

$$a - b = b + \frac{c}{d} - b$$

$$a - b = \frac{c}{d}$$

**Step 2:** Multiply or divide terms to isolate $d$.

$$a - b = \frac{c}{d}$$

$$(a - b)d = c$$

$$d = \frac{c}{(a - b)}$$

## Graphs

The equation of a straight line is $y = mx + c$, where $m$ is the gradient and $c$ is the $y$-intercept. The gradient and the intercept can be positive or negative. You can deduce information from straight-line graphs. If the line goes through the origin (no intercept) then $y$ is *directly proportional* to $x$, and $y = mx$. When you plot a graph you should think about the physical reason why a graph might (or might not) go through (0, 0).

**Hint**

Graphs such as the graph of force against extension for a spring will be a straight line through (0, 0). The equation of the line is $F = kx$, and this is the same as $y = mx + c$ with $y = F$, $m = k$, and $c = 0$.

You can test a relationship using a graph. You might collect data relating to the pressure and volume of a gas. Plotting the data produces a curve, but plotting $V$ against $\frac{1}{p}$ gives a straight line. This shows that $pV$ = constant (equal to the gradient).

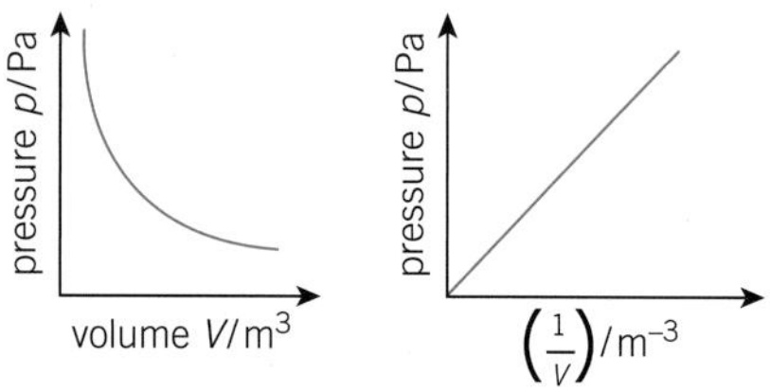


▲ **Figure 1** *The straight line shows that $p \propto \frac{1}{V}$, so $p$ = (gradient) × $\frac{1}{V}$*

### Worked example: Curve to straight line

You have collected data for two quantities, $t$ and $y$, which should be related by the equation $y = \frac{k}{t^2}$, where $k$ is a constant. State which graph you need to plot to show that this is correct. Explain how the shape of the graph would, or would not, confirm your belief.

**Step 1:** Change the equation into the form $y = mx$.

If $y = \frac{k}{t^2}$, then it becomes $y = mx$ if $k = m$ and $\frac{1}{t^2} = x$.

So you need to plot $y$ against $\frac{1}{t^2}$.

**Step 2:** Describe the shape you would get if the relationship was or was not true.

If the graph is a straight line through (0, 0), then this confirms that $y = \frac{k}{t^2}$ and the gradient = $k$.

If the graph is a curve then $y$ is not equal to $\frac{k}{t^2}$.

## Calculations using graphs

You use graphs to display data. When you draw a graph you choose a type of scale, and add numbers vertically and horizontally to the axes. The graph will show the relationship between physical quantities, so you should also include the units as well as the name of the quantity on each of the axes.

You often need to calculate **gradients** and areas from graphs. These may represent important physical quantities. Their units can give an important clue as to what they represent.

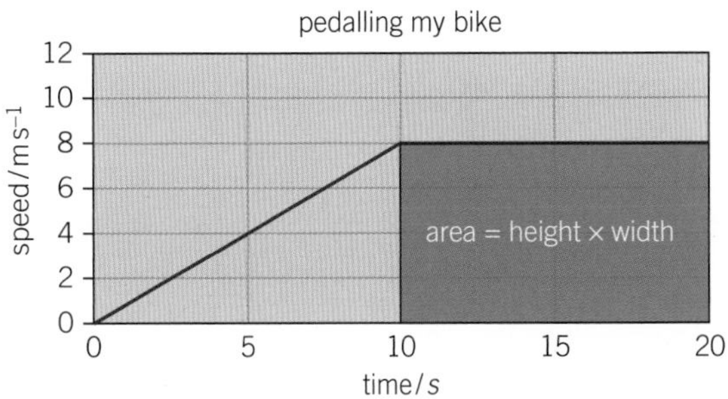


▲ **Figure 2** *A bike ride*

### Worked example: Gradients and area

Calculate the gradient of the graph in Figure 2 between 0 and 10 s, and the area under the graph between 10 and 20 s.

**Step 1:** Find the gradient using the equation gradient = $\frac{\text{change in } y}{\text{change in } x}$.

$$\text{gradient} = \frac{(8\,\text{m}\,\text{s}^{-1} - 0\,\text{m}\,\text{s}^{-1})}{(10\,\text{s} - 0\,\text{s})} = 0.8\,\text{m}\,\text{s}^{-2}$$

These are the units of acceleration. The gradient of a speed–time graph = acceleration.

**Step 2:** Substitute values into area = height × width.

Area = $8\,\text{m}\,\text{s}^{-1} \times 10\,\text{s} = 80\,\text{m}$. These are the units of distance.

The area under a velocity–time graph = distance travelled.

**Synoptic link**

For worked examples of finding gradients and areas for curved graphs, see Topic 8.1, Graphs of motion.

You need to make sure that the gradient that you are calculating is meaningful. The gradient of a velocity–time graph is equal to acceleration, but resistance is equal to the ratio of the potential difference and the current, not the gradient of a graph of potential

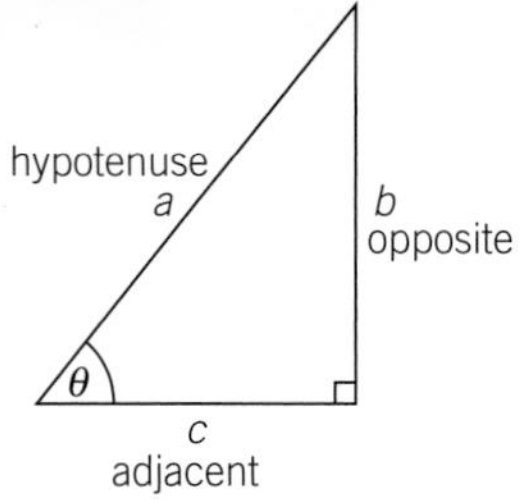


▲ **Figure 3** *You define sin, cos, and tan using the sides of a right-angled triangle*

difference and current (except in the special case of Ohm's law). The resistance is *not* the change of potential difference with respect to current, but acceleration *is* the change of velocity with respect to time.

## Trigonometric (trig) functions

You will use three common trigonometric functions during your physics course. For the triangle in Figure 3

$$\sin\theta = \frac{\text{opp}}{\text{hyp}} = \frac{b}{a},\ \cos\theta = \frac{\text{adj}}{\text{hyp}} = \frac{c}{a},\ \text{and}\ \tan\theta = \frac{\text{opp}}{\text{adj}} = \frac{b}{c}$$

Rearranging these equations means that $b = a\sin\theta$ or $c\tan\theta$, and $c = a\cos\theta$. You will use these ideas when you resolve vectors into two perpendicular components.

### Hint

You can use Pythagoras' theorem to calculate the length of one side from the other two. In Figure 10, $a^2 = b^2 + c^2$.

In general, $h^2 = x^2 + y^2$

where $h$ = hypotenuse (the longest side of a right-angled triangle), $x$ and $y$ are the other two sides.

## Small angle approximation

There may be situations where the angle is very small. Think about what happens to the triangle when this happens.

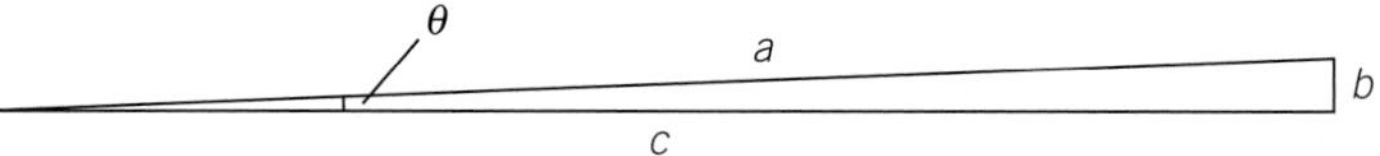


▲ **Figure 11** *When the angle is small the hypotenuse approximately equals the adjacent*

In this situation $a \approx c \Rightarrow \cos\theta = \frac{c}{a} \approx \frac{c}{c} = 1$. Also, $\sin\theta = \frac{b}{a} \approx \frac{b}{c} = \tan\theta$. The triangle in Figure 11 is almost identical to a tiny segment of a circle, where $a = c$ = radius, and $b$ = the length of the arc. The angle $\theta$ measured in radians $= \frac{\text{arc length}}{\text{radius}}$, which is also equal to $\frac{b}{a}$ and $\frac{b}{c}$.

So the small angle approximation says that for small angles $\cos\theta = 1$ *and* $\sin\theta = \tan\theta = \theta$ (measured in radians).

### Synoptic link

You have used trigonometry when resolving vectors in Topic 8.2, Vectors.

### Synoptic link

Small angle approximation is required for Young's double slit experiment in Topic 6.4, Interference and diffraction of light.

### Graphs of $\sin\theta$ and $\cos\theta$

The graphs of $y = \sin\theta$ and $y = \cos\theta$ are shown in Figure 12.

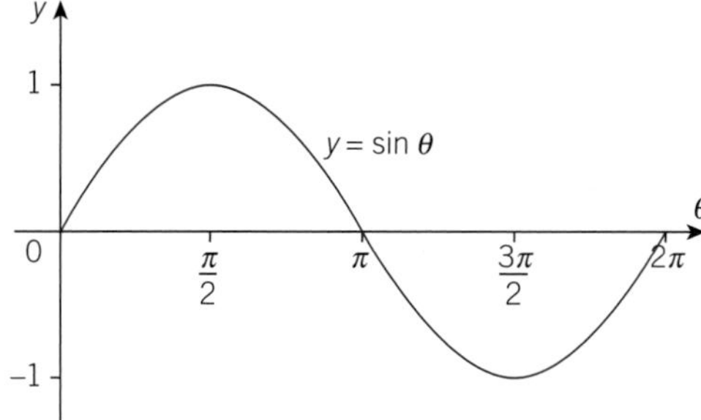


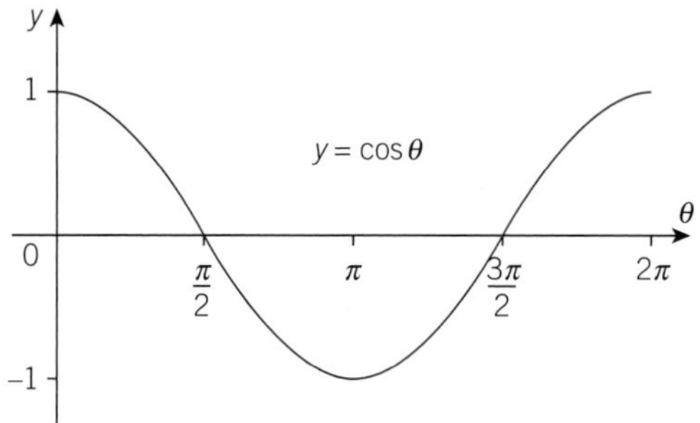


◀ **Figure 12** *Graphs of $y = \sin\theta$ and $y = \cos\theta$ are 90° or $\frac{\pi}{2}$ out of phase*

### Synoptic link

You have seen graphs showing waves in and out of phase in Topic 6.1, Superposition of waves.

## Logarithms

A **logarithm** is another name for a power (or exponent, or index). You can raise any number to a power, for example, $10^3$, $2^3$, $e^3$ ($e = 2.718...$). In all these examples the logarithm (or power) is 3.

The number which is raised to the power is called the **base**. In the example above the bases are 10, 2, and e, respectively. During your physics course you will meet all three of these bases. The logarithm of a number is the power to which you need to raise a base to get that number. For example:

- $10^6 = 1\,000\,000 \Rightarrow \log_{10}(1\,000\,000) = 6$, or $\log_{10}(10^6) = 6$
- $2^6 = 64$. The logarithm to the base 2 of the number 64 is 6
  We write this as $\log_2(64) = 6$.
- $e^{-2} = 0.135 \Rightarrow \log_e(0.135) = -2$. This can be written $\ln(0.135) = -2$.

**Hint**

Do not forget that $\ln = \log_e$.

### Worked example: Finding logs

Find $\log_e(2)$ using your calculator. You will need to use the $\log_e$ or ln button. Check that you have the correct answer by finding $e^{(ans)}$.

**Step 1:** Find $\log_e(2)$ by pressing the ln, then the 2 button.

$$\ln 2 = 0.693 \text{ (3 s.f.)}$$

**Step 2:** Find $e^{0.693....}$

$$e^{0.693...} = 2$$

## Logarithmic scales

When you use any scale (on a graph, on a number line) you can use a **linear scale** or a **logarithmic scale**.

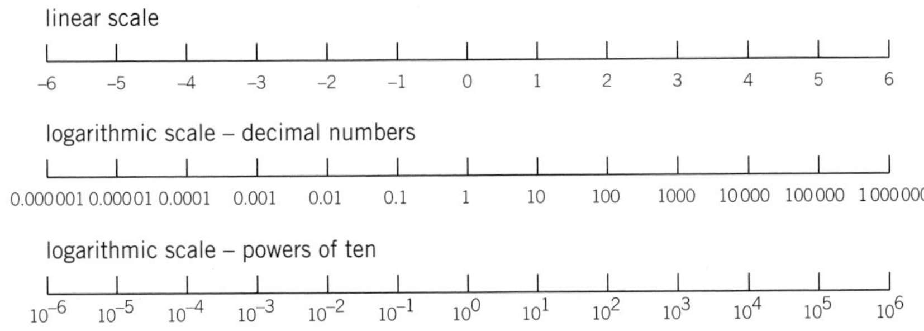


▲ **Figure 13** *A logarithmic scale is a 'times' scale*

You move from left to right on a linear scale by *adding* a fixed number. You move from left to right on a logarithmic scale by *multiplying* by a fixed number (usually, but not always, equal to 10). You need to use logarithmic scales when the values vary over many orders of magnitude, and you will meet many logarithmic scales in your physics course: distance, time, resistivity.

# Data, Formulae and Relationships

## Data

Values are given to three significant figures, except where more – or fewer – are useful.

## Physical constants

| | | |
|---|---|---|
| speed of light | $c$ | $3.00 \times 10^{8}\,\mathrm{m\,s^{-1}}$ |
| permittivity of free space | $\varepsilon_0$ | $8.85 \times 10^{-12}\,\mathrm{C^2\,N^{-1}\,m^{-2}}$ ($\mathrm{F\,m^{-1}}$) |
| electric force constant | $k = \dfrac{1}{4\pi\varepsilon_0}$ | $8.98 \times 10^{9}\,\mathrm{N\,m^2\,C^{-2}}$ ($\approx 9 \times 10^{9}\,\mathrm{N\,m^2\,C^{-2}}$) |
| permeability of free space | $\mu_0$ | $4\pi \times 10^{-7}\,\mathrm{N\,A^{-2}}$ (or $\mathrm{H\,m^{-1}}$) |
| charge on electron | $e$ | $-1.60 \times 10^{-19}\,\mathrm{C}$ |
| mass of electron | $m_e$ | $9.11 \times 10^{-31}\,\mathrm{kg} = 0.000\,55\,\mathrm{u}$ |
| mass of proton | $m_p$ | $1.673 \times 10^{-27}\,\mathrm{kg} = 1.007\,3\,\mathrm{u}$ |
| mass of neutron | $m_n$ | $1.675 \times 10^{-27}\,\mathrm{kg} = 1.008\,7\,\mathrm{u}$ |
| mass of alpha particle | $m_\alpha$ | $6.646 \times 10^{-27}\,\mathrm{kg} = 4.001\,5\,\mathrm{u}$ |
| Avogadro constant | $L$, $N_A$ | $6.02 \times 10^{23}\,\mathrm{mol^{-1}}$ |
| Planck constant | $h$ | $6.63 \times 10^{-34}\,\mathrm{J\,s}$ |
| Boltzmann constant | $k$ | $1.38 \times 10^{-23}\,\mathrm{J\,K^{-1}}$ |
| molar gas constant | $R$ | $8.31\,\mathrm{J\,mol^{-1}\,K^{-1}}$ |
| gravitational force constant | $G$ | $6.67 \times 10^{-11}\,\mathrm{N\,m^2\,kg^{-2}}$ |

## Other data

| | | |
|---|---|---|
| standard temperature and pressure (stp) | | $273\,\mathrm{K}$ ($0\,°\mathrm{C}$), $1.01 \times 10^{5}\,\mathrm{Pa}$ (1 atmosphere) |
| molar volume of a gas at stp | $V_m$ | $2.24 \times 10^{-2}\,\mathrm{m^3}$ |
| gravitational field strength at the Earth's surface in the UK | $g$ | $9.81\,\mathrm{N\,kg^{-1}}$ |

## Conversion factors

| | | |
|---|---|---|
| unified atomic mass unit | 1 u | $= 1.661 \times 10^{-27}\,\mathrm{kg}$ |
| | 1 day | $= 8.64 \times 10^{4}\,\mathrm{s}$ |
| | 1 year | $\approx 3.16 \times 10^{7}\,\mathrm{s}$ |
| | 1 light year | $\approx 10^{16}\,\mathrm{m}$ |

## Mathematical constants and equations

$e = 2.72$ $\pi = 3.14$ 1 radian = 57.3°

arc = $r\theta$

circumference of circle = $2\pi r$

$\sin\theta \approx \tan\theta \approx \theta$ and $\cos\theta \approx 1$ for small $\theta$

area of circle = $\pi r^2$

surface area of cylinder = $2\pi rh$

$\ln(x^n) = n \ln x$

volume of cylinder = $\pi r^2 h$

$\ln(e^{kx}) = kx$

surface area of sphere = $4\pi r^2$

volume of sphere = $\frac{4}{3}\pi r^3$

## Prefixes

| $10^{-12}$ | $10^{-9}$ | $10^{-6}$ | $10^{-3}$ | $10^3$ | $10^6$ | $10^9$ |
|---|---|---|---|---|---|---|
| p | n | μ | m | k | M | G |

## Formulae and relationships

### Imaging and signalling

focal length $\frac{1}{v} = \frac{1}{u} + \frac{1}{f}$

linear magnification $m = \frac{v}{u}$

refractive index $n = \frac{\sin i}{\sin r} = \frac{C_{1\text{st medium}}}{C_{2\text{nd medium}}}$

noise limitation on maximum bits per sample $b = \log_2\left(\frac{V_{\text{total}}}{V_{\text{noise}}}\right)$

alternatives, $N$, provided by $n$ bits $N = 2^b$, $b = \log_2 N$

### Electricity

current $I = \frac{\Delta Q}{\Delta t}$

potential difference $V = \frac{W}{Q}$

power and energy $P = IV = I^2R$, $W = VIt$

e.m.f and potential difference $V = \varepsilon - Ir$

conductors in series and parallel $\frac{1}{G} = \frac{1}{G_1} + \frac{1}{G_2} + \ldots$

$G = G_1 + G_2 + \ldots$

resistors in series and parallel $R = R_1 + R_2 + \ldots$

$\frac{1}{R} = \frac{1}{R_1} + \frac{1}{R_2} + \ldots$

potential divider $V_{\text{out}} = \frac{R_2}{R_1 + R_2} V_{\text{in}}$

conductivity and resistivity $G = \frac{\sigma A}{L}$

$R = \frac{\rho L}{A}$

capacitance $C = \frac{Q}{V}$

energy stored in a capacitor $E = \frac{1}{2}QV = \frac{1}{2}CV^2$

discharge of capacitor $\frac{dQ}{dt} = -\frac{Q}{RC}$

$Q = Q_0 e^{-t/RC}$

$\tau = RC$

### Materials

Hooke's law $F = kx$

elastic strain energy $\frac{1}{2}kx^2$

Young modulus $E = \frac{\text{stress}}{\text{strain}}$,

$\text{stress} = \frac{\text{tension}}{\text{cross – sectional area}}$,

$\text{strain} = \frac{\text{extension}}{\text{original length}}$

### Gases

kinetic theory of gases $pV = \frac{1}{3}Nm\overline{c^2}$

ideal gas equation $pV = nRT = NkT$

### Motion and forces

momentum $p = mv$

impulse $F\Delta t$

force $F = \frac{\Delta(mv)}{\Delta t}$

work done $W = Fx$ $\Delta E = F\Delta s$

power $P = Fv$, $P = \frac{\Delta E}{t}$

components of a vector in two perpendicular directions

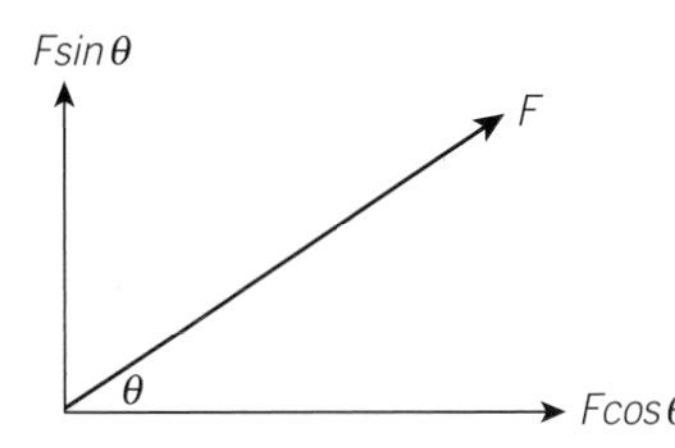

| | |
|---|---|
| equations for uniformly accelerated motion | $s = ut + \frac{1}{2}at^2$ <br> $v = u + at$ <br> $v^2 = u^2 + 2as$ |
| for circular motion | $a = \frac{v^2}{r}, F = \frac{mv^2}{r} = mr\omega^2$ |

## Energy and thermal effects

| | |
|---|---|
| energy | $\Delta E = mc\Delta\theta$ |
| average energy approximation | average energy ~ $kT$ |
| Boltzmann factor | $e^{-\frac{E}{kT}}$ |

## Waves

| | |
|---|---|
| wave formula | $v = f\lambda$ |
| frequency and period | $f = \frac{1}{T}$ |
| diffraction grating | $n\lambda = d\sin\theta$ |

## Oscillations

| | |
|---|---|
| simple harmonic motion | $\frac{d^2x}{dt^2} = a = -\left(\frac{k}{m}\right)x = -\omega^2 x$ <br> $x = A\cos(\omega t)$ <br> $x = A\sin(\omega t)$ <br> $\omega = 2\pi f$ |
| Periodic time | $T = 2\pi\sqrt{\frac{m}{k}}$ <br> $T = 2\pi\sqrt{\frac{L}{g}}$ |
| total energy | $E = \frac{1}{2}kA^2 = \frac{1}{2}mv^2 + \frac{1}{2}kx^2$ |

## Atomic and nuclear physics

| | |
|---|---|
| radioactive decay | $\frac{\Delta N}{\Delta t} = -\lambda N$ <br> $N = N_0 e^{-\lambda t}$ |
| half life | $T_{\frac{1}{2}} = \frac{\ln 2}{\lambda}$ |
| radioactive dose and risk | absorbed dose = energy deposited per unit mass <br> effective dose = absorbed dose × quality factor |
| mass–energy relationship | $E_{rest} = mc^2$ |
| relativistic factor | $\gamma = \sqrt{\frac{1}{1 - v^2/c^2}}$ |
| relativistic energy | $E_{total} = \gamma E_{rest}$ |

| | |
|---|---|
| energy–frequency relationship for photons | $E = hf$ |
| de Broglie | $\lambda = \frac{h}{p}$ |

## Field and potential

| | |
|---|---|
| for all fields | fields strength $= -\frac{dV}{dr} \approx -\frac{\Delta V}{\Delta r}$ |
| gravitational fields | $g = \frac{F}{m}, E_{grav} = -\frac{GmM}{r}$ <br> $V_{grav} = -\frac{GM}{r}, F = -\frac{GmM}{r^2}$ |
| electric fields | $E = \frac{F}{q} = \frac{V}{d},$ <br> electrical potential energy $= \frac{kQq}{r}$ <br> $V_{electric} = \frac{kQ}{r}, F = \frac{kQq}{r^2}$ |

## Electromagnetism

| | |
|---|---|
| magnetic flux | $\Phi = BA$ |
| force on a current carrying conductor | $F = ILB$ |
| force on a moving charge | $F = qvB$ |
| Induced e.m.f | $\varepsilon = -\frac{d(N\Phi)}{dt}$ |

# Glossary

**acceleration** A vector quantity — the rate of change of velocity, $a = \frac{\Delta v}{\Delta t}$.

**alloy** An alloy is a material composed of two or more metals, or a mixture of metals and other materials.

**amorphous** At the microscopic scale, an amorphous material has no long range order. The microscopic structure of glass is amorphous — the atoms in glass form strong bonds with one another to make up a rigid structure without any regularity.

**ampere (A)** The S.I. unit of electrical current. $1\,A = 1\,C\,s^{-1}$

**amplitude (wave)** The amplitude of a wave at a point is the maximum displacement from some equilibrium value at that point.

**antinode** An antinode is a position of maximum amplitude of oscillation on a standing wave.

**antiphase** Two oscillations are in antiphase when their phase difference is $\pi$ radians, or 180°.

**bit** A bit is the smallest unit of digital information, represented as a 0 or a 1 corresponding to low voltage or high voltage in a digital circuit.

**brittle** A brittle material breaks by snapping cleanly. It undergoes little or no plastic deformation before fracture.

**byte** A byte is a sequence of eight bits coded to represent one of 256 alternatives.

**conductance, *G*** Conductance $G$ is the ratio $\frac{I}{V}$ for a circuit component, and is measured in siemens (S) for current in A and p.d. in V.

**coherence** Two sources of waves are coherent if they emit waves with a constant phase difference and have the same frequency. Two waves arriving at a point are said to be coherent if there is a constant phase difference between them as they pass that point.

**component of a vector** One of two vectors (in two perpendicular dimensions) into which a vector may be split, e.g. horizontally and vertically.

**compression** Compressive forces are squashing forces. An object is in compression when two forces act on it in opposite directions to make the object compress (squash) along the line of action of the forces.

**conductor (electrical)** A material which conducts electricity well on account of having many free charge carriers.

**conservation of energy** During any interaction, the total energy before the interaction is the same as the total energy after the interaction.

**conservation of momentum** During any interaction, the total momentum before the interaction is the same as the total momentum after the interaction (see Newton's second law).

**coulomb (C)** The S.I. unit of electrical charge. 1 C is the charge flowing through a point in 1 s where there is a current of 1 A.

**current, *I*** The rate at which charge flows through a point in an electrical circuit

**density** Density is mass per unit volume. The units of density are $kg\,m^{-3}$. Density is often represented by the Greek letter rho, $\rho$.

**diffraction** Diffraction is the spreading of waves after passing through a gap or past the edge of an obstacle. The spreading increases if the gap is made narrower or if the wavelength of the waves is increased.

**dislocation** A dislocation is a defect in the regular structure of a crystal or crystalline region of a material. Dislocations in metals are mobile and make metals ductile.

**displacement** A vector quantity — the distance travelled in a specified direction.

**dissipation** A thermal energy transfer resulting in an increase in internal energy, often of the surroundings.

**distance** A scalar quantity — the separation between two points in space, possibly along a curved path, with no reference to direction.

**drift velocity** The mean velocity of charge carriers in a conductor carrying an electrical current.

**ductile** A ductile material can be easily drawn into a wire (e.g. copper is easier to draw into a wire than tungsten). Metals are ductile because the non-directional metallic bonds allow ions to slide past one another.

**e.m.f., $\varepsilon$** The energy per unit charge given by any source of electrical supply to the charges set in motion.

**elastic deformation** When a material deforms elastically it regains its original shape after deformation.

**elastic limit** The elastic limit is the maximum stress at which an object returns to its original shape after the deforming stress is removed.

**electrical conductivity, $\sigma$** A constant for electrical conductors given by the equation $G = \frac{\sigma A}{L}$

**electrical resistivity, $\rho$** A constant for electrical conductors given by the equation $R = \frac{\rho L}{A}$

**electronvolt** One electronvolt (eV) is the work done when an electron is moved through a potential difference of 1 volt. $1\,eV = 1.6 \times 10^{-19}\,J$.

**energy level** Electrons in atoms can only take certain fixed energies. These are the energy levels of the electrons in the atom. When electrons gain or lose energy they move between energy levels.

**focal length** The focal length $f$ of a thin lens is the distance from the centre of the lens to the focal point $F$.

**focal point (focus)** The focal point $F$ of a converging lens is the point where light from a very distant object on the axis of the lens is brought to a focus by the lens. This point is also called the focus.

**force, $F$** The 'push' or 'pull' acting on an object associated with a change in its momentum: $F = \frac{\Delta p}{\Delta t}$ (see Newton's second law).

**fracture** An object fractures when it breaks into two or more pieces when placed under stress.

**fracture stress** Fracture stress is the stress at which fracture occurs. The fracture stress of a material in tension is sometimes called its tensile strength.

**grain boundary** A grain boundary is the line along which grains meet in a crystalline material.

**gravitational potential energy, $E_{grav}$** Potential energy of a mass due to its position in a gravitational field — the gravitational potential energy difference between two points in a uniform field $\Delta E_{grav} = mgh$, where $h$ is the vertical separation between the two points.

**hard** A material is hard if it is difficult to dent its surface. Many ceramics are very hard.

**impulse** The product $F\Delta t$ summed over the whole action of a force on an object resulting in a momentum change $\Delta p$ = impulse

**in parallel** Components joined alongside which share current and have the same p.d.

**in series** Components joined end to end which share the p.d. and have the same current

**inertia** The tendency of any stationary object to remain stationary or of any moving object to continue with the same momentum (see Newton's First Law)

**input voltage, $V_{in}$** The p.d. applied to a sensor circuit (frequently a potential divider)

**insulator (electrical)** A material which conducts electricity poorly on account of having very few charges which are free to move

**intensity** The intensity of a wave is the energy per unit time carried by the waves and incident normally per unit area of surface.

**interference** Interference arises from the superposition of waves on top of one another. When waves overlap, the resultant displacement will be equal to the sum of the individual displacements at that point and at that instant (if the waves superpose linearly).Interference is produced if waves from two coherent sources overlap or if waves from a single source are divided and then reunited.

**internal resistance, $r$** Resistance within a source of e.m.f. resulting in a drop of terminal p.d. when a current is drawn from the source.

**iterative model** A mathematical treatment, often using a computer, whereby small step-wise changes to variables such as displacement and velocity are made at regular time intervals $\Delta t$ with the assumption that those variables change only at the end of each time interval.

**kinematic equations** Four equations for motion in a straight line under uniform acceleration (also called suvat equations).

**kinetic energy, $E_k$** The energy possessed by an object by virtue of its motion: $E_k = \frac{1}{2}mv^2$.

**Kirchhoff's first law** At any electrical junction, the total current into the junction = the total current out of the junction.

**Kirchhoff's second law** Around any electrical circuit, the sum of all e.m.f.s = the sum of all p.d.s.

**magnification** Linear magnification $= \frac{\text{image distance}}{\text{object distance}} = \frac{v}{u}$.

**malleable** A material is malleable if it is easy to hammer or press a sheet of material into a required shape.

**mass, $m$** The amount of matter in any object.

**momentum, $p$** A vector quantity: the product mass × velocity.

**Newton's first law** A stationary object will remain stationary, and a moving object will continue moving with the same momentum, unless an external force acts upon it.

**Newton's second law** When an external force acts on an object, it produces a momentum change according to $F$ = rate of change of momentum = $\frac{\Delta p}{\Delta t} = ma$ if there is no change in mass.

**Newton's third law** When an object A exerts a force $F$ upon an object B, then the object B exerts a force $(-F)$ on object A.

**node** A node is a position of minimum amplitude of oscillation on a standing wave.

**normal line** The normal line is line drawn at ninety degrees to the surface of a transparent material. When light is represented as a ray the angle of incidence at the surface is the angle the ray makes to the normal line.

**number density, $n$** The number of mobile charge carriers per unit volume of the conductor, measured in $m^{-3}$.

**ohm ($\Omega$)** The S.I. unit of electrical resistance, $R$: $1\,\Omega = \frac{1\,V}{1\,A}$.

**output voltage, $V_{out}$** The p.d. measured in a sensor circuit (frequently a potential divider).

**path difference** When waves travel from one point to another by two or more routes, the difference in the distance travelled by each wave is the path difference. The importance of a path difference is that it introduces a time delay, so that the phases of the waves differ when they meet. It is the difference in phase that generates interference effects.

**phase and phase difference** 'Phase' refers to stages in a repeating change, as in 'phases of the Moon'. The phase difference between two objects vibrating at the same frequency is the fraction of a cycle that passes between one object being at maximum displacement in a certain direction and the other object being at maximum displacement in the same direction.

**phase angle** Phase difference is expressed as a fraction of one cycle, or of $2\pi$ radians, or of 360°. This is known as the phase angle. For example, two waves which are half a cycle apart have a phase angle of $\pi$ radians or 180°.

**phasor** Phasors are used to represent amplitude and phase in a wave. A phasor is a rotating arrow used to represent a sinusoidally changing quantity.

**photons** Electromagnetic waves of frequency $f$ are emitted and absorbed in quanta of energy $E = hf$, called photons.

**pixel** A pixel is a single 'picture element'. In a digital camera, a lens is used to form a real image on a chip with an array of the order of a million very small light-sensitive detectors. Each detector corresponds to one pixel in the final image.

**plane wavefronts** A plane wavefront is one with zero curvature. Waves from distant sources have plane wavefronts.

**plastic deformation** When a material deforms plastically it undergoes permanent stretching or distortion before breaking.

**polarisation** Transverse waves are linearly polarised if they vibrate in one plane only. Unpolarised transverse waves vibrate in a randomly changing plane. Longitudinal waves cannot be polarised.

**polycrystalline** Polycrystalline materials are composed of tiny crystal grains. Within each grain the material shows an ordered structure but the orientation of each individual grain is random.

**potential difference (p.d.),** $V$ The energy transfer per unit charge moving between the two points in question.

**power (of a lens)** The power of a lens in dioptres (D) $= \frac{1}{f}$, where $f$ is the focal length in metres. The shorter the focal length the more powerful the lens.

**probability** Probability has to do with uncertainty, randomness and quantum effects. Probability is a measure of the chance of one of a number of possible things happening.

**projectile** An object thrown or fired so that its subsequent motion is affected by a gravitational field.

**refractive index** The refractive index of a transparent material is the ratio of the speed of light in a vacuum to the speed of light in the material. The refractive index is also given (in Snell's Law) as the ratio of the sine of the angle of incidence to the sine of the angle of refraction as the light enters the transparent material.

$$\text{refractive index} = \frac{\sin\text{(angle of incidence)}}{\sin\text{(angle of refraction)}}$$

**relative velocity** The velocity a moving object appears to have when viewed from another moving object.

**resistance,** $R$ Resistance $R$ is the ratio p.d. $V$/current $I$ for a circuit component, and is measured in ohms ($\Omega$) for p.d. in V and current in A.

**resolution** Splitting up a vector quantity into two components.

**resolution (of a digital image)** The resolution of an image is the scale of the smallest detail that can be distinguished.

**scalar** A quantity (e.g. speed, mass, energy) with magnitude but no direction.

**semiconductor** A material midway in electrical conductivity and resistivity between conductors and insulators.

**sensor circuit** A circuit whose electrical properties depend upon environmental variables and which can be used to monitor or measure those variables.

**siemen (S)** The S.I. unit of electrical conductance, $G$: $1\,\text{S} = \frac{1\,\text{A}}{1\,\text{V}}$.

**speed** A scalar quantity — the distance per unit time travelled with no account taken of the direction of movement.

**stiff** A stiff material has a small extension per unit force. The stiffness is indicated by the Young modulus.

**strain** Strain is the change of length per unit length. Strain is a ratio of two lengths and therefore has no unit. Strain is often represented by the Greek letter epsilon, $\varepsilon$.

**stress** Stress (tensile and compressive) is the force per unit area acting at right angles to a surface. The units of stress are Pa or $\text{N}\,\text{m}^{-2}$. Stress is often represented by the Greek letter sigma, $\sigma$.

**superposition** When two or more waves meet, their displacements can superpose, that is, be added together at every point. The principle of superposition states that when two or more waves overlap, the resultant displacement at a given instance and position is equal to the sum of the individual displacements at that instance and position.

**tension** Tensile forces are stretching forces. An object is in tension when two forces act on it in opposite directions to make the object stretch along the line of action of the forces.

**threshold frequency** The threshold frequency is the minimum frequency of light that will eject photoelectrons from a given surface. The threshold frequency varies with material.

**tough** A material is tough if it does not break by snapping cleanly. A tough material is resistant to the propagation of cracks. Toughness is the opposite of brittleness.

**vector** A quantity (e.g. velocity, force, momentum) with both magnitude and direction.

**velocity** A vector quantity — the distance per unit time travelled in a specified direction.

**volt (V)** The S.I. unit of electrical potential and potential difference, $V$: $1\,\mathrm{V} = \frac{1\,\mathrm{J}}{1\,\mathrm{C}}$

**watt (W)** The S.I. unit of power, $P = \frac{\Delta E}{\Delta t}$. One Watt is equal to one joule per second.

**wavefront** A wavefront is an imaginary line or surface that moves along with a wave. All points on the wavefront have the same phase. In water waves, the line of the crest of the wave can be thought of as a wavefront.

**work function** The work function $\Phi$ is the minimum energy required to eject photoelectrons from a given surface. The work function for a given material is found by multiplying the threshold frequency for that material by the Planck constant.

**work, $W$** An energy transfer when a force moves its point of application: $W$ = component of force in the direction of the diplacement × displacement = $Fs\cos\theta$.

**yield stress** Yield stress is the stress at which a specimen begins to yield (where plastic deformation begins).

**Young modulus** The Young modulus is the ratio of stress over strain, $\frac{\sigma}{\varepsilon}$. The units of the Young modulus are Pa or $\mathrm{N\,m^{-2}}$.

# Answers

## 1.1

1 a curvature = 3.1 D

2 a power = (+) 20 D

b curvature added = 20 D

3 focal length = 140 m

4 a Note that the wavelength is the same either side of the lens.

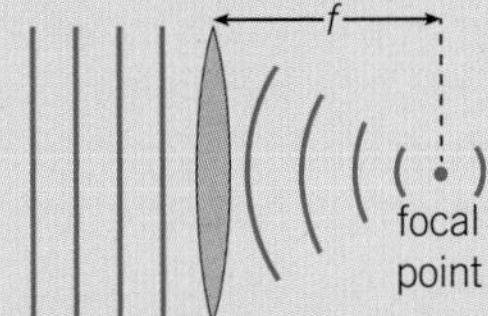


b The focal length is twice that in **4 (a)**.

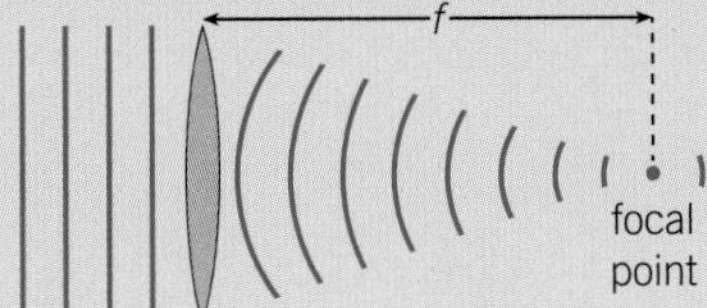


## 1.2

1 image distance = 53 mm

2 magnification = 1.7

3 Magnification is defined as $\frac{\text{image distance}}{\text{object distance}}$. If the image distance is less than the object distance the magnification will be less than one — the image size will be smaller than the object size.

4 a power of lens = +5 D

b To form an image further from the lens the curvature of waves leaving the lens must be less. Moving the lamp nearer the lens means that the curvature of the waves incident on the lens will be more negative. As the lens adds a constant (positive) curvature to waves passing through it, the (positive) curvature of the waves leaving will be less.

Alternatively, you can argue using the lens equation. If $\frac{1}{u}$ becomes more negative (as it will when u decreases) $\frac{1}{v}$ will decrease. If $\frac{1}{v}$ decreases, $v$ must increase.

## 1.3

1 a mean = 111

b median = 100

c Replacing the pixel with the median is the best method. A 'noisy' pixel will have a very different value from the surrounding pixels. Using the mean would result in a pixel which still has a different value from the surrounding pixels and so could still show up as noise.

2 number of bits = 12
number of bytes = 1.5

3 a Resolution is the distance on the object that is represented by one pixel in the image.

b number of pixels = $1 \times 10^6$

c memory required = 3 Mbyte

## 1.4

1 wavelength = 3.3 m. Same number of significant figures in the answer as the data provided.

2

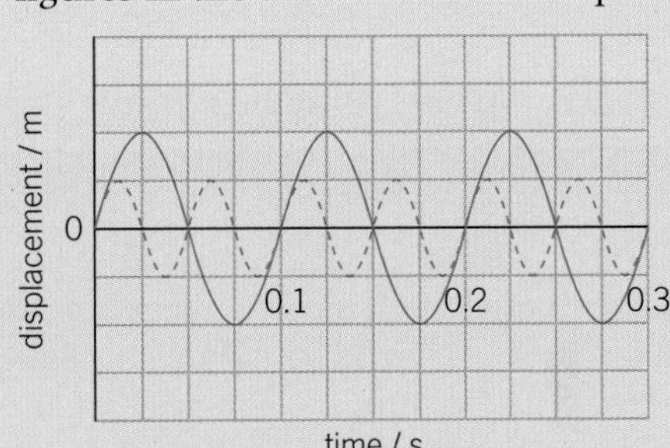


The solid line represents a wave of time period 0.1 s. The dashed line represents a wave of frequency 20 Hz (time period = 0.05 s).

3 Rotate the receiving aerial in a plane at right angles to the direction of travel of the waves. If the radio waves are polarised the received signal will vary in amplitude from maximum to minimum as the aerial is rotated through 90°. If there is no change in amplitude on rotation of the aerial the radio emission is unpolarised.

## 2.1

1 1016 bits

2 a 64 levels

b 0.19 V

3 Many possible advantages including: ability to edit, error correction, ease of storage, data compression, digital files are playable on a number of devices. These are all possible because the data is numeric and can be mathematically processed easily and quickly.

4 $\log_2 75 = 6.2$ so 7 bits needed

5 a See Figure 4 on main content page.

b Quantisation error is the difference between the signal value and the quantisation level that represents the signal value.

## 2.2

1 256 kHz

2 3.6 ms

3 For accurate recording the sample rate must be greater than twice the highest frequency present in the signal. If the sample rate is lower than this, aliasing can occur producing spurious low frequency signals. If the original signal has a maximum frequency of 20 000 Hz, the sample rate should be greater than twice this, that is, 40 000 Hz.

4 Aliasing could produce lower frequency signals that are within the range of human hearing from frequencies above the range of hearing.

5 a For example, ease of manipulation can lead to false images and ease of communication can lead to security lapses.

b For example, digital images can be altered easily. Techniques such as edge detection and others can be very useful in medical imaging and are of great benefit, but images can also be altered with malicious intent.

## 3.1

1 a 5A

b 60 W

2 $V_1 = 3.4$ V, $V_2 = 2.6$ V and $I_1 = 0.3$ A (one mark each)

3 $1.5 \times 10^{18}$

4 copper ions: $1.3 \times 10^{20}$ ions
chloride ions: $2.6 \times 10^{20}$ ions

## 3.2

1 26 V

2 a As current increases, the wire heats and resistance increases, so $\frac{I}{V}$ gets smaller.

b If $R$ and $G$ were constant, the $I$-$V$ graph would be linear with gradient $G$ and the $R$-$V$ and $G$-$V$ graphs would be horizontal straight lines. The $I$-$V$ graph starts linear but drops more and more below that line as $V$ and $I$ increase. This shows that $R$ is increasing and $G$ decreasing, as shown by the $G$-$V$ and $R$-$V$ graphs.

3 15.9 °C (3 s.f.)
if resistance not measured to high precision, $R_\theta - R_0$ will not have 3 s.f.

4 $R$ increases (from a non-zero value) at 0 °C to 100 °C; $G$ decreases (from a non-zero value) over that range (ignore curvature in either)

5 4.1 V

## 3.3

1 $\rho = \frac{RA}{L} = \Omega\ \text{m}^2/\text{m} = \Omega$ m; $\sigma = \frac{1}{\rho} = (\Omega\ \text{m})^{-1}$
$= (\Omega)^{-1}\ (\text{m})^{-1} = \text{S m}^{-1}$

2 1.3 m

3 $G \approx 7 \times 10^{-10}$ S $\Rightarrow I \approx 3 \times 10^{-6}$ A so the experiment should be possible.

## 3.4

1 Any two points from: silver has more atoms per unit volume, so more free electrons; crystal structure is more closely packed in iron so that electrons are obstructed more; iron has larger ion cores so electrons are obstructed more.

2 $v_B = 10 \times v_A$ because total charge flowing through per second is the same so $(10n_B)\ Av_Ae = (n_B)\ Av_Be$.

3 $I = \frac{\Delta Q}{\Delta t}$ so $\Delta Q = I\Delta t$ $\quad N = \frac{\Delta Q}{e} = \frac{I\Delta t}{e}$

$n = \frac{N}{V} = \frac{N}{LA} = \frac{I\Delta t}{eLA}$ so $L = \frac{I\Delta t}{nAe}$

$v = \frac{L}{\Delta t} = \frac{\left(\frac{I\ t}{nAe}\right)}{t} = \frac{I}{nAe}$

## 3.5

1 5.0 V

2 Suitable fixed resistor $R$ in range 5.4 kΩ to 1.0 MΩ, although 200 kΩ to 600 kΩ is better.

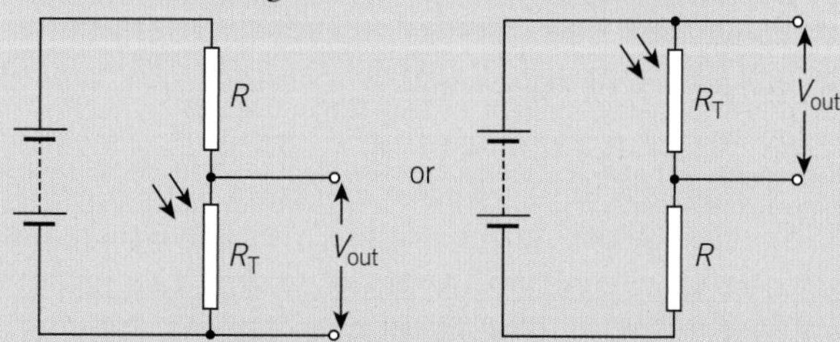


3 The voltmeter has too low a resistance so that when it is in parallel with resistor **A**, the value of parallel resistance drops to 5000 Ω and, as it is $\frac{1}{3}$ of the total resistance, it takes $\frac{1}{3}$ of the total p.d. = 2.0 V; the same happens when the voltmeter is connected in parallel with resistor **B**.

## 3.6

1 The p.d. across the terminals $V = \varepsilon - Ir$ and so is less than the 'expected' p.d. $\varepsilon$ by the amount $Ir$, which is the p.d. across the internal resistance. This p.d. seems to have vanished: it is 'lost' inside the cell.

2 3.6 V, 0.41 W

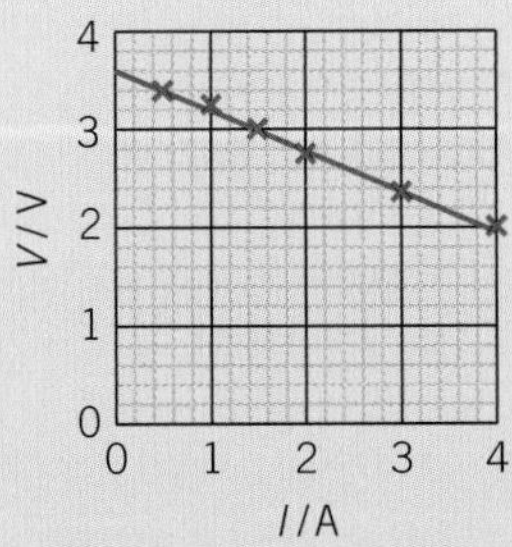


3 The starter motor draws a (very) large current, so the total current through the internal resistance increases. This means the 'lost volts' *Ir* increase, so that the p.d. across the terminals drops. This p.d. is not enough now to light the car lights to full brightness.

## 4.1

1 a Steel is a metal, china is a ceramic, rubber and acrylic are polymers.

b Link the properties of the material to its classification. For example, steel is ductile, china is hard and brittle, and the polymers are both tough.

2 For example, keyboard plastic is a tough material, window glass is brittle.

3 Scalpels need to be hard and stiff. Hardness is required to ensure the blade is not blunted. Stiffness is required because the scalpel shouldn't flex.

4 a For example, strong, stiff, hard. (The outer casing of the phone protects the sensitive electronics.)

b The outer sleeve or case needs to be tough to dissipate energy on impact.

c Casing: often steel (metal) with a glass face. Sleeve: often made from a polymer (leather, rubber, or others).

## 4.2

1 See Figure 5 on main content page.

2 $5500\,\mathrm{N\,m^{-1}}$

3 a The graph is a straight line through the origin.

b $83\,\mathrm{N\,m^{-1}}$

c 0.20 J

4 The student realises that the graph shows proportional behaviour and assumes that this behaviour is maintained up to loads of 35.0 N. Using $F = kx$ and the calculated value of $k$ in **3 (b)** gives an extension of 42.0 cm at a force of 35.0 N. This may be incorrect as the spring may begin extending non-elastically at a lower force than 35.0 N, making the Hooke's Law equation inapplicable.

## 4.3

1 a Fracture stress is the stress at which a material breaks (fractures). Yield stress is the stress at which plastic deformation first occurs.

b $10^2\,\mathrm{MN\,m^{-2}}$, $10^2\,\mathrm{N\,mm^{-2}}$

2 a 0.12 or 12%

b 6.3 cm. Assume elastic behaviour.

3 extension (assuming circular cross-section of radius 25 mm) = 7.6 mm

4 a percentage uncertainty in diameter is ± 6%

b $5.8 \times 10^{10}\,\mathrm{N\,m^{-2}}$

c range: $5.5 - 6.1 \times 10^{10}\,\mathrm{N\,m^{-2}}$

d If the mass and density are known, the volume of the wire can be calculated using the equation volume = mass/density. Measuring the length of the wire allows a calculation of cross-sectional area as cross-sectional area = volume/length. If the mass, density, and length are known to small percentage uncertainties, the percentage uncertainty in cross-sectional area may be smaller than obtained by measuring the diameter of the wire.

e The shape of the curve will be similar to that in Figure 5, Topic 4.3. The yield stress of copper is about 100 MPa. The fracture stress of copper is about 200 MPa.

## 4.4

1 From Figure 4 you can see that the stiffest metal has a Young Modulus of about 400 GPa. The least stiff metal has a Young Modulus of about 12 GPa (it is difficult to see exactly). The ratio is $\frac{400}{12}$, which is 30 (to 1 s.f.).

2 between 700 and 800

3 a 10.2 kN

b 3.6 mm

c The wire will yield at lower stress. The designers have to include margin for error to cover, for example, the lift being used by eight people of greater average weight than 650 N.

## 5.1

1 Information required: number of atoms along one side of the image and length represented by one side of the image. The diameter of

one atom can be estimated from the ratio $\frac{\text{length of one side of the image}}{\text{number of atoms along one side}}$

2 The oil patch will be at least one molecule thick so one molecule cannot be longer than the calculated value. If the patch is more than one molecule thick, the length of the molecules will be less than that calculated.

3 order of magnitude = $10^{-9}$ m (calculated value = $2.1 \times 10^{-9}$ m)

4 The density of gold is found by measuring the volume of a known mass of gold. To apply this density value to a single atom assumes that there are no spaces between the atoms in a macroscopic specimen of gold.

## 5.2

1 Amorphous materials are disordered at the microscopic scale. Polycrystalline materials have grains oriented randomly but with an ordered (crystal) structure within the grains.

2 Weakness is a description of breaking stress. Glass (the material) does not have a low breaking stress. Strong materials have a high breaking stress. Brittleness is a description of the fracture process in a material. The fracture process is independent of the strength of the material.

3 Metals may have mobile dislocations. These allow planes of atoms to slip over one another by breaking the bonds one at a time. Plastic deformation can occur at a lower stress if dislocations are present. Alloying atoms can stop the dislocations moving through the metal, making it less plastic. This increases the hardness and brittleness of the material. See Figure 5, Topic 5.2.

4 See explanation under *Stress concentration and crack propagation*, Topic 5.2. The stress is concentrated at the tip of the crack as the area at the tip is small.

5 Dislocations in ceramics are not mobile and so cannot move through the material.

## 5.3

1 The material may become stiffer because the chains will not unravel so easily.

2 The bonds in metals are strong but non-directional. This allows the ions to slip. Ions in ceramics are held together by directional bonds that lock them into place.

3 Metals deform elastically when the distance between positive ions increases. Up to strains of around 0.1% the ions will be pulled back into position when the deforming force is removed. At strains greater than this the force between the ions grows weaker. Polymer elasticity is explained by the rotation of the bonds in the long chain molecule, allowing it to unravel.

4 Cross-linking joins polymer chains together at points along the chains. This reduces the flexibility of the polymer because the chains cannot unfold to the same degree.

## 6.1

1 a The two waves are on top of each other.

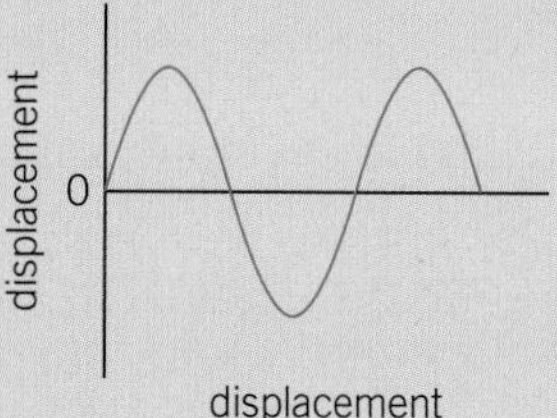


b The two solid lines shown.

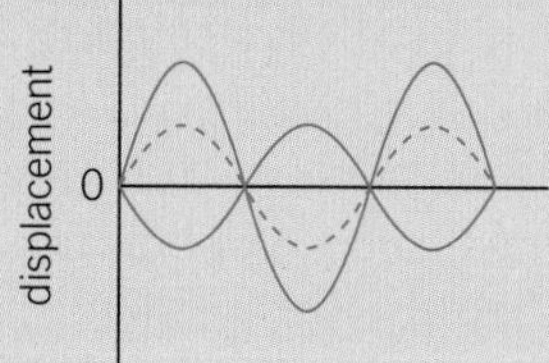


c

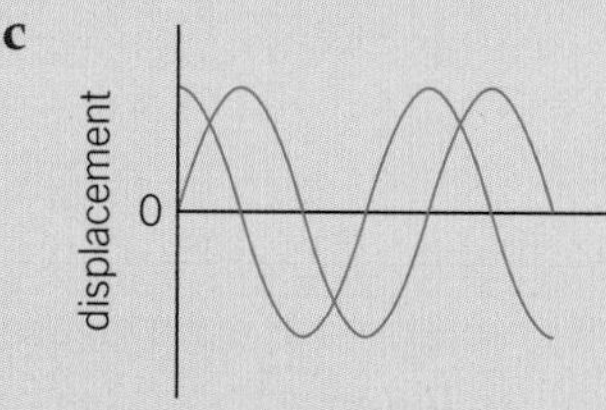


d See dashed line in **1(b)**.

2 a A progressive wave travels along a string or tube. The wave reflects at the end of the string or tube. Waves travelling in opposite directions superpose. At points where the waves always meet in phase an antinode of maximum displacement forms. Where waves always meet in antiphase a node of minimum displacement is formed.

b distance between nodes = 0.25 m

3 2.8 m

4 a 15 GHz

b 0.02 m

c

Place the reflector about 1.5 m from the transmitter. Place the detector between the transmitter and receiver. A minimum signal shows at the point at which the waves from the transmitter and reflector are in antiphase. Measure the position of such a minimum signal. Slowly move the detector towards the reflector, observing the amplitude of the detected signal. Find the next minimum signal position. The distance between one minimum and the next is half a wavelength.

## 6.2

1 $1.2 \times 10^8\,\mathrm{m\,s^{-1}}$. Number of significant figures is that of the least precise value in the data given.

2 27°

3 a

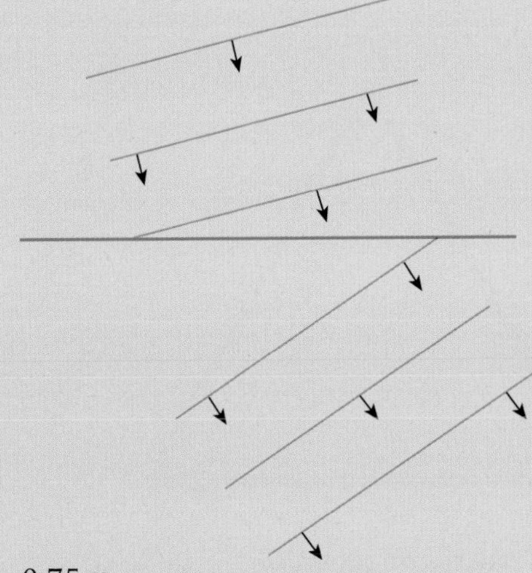

b 0.75 m

4 If the refractive index of the two materials is the same the light will not deviate when it moves from one material to the next. Glycerol and Pyrex are both transparent so the tube will only be detected by the distortion it produces. As there is no deviation of light the observer will not see any distortion and so the tube is invisible.

5 a Refractive index gives a measure of the change of speed when light enters the material.

b Different wavelengths are deviated by different amounts. As the deviation of light in the glass is due to the speed changing on entering the glass it follows that different wavelengths travel at different speeds in glass.

c In this case, refractive index is a measure of the amount of deviation when light enters the glass from air. As different wavelengths are deviated by different amounts the refractive index of the glass is wavelength-dependent.

6 a In this case, refractive index is a measure of the amount of deviation when light enters the glass from air. As different wavelengths are deviated by different amounts the refractive index of the glass is wavelength-dependent.

| $\sin i$ | $\sin r$ +/– 0.04 |
|---|---|
| 0 | 0 |
| 0.17 | 0.09 |
| 0.34 | 0.24 |
| 0.5 | 0.34 |
| 0.64 | 0.44 |
| 0.77 | 0.53 |
| 0.87 | 0.66 |

b

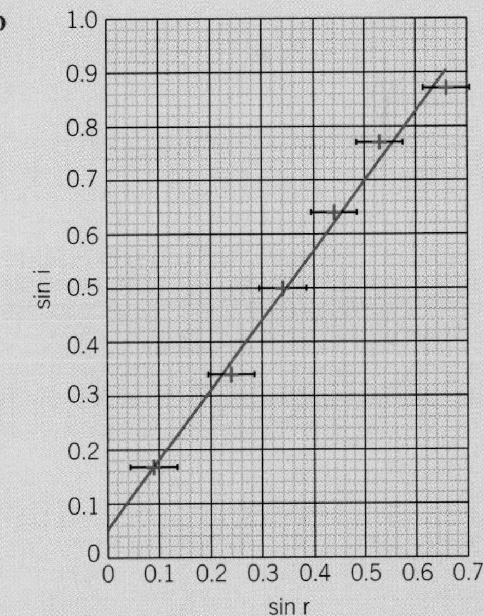


(The uncertainty of +/– 0.04 is a pessimistic estimate based on calculating the range of sin *r* values when *r* has an uncertainty of +/– 2°.)

Gradient of graph = 1.3 +/– 0.1

## 6.3

1 Path difference is the difference in distance between two (or more) sources of waves and a receiver. Phase difference is the difference in the position in the wave cycle of two (or more) waves at a given instant.

2 Coherent waves will show a constant phase difference

3 a Path difference = $n\lambda$

b 0.75 m

c Path difference has changed by 0.25 m. This represents a change of path difference of one wavelength. Therefore, the wavelength of the sound is 0.25 m.

## 6.4

1 a Diffraction is the spreading of waves when they pass through a gap or around an object. There is no change to speed, wavelength, or frequency on diffraction.

b $4.8 \times 10^{-7}$ m

2 a 19°

b 6 orders

3 a about 0.03°

b It is too small to be easily noticed. For example, a first order minimum of 0.09° corresponds to width of about 1.6 mm on a screen 1 m from the slit. This is a very small amount of spreading.

4 a Largest value is about 1600 nm, smallest value about 800 nm.

b The measurement of the gap between the slits contributes most uncertainty. The percentage uncertainty in this measurement is 20% compared to 14% for the fringe separation and less than 1% for the slit-screen distance.

## 7.1

1 $2.0 \times 10^{-24}$ J, $1.3 \times 10^{-5}$ eV; $4.0 \times 10^{-19}$ J, 2.5 eV; $7.9 \times 10^{-16}$ J, 5.0 keV

2 Use the ideas in the chapter, remembering to describe the phenomenon carefully and then explain it using energy quanta.

3 work function = $4.8 \times 10^{-19}$ J, threshold frequency = $7.3 \times 10^{14}$ Hz

4 a Energy transfer when an electron of charge $e$ passes through a potential difference $V = eV$. The energy is released as a photon.

b

| wavelength of light / nm | frequency of photon / 1014 Hz | striking potential / V | photon energy / 10–19 J |
|---|---|---|---|
| 470 +/– 30 | 6.4 +/– 0.4 | 2.6 +/– 0.2 | 4.2 +/– 0.4 |
| 503 +/– 30 | 6.0 +/– 0.4 | 2.5 +/– 0.2 | 4.0 +/– 0.4 |
| 585 +/– 30 | 5.1 +/– 0.4 | 2.1 +/– 0.2 | 3.4 +/– 0.4 |
| 620 +/– 30 | 4.8 +/– 0.4 | 2.0 +/–0.2 | 3.2 +/– 0.4 |

c Correctly plotted graph using the values given in **4(b)**.

d gradient from graph gives a value of about $7 \times 10^{-34}$ J s. Uncertainty can be estimated by considering the variation of gradients possible within the uncertainty bars. These gradients give a wide range of values. The uncertainty can also be estimated by considering the percentage uncertainty in each variable.

## 7.2

1 rotations per second = $5.0 \times 10^{14}$

2 a $2.0 \times 10^{-8}$ s

b $1.2 \times 10^{7}$ rotations

3

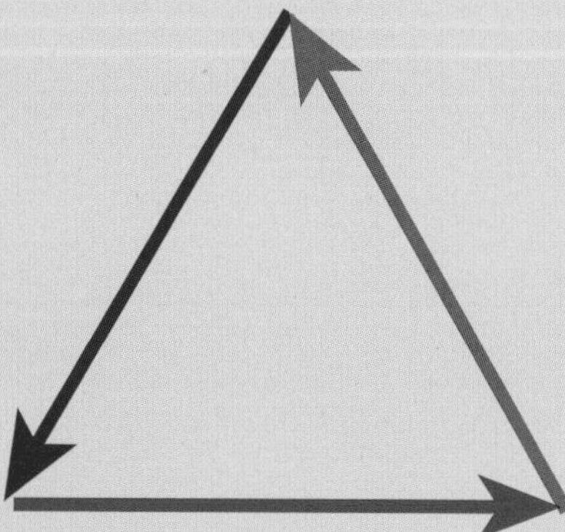

4 a All these trip times are the same.

b Photons from all three paths meet in phase (phasor arrows are all in the same direction).

c Phasor amplitude will be greatest at F, therefore the probability of the arrival of a photon will be greatest at F. As intensity is proportional to probability of arrival, this will also be greatest at F and lower elsewhere where the path lengths are different and the phasors do not meet in phase.

5 a $3 \times 10^{11}$ mm s$^{-1}$

b 1 ns

c $5 \times 10^{-11}$ s

d $2 \times 10^{11}$ mm s$^{-1}$

e 1.5

f They have the same trip time because the phasors following route ADEB slow down in the glass but their frequency of rotation remains the same.

## 7.3

1 $1 \times 10^{-10}$ m

2

| Momentum $mv$ | Speed $v$ | Wavelength |
|---|---|---|
| $6.6 \times 10^{-26}$ kg m s$^{-1}$ | $7.3 \times 10^{4}$ m s$^{-1}$ | 10 nm |
| $1.0 \times 10^{-24}$ kg m s$^{-1}$ | $1.1 \times 10^{6}$ m s$^{-1}$ | 0.66 nm |
| $1.8 \times 10^{-24}$ kg m s$^{-1}$ | $2.0 \times 10^{6}$ m s$^{-1}$ | 0.37 nm |

3 The wavelength of an electron of momentum $7 \times 10^{-24}$ kg m s$^{-1}$ is approximately $9 \times 10^{-11}$ m. This is far smaller than the wavelength of a photon of equivalent energy. As interference effects only become noticeable when the gap is of the order of magnitude of the wavelength, electrons will require smaller gaps for interference effects to be observed.

4 a A spot is formed when a discrete particle strikes the image in a particular place. The energy of the particle is transferred over a small area. This would not happen with a wave.

b This is a superposition effect. Such effects are usually associated with waves.

c Phasors explore all possible paths. Where the phasor arrows combine to form a large resultant there is a larger probability of an electron being found at that spot. The probability of detection of an electron can be calculated from the square of the resultant phasor.

## 8.1

1

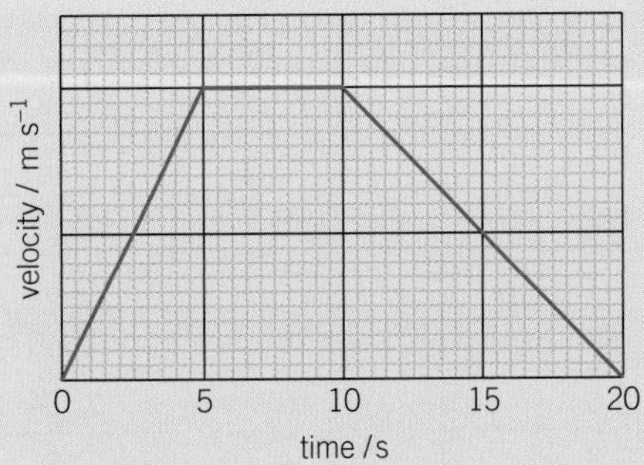


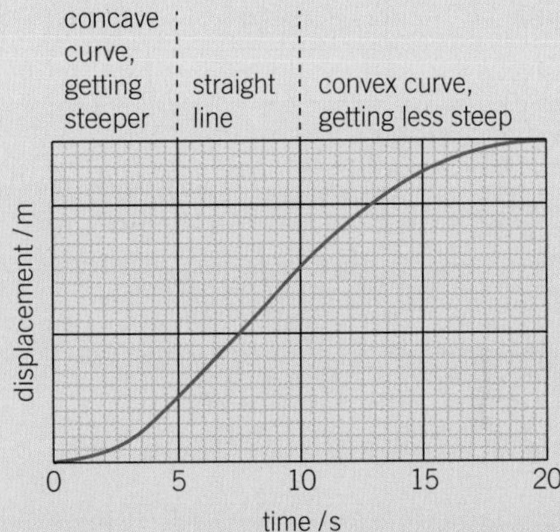


2 Answer in range 1950 – 2150. Displacement is not distance travelled as direction changes, so the car is not always heading directly away from the starting place.

3 $a = 3.5$ m s$^{-2}$
distance = area under curve = 125 m (± 8 m)

## 8.2

1 magnitude of displacement = 21 paces
direction = N 17° E / bearing 017°

2 Diagram has boat pointing upstream at angle $\theta$ so that resultant velocity is straight across. $\theta = 24°$.

3 velocity = 240 m s$^{-1}$ in a direction N 54° W / bearing 306°

## 8.3

1

| t/s | v/m s$^{-1}$ | s/m |
|---|---|---|
| 0 | 0.00 | 0.00 |
| 0.1 | 0.34 | 0.00 |
| 0.2 | 0.68 | 0.03 |
| 0.3 | 1.02 | 0.10 |
| 0.4 | 1.36 | 0.20 |
| 0.5 | 1.70 | 0.34 |
| 0.6 | 2.04 | 0.51 |
| 0.7 | 2.38 | 0.71 |
| 0.8 | 2.72 | 0.95 |

2 The increments in velocity are calculated only from the time interval (which is constant) and the previous value of acceleration (which equals zero at terminal velocity), that is, by $a\Delta t$. Therefore, when terminal velocity is reached, $a = 0$, so $a\Delta t = 0$, regardless of the time interval, $\Delta t$.

3 On each interval, $\Delta s$ is 0.8 m downstream. In the diagram, scale = 10 divisions = 2 m.

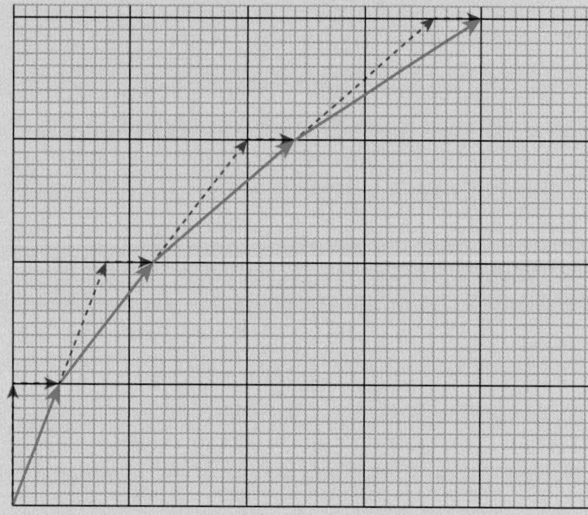

## 8.4

1 $u = 26.8\,\text{m s}^{-1}$ and $s = 55\,\text{m}$ gives $a = -6.5\,\text{m s}^{-2}$; $u = 13.4\,\text{m s}^{-1}$ and $s = 14\,\text{m}$ gives $a = -6.4\,\text{m s}^{-2}$

2 160 m

3 0.86 s

4 0.53 s or 1.9 s: the stone can hit the kite on the way up, or pass the kite and hit it on the way back down again.

## 9.1

1 a 27 000 kg m s$^{-1}$

b E.g. 98 kg m s$^{-1}$, using 70 kg for the mass and 1.4 m s$^{-1}$ for the walking pace.

2 260 000 m s$^{-1}$ assuming no other interactions with the atom.

3 $v_{min}$ car = 15.9 m s$^{-1}$ for the sum of momenta to be forwards, so yes, the car was going too fast.

## 9.2

1 (−)170 000 N

2 a no momentum before firing, so forwards momentum of bullet = backwards momentum of gun

b gun pushes bullet forwards; bullet pushes gun backwards.

3 By Newton's first law, you carry on going forwards until a force (seatbelt/airbag/ windscreen) stops you. The car is being decelerated by a force so, relative to the car, you are still moving forwards.

4 (−)4000 N (2 s.f.)

5 $\Delta p = 0.5\,\text{kg} \times 250\,\text{m s}^{-1} = 125\,\text{kg m s}^{-1}$ and $\Delta t = 1\,\text{s}$

$F = \dfrac{\Delta p}{\Delta t} = \dfrac{125\,\text{kg m s}^{-1}}{1\,\text{s}} = 100\,\text{N}$ (1 s.f.)

At the end of the 3 s, mass lost by the rocket = $0.5\,\text{kg s}^{-1} \times 3\,\text{s} = 1.5\,\text{kg}$.

The rocket mass drops from 2.5 kg tp 1.0 kg during the flight.

$a = \dfrac{F}{m}$, so with constant $F$ from the exhaust gases and decreasing $m$ of the rocket, $a$ will increase.

## 9.3

1 E.g. 530 000 J

2 0.65 m, or 0.85 m from the lowest point.

3 $\frac{1}{2}mv^2 \times 2m = m^2v^2 = p^2$

4 3 m s$^{-1}$

## 9.4

1 Carries on in a straight line until it 'runs out of impetus', and then falls vertically. A real projectile is falling all the time from leaving the mouth of the cannon, and follows a parabolic trajectory.

2 64 m
working to 4 s.f., 45° gives 63.78 m while both 44° and 46° give 63.74 m.

3 $t = 0.79\,\text{s}$ when $s_h = 4\,\text{m}$, which gives $s_v = 1\,\text{m}$ (the net is 1 m above the thrower).

## 9.5

1 81 000 W

2 4.8 J

3 110 000 J

# Index

**Acknowledgements**

**pX**: Rook76/Shutterstock; **pXI**: Esa/Rosetta/Philae/Civa/Science Photo Library; **p7**: Vadim Sadovski/Shutterstock; **p8**: Marques/Shutterstock; **p9**: Roman White/Shutterstock; **p15**: Alexander Mak/Shutterstock; **p16-17**: Giphotostock/ Science Photo Library; **p18**: Hikrcn/Shutterstock; **p24**: Johannes Poetzsch; **p25** (R): Andy Crump/Science Photo Library; **p25** (L): Asharkyu/Shutterstock; **p27**: Nasa/Science Photo Library; **p29**: Yanlev/Shutterstock; **p34**: Twin Design/ Shutterstock; **p42** (T): Shin Okamoto/Shutterstock; **p42** (B): European Space Agency,J. Huart/Science Photo Library; **p46**: Mino Surkala/Shutterstock; **p55**: Alfred Pasieka/Science Photo Library; **p58**: Andrew Lambert Photography/Science Photo Library; **p59**: Fanfo/Shutterstock; **p63**: Science Photo Library; **p65**: Science Photo Library; **p72-73**: Andrei Seleznev/ Shutterstock; **p74**: Trekandshoot/Shutterstock; **p75**: Sebastian Kaulitzki/Shutterstock; **p76**: Tongo51/Shutterstock; **p78**: Bunnyphoto/Shutterstock; **p80** (T): Mycteria/Shutterstock; **p80** (B): Michael Chamberlin/Shutterstock; **p86** (L): Africa Studio/Shutterstock; **p86** (R): Lewis Tse Pui Lung/Shutterstock; **p88** (R): Alan Jeffery/Shutterstock; **p88** (L): Dutourdumonde Photography/Shutterstock; **p92**: Stephan Raats/Shutterstock; **p94**: Philippe Plailly/Science Photo Library; **p96** (L): Science Photo Library; **p96** (C): Power And Syred/Science Photo Library; **p96** (R): Nick Pavlakis/Shutterstock; **p98**: G. Muller, Struers Gmbh/Science Photo Library; **p100**: D. Kucharski K. Kucharska/ Shutterstock; **p104** (TL): Bagdan/Shutterstock; **p104** (TR): Antoniomas/Shutterstock; **p104** (C): Boris15/Shutterstock; **p104** (B): Magnetix/Shutterstock; **p105**: Ornl/Science Photo Library; **p108-109**: Richard Peterson/Shutterstock; **p110** (T): Jeng_ Niamwhan/Shutterstock; **p110** (B): Nicholas Toh/Shutterstock; **p114**: Andrew Lambert Photography/Science Photo Library; **p117**: Pat_Hastings/Shutterstock; **p119**: Morphart Creation/ Shutterstock; **p122**: Mopic/Shutterstock; **p129**: DigitalGlobe; **p130**: Giphotostock/Science Photo Library; **p136**: Marty Pitcairn/Shutterstock; **p137**: David Parker/Science Photo Library; **p138**: Royal Institution Of Great Britain/Science Photo Library; **p144**: A Rose Advances In Biological And Medical Physics 5 211 (1957); **p146**: Science Photo Library; **p149**: OUP; **p150**: Krasowit/Shutterstock; **p152**: Rawpixel/Shutterstock; **p157**: Andrew Lambert Photography/Science Photo Library; **p158**: Clouds Hill Imaging LTD/Science Photo Library; **p160**: American Institute Of Physics/Science Photo Library; **p161**: Molekuul.Be/Shutterstock; **p164-165**: Denis Kuvaev/ Shutterstock; **p173**: © XCWeather 2015. All rights reserved.; **p175**: Nicku/Microstock; **p178**: Harvepino/Microstock; **p190**: New York Public Library/Science Photo Library; **p194** (T): 36Clicks/iStockphoto; **p194** (B): Jose Gil/Shutterstock; **p198**: Tonybaggett/iStockphoto; **p203**: Joe Munroe/Science Photo Library; **p207**: Science Photo Library; **p208**: Kristina Postnikova/Shutterstock; **p211**: Pal2Iyawit/Shutterstock; **p213**: 2Happy/Shutterstock; **p216** (T): Library Of Congress/Science Photo Library; **p216** (B): Edstock/iStockphoto; **p217**: Ssuaphoto/ iStockphoto;

Artwork by Q2A Media